MONMOUTHSHIRE LEPIDOPTERA

*The Butterflies and Moths
of
Gwent*

MONMOUTHSHIRE LEPIDOPTERA

The Butterflies and Moths of Gwent

Dr. G. A. Neil Horton,
MA, MRCS, LRCP, FLS, FRES

Published with the support of the Countryside Council for Wales

Published in 1994 by
Comma International Biological Systems.

© Dr G.A. Neil Horton 1994

The right of Dr G.A. Neil Horton to be identified as the author of this work has been asserted by him in accordance with the Copyright, Designs and Patents Act 1988.

This book is copyright. No part of it may be reproduced in any form without permission in writing from the publishers except by a reviewer who wishes to quote brief passages in connection with a review written for inclusion in a newspaper, magazine, radio or television broadcast.

British Library Cataloguing in Publication Data

A catalogue record for this book is available
from the British Library

ISBN 0 9513977 5 3

Designed and Produced by Images Publishing (Malvern) Ltd.
Printed and Bound by Bookcraft Ltd, Midsomer Norton, Bath, Avon.

Contents

Symbols and Abbreviations used in the text	7
Alphabetical List of Recorders etc	8
Acknowledgements	9
Map of Monmouthshire	10
Introduction	11
Monmouthshire – Boundaries, Geology and Topography	12
Rainfall and Climate	15
Monmouthshire Woodlands and their Lepidoptera	15
Bogs and Heathland	18
Lepidoptera of the Gwent Levels and the Severn Littoral	20
Lepidoptera of the Northern Hills and Moorland	21
Migrants	22
Vagrants	23
Species now Extinct	24
Lepidoptera Recording in Monmouthshire	24
(concluding remarks)	26
The Systematic List and Records	27
(Introductory Remarks)	27
Part One Butterflies	29
Part Two Moths	59
Section 1. The Larger Moths (macrolepidoptera)	61
Section 2. The Smaller Moths (microlepidoptera)	225
Bibliography and References	312
National Grid References	318
Abbreviations of Authors' names	321
Reference to Plates 23-27	322
Index of English names	326
Index of Scientific names	333

To the memory of my mother who throughout her 96 years detested moths, and to my wife Shelia who has learned to live with them.

Symbols and Abbreviations used in the text:

ab.	aberration	ms.	manuscript
abdt.	abundant	m.v.l.	mercury vapour light
act. l.	actinic light		
		NNR	National Nature Reserve
BE & NHS	British Entomological and Natural History Society		
		pers. comm.	personal communication
BMNH	British Museum (Natural History)	pers. obs.	personal observation
		plfl.	plentiful
c.	circa (about)		
coll.	collected	rec.	record
colln.	collection	ref.	reference
det.	determined	spp.	species
detn.	determination	Sq.	10 Km. square of National grid
dist.	district		
d°	ditto	ssp.	sub-species
		SSSI	Site of Special Scientific Interest
Ed.	earliest date/dates		
		svl.	several
f.	form		
frqt.	frequent	u.v.l.	ultra violet light
genit. exam	genitalia examination	var.	variety
		v-c.	Watsonian vice-county
ha.	hectare		
		Wds	Woods
in cop.	in copula (mating)		
in litt.	in correspondence		
inc.	including		
Km.	kilometre		
Ld.	latest date/dates		
		♀ female	♀♀ females
		♂ male	♂♂ males
macros.	macrolepidoptera		
micros.	microlepidoptera		

Alphabetical List of Recorders etc

Canon D.J.L. Agassiz (DJLA); Dr. M.E. Anthoney (MEA); the late J. Armitage Esq. (JA); Mrs. H. Bosanquet (HB); Dr. J.D. Bradley (JDB); J.M. Chalmers-Hunt Esq. (JMC-H); P. Clarkson-Webb Esq. (PC-W); S. Coxey Esq. (SC); the late Brig. A. Daniel (AD); Mrs. J.M. Dean (JMD); R. Dean Esq. (RD); the late R.P. Demuth Esq (RPD); the late E.W. Ecutt Esq. (EWE); C. Elliott Esq. (CE); T.G. Evans Esq. (TGE); Lieut.- Col. A. Maitland Emmet (AME); the late D.W.H. Ffennel Esq. (DWHF); Mrs. M.A. Finch (MAF); D.S. Fletcher Esq. (DSF); R. Gaunt Esq. (RG); Dr. P. Glading (PG); Barry Goater Esq. (BG); G.M. Haggett Esq. (GMH); E.F. Hancock Esq. (EFH); Dr. M.W. Harper (MWH); Dr. A.S. Henderson (ASH); G.A. Hill Esq. (GAH); Tony Hopkins Esq. (TH); Dr. G.A.N. Horton (GANH); S.W.N. Horton Esq. (SWNH); Mrs. S.Y. Horton (SYH); the late W.T. Horton Esq. (WTH); P.N. Humphreys Esq. (PNH); the late Lieut.- Col R.B. Humphreys (RBH); the late Rex A. Jackson Capt. R.N. (RAJ); C. Jones Esq. (CJ); M.E. Jones Esq. (MEJ). W.E. Keen Esq. (WEK); Miss S. Kerry (SK); Dr. J.R. Langmaid (JRL); M.J. Leech Esq. (MJL); David Lewis Esq. (DL); R.F. Mc Cormack Esq. (RFMcC); the late J.L. Messenger Esq. (JLM); H.W.N. Michaelis Esq. (HWNM); the late E.C. Pelham-Clinton, 10th Duke of Newcastle (ECP-C); the late R.E.M. Pilcher Esq. (REMP); I. Rabjohns Esq. (IR); C.J. Renshaw Esq. (CJR); the late Austin Richardson Esq. (AR); A.M. Riley Esq. (AMR); the late Mrs. S.H. Robbins (SHR); J.P. Sankey-Barker Esq. (JPS-B); Mrs. C.M. Sankey-Barker (CMS-B); Miss S. Sankey-Barker (SS-B); P. Scource Esq. (PS); Bernard Skinner Esq. (BS); R. Smith Esq. (RS); P.A. Sokoloff Esq. (PAS); D.R. Stephenson Esq. (DRS); Lieut.- Col. D.H. Sterling Esq (DHS); the late W.H.T. Tams Esq. (WHTT); V.C. Tidley Esq (VCT); Colin Titcombe Esq. (CT); D.G. Turner Esq. (DGT); Mrs. Stephanie Tyler (SJT); Dr. R.W.J. Uffen (RWJU); D.J. Upton Esq. (DJU); J.J. Wood Esq. (JJW); the late Baron C.G.M. de Worms (CGMdW); the late J.W. York Esq. (JWY).

Acknowledgements

Without the considerable help afforded me by many kind individuals this book could not have been published or even written.

I am indebted not only to those recorders named on a previous page who have allowed me to use their records but also to many who have given great assistance in identifying material I have supplied.

In particular I would like to thank Dr. J.D. Bradley of the Commonwealth Institute of Entomology. Not only has he accompanied me on many local moth-hunting expeditions but has devoted much time to identifying many of the micros I have turned up.

Similarly, Col. Maitland Emmet has also been of great assistance, by kindly confirming the identity of moths I have sent him and also by supplying me with his own Monmouthshire records of leaf-mining moths, and there are many such records.

Likewise Michael Chalmers-Hunt and others have placed their records at my disposal or otherwise provided assistance.

Many local landowners must be thanked for allowing me access to their properties, so also must the Forestry Commission for permitting me to visit their woodlands. I am grateful to the Gwent Wildlife Trust not only for allowing me entry to its reserves to collect and record insects but also for its help in arranging financial assistance.

Also must be thanked are those good friends who have accompanied me on my nocturnal forays in remote woods where we have encountered police, forestry officials, gamekeepers and poachers as well as moths. Names of these friends which come readily to mind are Jack York, Wayne Davies, both now deceased and Ivor Grindle a great countryman originally from the Forest of Dean, and there were many others.

Finally, the production of this book has been assisted by the Countryside Council for Wales and by two other main benefactors P.N. Humphreys Esq. and Dr. M.E. Anthoney respectively President and Vice-chairman of the Gwent Wildlife Trust and to them I express my sincerest thanks.

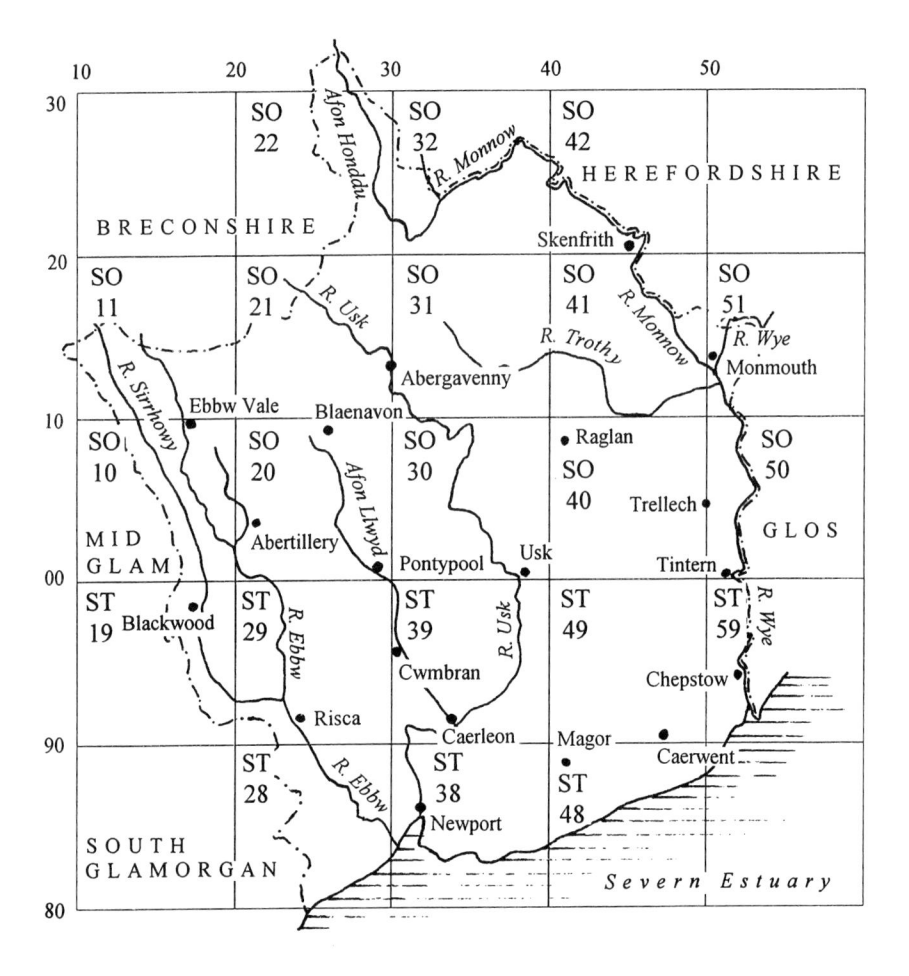

MAP OF MONMOUTHSHIRE (Gwent) showing the county boundaries, adjoining counties and the 10 km squares of the National Grid. Shown also are the chief rivers and some of the towns and villages referred to in the text.

Introduction

No comprehensive list of Monmouthshire Lepidoptera has hitherto been published nor have any local lists other than Charles Conway's 1833 list of the butterflies of the Pontnewydd district in Monmouthshire's Eastern Valley and several lists compiled by the author in the 1960s of the macrolepidoptera of Wentwood, Magor Marsh and the Usk and Wye Valleys.

When invited in 1968 by John Heath of the Biological Records Centre, Monkswood to be the Monmouthshire County Referee for the Lepidoptera Distribution Maps Scheme he was organising it soon became apparent that, with the exception of the Wye Valley area, this county had in the past been largely neglected and very under-recorded with local records being meagre or non-existent. Furthermore recorders for the Scheme were few here in Gwent and their activities were mainly confined to the butterflies.

As Monmouthshire remained one of the few counties for which no list was available I decided to fill this hiatus as far as possible in order to provide a guide to the distribution of its Lepidoptera, to gather together such information as could be found on the past history of its butterflies and moths and also to present a picture of its Lepidoptera at the present time so that future changes might be assessed.

I apprised Michael Chalmers-Hunt of my intentions and encouragingly he immediately placed at my disposal his list of Monmouthshire records gleaned mainly from the entomological literature.

Some twenty-five years have now elapsed and my self-imposed, though not onerous, task is virtually completed. I can but hope that this work will prove of interest and assistance to future local lepidopterists and that some will be stimulated to continue where I have left off. Needless to say, much work still remains to be done especially on the distribution of Monmouthshire's microlepidoptera.

MONMOUTHSHIRE – Boundaries, Geology and Topography

Monmouthshire, the southernmost county of the Welsh Marches derived its name from Monmouth, the Roman town of Blestium, situated at the confluence of the rivers Wye, Monnow and Trothy but on the political revision of administrative areas and boundaries in 1974 the county was renamed "Gwent".

Predominantly mountainous and hilly it is one of our smaller counties having an area of 138,000 hectares, approximately 530 square miles. The 1974 alterations to the county boundary were minimal and were of little moment to the lepidopterist as the limits of present-day Gwent still correspond fairly closely to those of Watsonian Vice-county 35 (Monmouthshire).

Monmouthshire's southern boundary formed by the Severn Estuary extends south-westwards from the mouth of the River Wye to the vicinity of Cardiff in the west. In the east, the Wye separates the county from Gloucestershire as far north as Redbrook where, several miles below Monmouth, the county boundary deviates from the course of the Wye taking in a small part of the Forest of Dean including Redding's Inclosure and Lady Park Wood.

North-west of Monmouth the R. Monnow, a tributary of the Wye, forms as far as its confluence with the Honddu, part of the boundary with Herefordshire. From here, a narrow northerly extension of Monmouthshire penetrates deep into the Black Mountains, enclosing the Llanthony Valley or Vale of Ewyas.

On the eastern side of the Vale, along the Herefordshire border, the Hatterrall Ridge rises to about 2,100 ft., while on the west, adjoining Breconshire (now part of Powys), is Chwarel y Fan which at 2,228 ft. is the highest point in Gwent. Southwards from here the boundary with Breconshire is formed by a river, the Grwyne Fawr, but then, from the vicinity of Gilwern in the Usk Valley, it takes an irregular and arbitrary course westwards, crossing the moorland plateau of Llangattwg and Llangynidr at a height of some 1,500 to 1,600 ft. to the headwaters of the R. Rhymney.

From here the county boundary, continuing now in a southerly direction, mainly along the ridge between the Rhymney and Sirhowy rivers, separates Gwent from Mid Glamorgan to the west. South of Risca this western boundary follows a more or less arbitrary line meeting the Bristol Channel at Peterstone Wentlooge separating the county from Mid and finally, South Glamorgan.

From its source in the Brecon Beacons the River Usk, flowing south past Abergavenny and Usk to enter the Bristol Channel at Newport, roughly bisects the county from north to south. West of a line from Abergavenny to Newport and extending into Glamorgan lie the Coal Measures, their north-south orientated industrial, and formerly mining, valleys drained by the rivers Rhymney, Sirhowy, Ebbw Fawr, Ebbw Fach and Afon Llwyd with their intervening ridges of mountain and moorland reaching a height of around 1,500 ft.. In places however, this altitude is exceeded, as at the Blorenge (1,833 ft.) and the Coity Mountain (1,905 ft.).

Most of the coalfield surface is formed by the Pennant Grit or Sandstone, but on its eastern limits and at a few places in the north, outcrops of Carboniferous Limestone occur as in the Clydach Gorge or are exposed by quarrying, while a small area of limestone pavement can be seen on the Blorenge. A line from Cwmbran to Risca approximately marks the southern limit of the Coal Measures in Gwent.

Much of central and eastern Monmouthshire, chiefly to the east of the R. Usk, is formed by the Old Red Sandstone of the Devonian era giving rise, north of Abergavenny, to the Black Mountains with their two southern outlying hills the Sugar Loaf (1955 ft.) and the Skirrid Fawr or Holy Mountain (1596 ft.) while in the east near Grosmont an isolated hill, Graig Syfyrddin, rises to about 1375 ft. The Old Red Sandstone also gives rise to the marls of the undulating farmland region lying to the north-east of a line running from Abergavenny through Raglan to Monmouth and to a similar farming area between the River Usk and the Afon Llwyd. The Blorenge Mountain, also to the west of the River Usk, is largely Old Red Sandstone but its northern and eastern escarpments are of Carboniferous Limestone.

The Devonian era also provided the sandstones and quartz conglomerates of the elevated Wentwood and Trellech districts. This hilly well-wooded region, occupies the area between Monmouth, Raglan, Chepstow and Caerleon reaching as far as the Wye, and rising in general to between 400 and 900 ft., although at Trellech Beacon it reaches 1,005 ft. while Wentwood's highest point is 1,013 ft..

In the south-east of the county extending from Caerwent through Shirenewton and Chepstow to Tintern is an area of Carboniferous Limestone and here a little limestone grassland is to be found as in the neighbourhood of Rogiet and Dinham but little of it remains "unimproved". In places, as at the Wyndcliff between Tintern and Chepstow, in an area well-known to lepidopterists, the high, wooded, limestone cliffs lend great scenic beauty and interest to the lower Wye Valley. Carboniferous Limestone also occurs in a limited area at Lady Park Wood north of Monmouth.

In central Monmouthshire is found a roughly elliptical area of Silurian rocks measuring approximately nine by four miles with its long axis running south-westwards from Clytha in the north to the vicinity of Llanfrechfa in the south. These rocks consist mainly of mudstones of the Ludlow series while a narrow band of harder Wenlock Limestone is exposed here and there in the Usk district where, in the recent past, it was quarried to produce lime for local agricultural use, its presence now indicated by small quarries, some below ground, by the ruins of small wayside lime-kilns and by small pockets of calcicole plants. The River Usk bisecting this Silurian Inlier enters these rocks at Chain Bridge near Kemeys Commander.

In the extreme south-east corner of the county at St Pierre, Caldicot and Mathern is a small area of New Red Sandstone laid down in the Triassic era. The Jurassic system is represented by small areas of Lias formation found on the south-eastern outskirts of Newport near Llanwern and Bishton.

The Gwent Levels, the flat low-lying area of land which borders the Severn Estuary, are alluvial and are protected from tidal flooding by a sea-wall. The Caldicot Level stretches from the mouth of the Wye to Uskmouth while the Wentlooge Level lies to the west of Newport. Nearly four miles in width at their widest part, the Levels constitute, together with the flood-plains of the Usk and its tributary the Olway, the only really flat areas of land in the county. The Gwent Levels like the Somerset and Pevensey Levels are an extensive area of reclaimed pasture drained by a network of drainage ditches known locally as "reens" and form the largest area of its kind in Wales. The Levels were probably originally reclaimed by the Romans for wheat-growing at the end of the 3rd century, the sea being excluded by earthen and stone banks, though in time these were washed away. However, the Monasteries carried out further reclamation work in the 13th century but, through neglect, these sea defences fell into disrepair and led to the disastrous floods of 1606 when many people drowned.

The sea-walls were eventually repaired but, prior to the 1939-45 war, these coastal flats were subject to considerable seasonal flooding and there was much permanent marsh and fen as the land lay well below high tide level and was drained by the reens emptied by sea-wall sluices opened at low tide. The tidal rise and fall along this shore is as much as 45 ft. a figure which is exceeded only in the Bay of Fundy in Nova Scotia. Since the war the sea-walls have been rebuilt, raised and realigned and the water-table has been drastically lowered by increased drainage from the reen system through deepening of the ditches and, in the case of the Caldicot Level, also by pumping.

There is little left now of this botanically and entomologically-rich marshy terrain and it is represented only by a few remaining reed-beds as at Newport Docks, Uskmouth Power Station, Llanwern Steelworks and by Magor Marsh a sedge-peat bog and marshland area extending over 60 acres (25 ha.), now a Gwent Wildlife Trust Reserve, and the largest remaining fragment of the once extensive estuarine fen. The reens too, with their formerly rich aquatic flora and dependent richly diverse invertebrate fauna, especially of moths and beetles, including very many nationally local and rare species, are now suffering not only from this increased drainage but also by the continued spraying of their banks with herbicides.

The Levels between Magor and Cardiff are now (July 1993) very seriously threatened by a proposed new motorway system to serve the Second Severn Crossing, already under construction, and several sites with existing SSSI status in this sensitive area may well suffer disastrously.

Formerly regarded as a "maritime" county, especially in the days when

Newport was a major coal-shipping port, and boasting a littoral of some twenty-three miles, Monmouthshire sadly lacks sea cliffs, sand-dunes and sandy or shingle beaches its shore consisting of a narrow strip of mud between the sea-wall and the water line. At low tide large expanses of mud-flats are exposed which are of great importance to over-wintering and migratory waders and water-fowl. There are, however, small areas of saltmarsh but these are becoming increasingly invaded by Spartina grass. It is thus evident that many of our coastal species of plants and insects will not be found in Gwent.

However, all this having been said, Monmouthshire small though it is, but with a very varied topography and many contrasting and varying habitats sustains a wealth of plant and insect life and can claim many interesting and scarce species of Lepidoptera.

RAINFALL and CLIMATE

The mean annual rainfall varies considerably in different parts of Monmouthshire the differences depending mainly on altitude. It can be as little as 30" in the eastern district between Abergavenny and Monmouth but in the low central Raglan – Usk – Newport area it is in the region of 40" rising to 50" in the more hilly Wentwood – Trellech district. The wettest part of the county is the mountainous northern and north-western area where an annual rainfall of 60" is recorded.

Lying as it does on the northern shores of the tidal Severn Estuary Monmouthshire enjoys, on the whole, a more equable climate than some of its neighbouring inland counties such as Powys and Herefordshire being milder in general and suffering less extreme winter temperatures.

MONMOUTHSHIRE WOODLANDS and their LEPIDOPTERA

Monmouthshire is among the most wooded of the Welsh counties with about one eighth of its total area being covered by woodland, more than twice the national average. In the Chepstow district the proportion is one sixth. The 1979 – 1982 Forestry Commission survey of woods larger than 0.25 ha. gave a total woodland area of 16,900 ha. or 12.3 %, while if small brakes and copses were included, the figure rose to 20,706 ha. or 15%, of which a half were broad-leaved. The Nature Conservancy Council however estimates that only some 3,249 ha. consist of ancient semi-natural woodland and believes that 67 % of such woodland has been lost to the county since the 1939 – 45 war.

Nowhere is this more apparent than in the Wye Valley and in Wentwood. I have known both these districts and their woods for well over sixty years and during this time the character of these woodlands, together with their flora and insect life, has changed almost beyond recognition. The Forestry Commission took over the management of the Crown Woods in the Wye Valley and much of these former broad-leaved woodlands has been replaced by conifers. In addition, the Hendre Woods near Monmouth suffered severely. When first I knew them they were fine mixed deciduous woods containing also very large numbers of ancient yews. When the Forestry Commission took over, the fine oaks and ash were gradually felled and largely replaced with the inevitable conifers among which were to be seen scores of dead and leafless yews which had been poisoned and allowed to stand where they died. The Trellech ridge and Wentwood also have large conifer plantations replacing much of the former deciduous woodland. Largely coniferised too, has been Redding's Inclosure, part of the Dean Forest within Monmouthshire and noted for its fine oaks and for being a former haunt of the Purple Emperor butterfly.

Apart from the large forested expanse of Wentwood, the central and north-eastern rural areas associated with the marls of the Old Red Sandstone were characterised by well-developed hedges with scattered ash-oak woods but many of these woods have disappeared over the last four decades and many hedges have been uprooted and large hedgerow trees felled. In the wetter parts alder-willow woods were common as were the riverside growths of these trees but even these have not been immune to the activities of the water and river authorities working in the interests of drainage and "flood prevention". Fortunately, a few areas of alder carr and alder woods still remain in the county as in the north at Cwm Coed-y-cerrig which was declared a Local Nature Reserve by the Brecon Beacons National Park in 1977. This is the only district in Monmouthshire where I have found (in 1973) the Devon Carpet moth (*Lampropteryx otregiata* Metc.).

Examples of "ancient" woodland can be seen in several parts of Monmouthshire. Such old native woods are to be found surviving on the Carboniferous Limestone cliffs of the Lower Wye Gorge at Piercefield Park, Wyndcliff and Blackcliff and also again beside the Wye on the limestone in the far north-east corner of the county at the old Hadnock Quarries and Lady Park Wood now a National Nature Reserve where the woods are sessile oak-beech-ash and oak-lime.

In the lower Wye Valley the native woods on the steep stony slopes and limestone cliffs are dominated by beech, ash, sessile oak (*Quercus petraea*), pedunculate oak (*Q. robur*), small-leaved lime (*Tilia cordata*), and Yew, while such trees as birch (*Betula pubescens*), gean (*Prunus avium*), wych elm (*Ulmus glabra*), whitebeam (*Sorbus aria*), wild service tree (*S. torminalis*) and others are also here. These woods support a great diversity of shrubs and plants many of them scarce or locally rare as I well remember from botanising here in my younger days in the late 1920s and early 1930s. It is not surprising that with such a wealth and variety of plant species this region should prove such an ideal area for entomologists and for lepidopterists in particular.

Many other pockets of ancient woodland occur throughout the county as at Mounton, the Cwm near Shirenewton and in the Usk and Ebbw Valleys. Some natural woodlands in the west of the county are dominated by mature beech and are the most westerly self-regenerating native beechwoods in Britain. They are mainly montane occurring up to about 1500 ft. in the region of high rainfall at the heads of the valleys, as in the vicinity of Ebbw Vale, (eg. Cwm Merddog), and in the Blaenavon and Abergavenny districts including Clydach Gorge, the northern and eastern slopes of the Blorenge Mountain and the small valleys descending from the Sugar Loaf towards Abergavenny. Although often planted in Wales beech is only native to the east of the R. Rhymney.

In the past the mountainous ridges between the river valleys of western Gwent had supported woods of sessile oak (*Quercus petraea*) and birch up to about 1,400 ft. but much of this woodland was felled after the First World War and in places replanted with conifers so that only scattered fragments of the sessile oakwood remain and these ridges are now largely bracken-covered and grazed by sheep. Also some commercial conifer planting has taken place. The tops are open moorland with heather and bilberry being dominant in the drier areas while, in wetter parts, purple moor-grass (*Molinia caerulea*) and heath rush (*Juncus squarrosus*) are the domninant species.

When one considers the variety and extent of its woods it is little wonder that Monmouthshire can claim a rich and varied lepidopterous woodland fauna.

Although no longer what it used to be and with the numbers of many of its species of moths and butterflies declining markedly in the preceeding three decades, Wentwood, the large forested district in the central region of southern Monmouthshire, could still present in 1974 an impressive range of macrolepidoptera. My Wentwood list recorded over the previous ten years, contained 33 species of butterflies and 300 species of the larger moths. In each case these numbers approximated to one third of the numbers of all such species on the British list.

Butterflies included the Grizzled Skipper, Purple Hairstreak, White-letter Hairstreak, Small Pearl-bordered Fritillary, Pearl-bordered Fritillary and Marbled White.

The moths, to name but a few, included the Barred Hook-tip (*Drepana cultraria* Fabr.), Satin Lutestring (*Tetheella fluctuosa* Hb.), Small Waved Umber (*Horisme vitalbata* D.& S.), Little Thorn (*Cepphis advenaria* Hb.), Speckled Yellow (*Pseudopanthera macularia* Linn.), Lilac Beauty (*Apeira syringaria* Linn.), Satin Beauty (*Deileptenia ribeata* Cl.), Privet Hawk-moth (*Sphinx ligustri* Linn.), Lobster Moth (*Stauropus fagi* Linn), Scarce Prominent (*Odontosia carmelita* Esp.), Four-dotted Footman (*Cybosia mesomella* Linn.), Clouded Buff (*Diacrisia sannio* Linn.), White-marked (*Cerastis leucographa* D.& S.), Light Brocade (*Lacanobia w-latinum* Hufn.), Sprawler (*Brachionycha sphinx* Hufn.), Dusky-lemon Sallow (*Xanthia gilvago* D.& S.), Alder Moth (*Acronicta alni* Linn.), Slender Brindle (*Apamea scolopacina* Esp.) and Beautiful Snout (*Hypena crassalis* Fabr).

The semi-natural ancient woods of eastern Monmouthshire and the Wye

Valley also support a very large number of species and harbour many locally scarce and rare moths including the Oak Lutestring (*Cymatophorima diluta* D.& S.), The Mocha (*Cyclophora annulata* Schulze), Ruddy Carpet (*Catarhoe rubidata* D.& S.), Marbled Pug (*Eupithecia irriguata* Hb.), Blomer's Rivulet (*Discoloxia blomeri* Curt.), Waved Carpet (*Hydrelia sylvata* D.& S.), Peacock Moth (*Semiothisa notata* Linn.), Great Oak Beauty (*Boarmia roboraria* D.& S.), Brindled White-spot (*Ectropis extersaria* Hb.), Alder Kitten (*Furcula bicuspis* Borkh.), Great Prominent (*Peridea anceps* Goeze), Black Arches (*Lymantria monacha* Linn.), Cream-spot Tiger (*Arctia villica* Linn.), The Coronet (*Craniophora ligustri* D.& S.), Scarce Silver-lines (*Bena prasinana* Linn.) and others.

In the deep woods of the Wye Valley no less than ten species of Lithosiinae (Footman moths) are to be found including the Rosy Footman (*Miltochrista miniata* Forst.), Muslin Footman (*Nudaria mundana* Linn.), Red-necked Footman (*Atolmis rubricollis* Linn.), Orange Footman (*Eilema sororcula* Hufn.) and, surprisingly, the Hoary Footman (*F caniola* Hb).

One locality in particular supports three or four of our national rarities including the Barred Carpet (*Perizoma taeniatum* Steph.), and Fletcher's Pug (*Eupithecia egenaria* H.-S.). This Pug was discovered here in 1962 but elsewhere in Britain it occurs only at one very limited site in Norfolk. The Scarce Hook-tip (*Sabre harpagula* Esp.), a rare species previously known from the Avon Gorge and thought to be extinct, was rediscovered here in the lower Wye valley in 1961. Here too is the only known British station for the pyralid moth *Salebriopsis albicilla* H.-S. a member of the Phycitinae first found here in 1964.

Butterflies of the Wye Valley woodlands include the Wood White and the White Admiral. During the 1950's and early 1960's at a wood in the Angidy Valley six species of Fritillary were to be found including the High Brown, Dark Green and the Marsh Fritillary but sadly now, in the early 1990s, only the Silver-washed Fritillary remains and that in much reduced numbers.

The smaller moths (microlepidoptera) are well-represented in these old woodlands and include nationally scarce and rare species such as *Oecophora bractella* Linn., *Celypha rurestrana* Dup., and *Eucosmomorpha albersana* Hb. as well as *Salebriopsis albicilla*.

BOGS and HEATHLAND

Although an area of sub-montane heath and calcareous grassland is found on the Blorenge, Monmouthshire now has little remaining heath but until about sixty years ago much semi-heath was still to be seen in the hills to the west of the Wye in the Trellech and Whitelye districts but this has largely gone. In Wentwood also there were areas of semi-heath as at Bica Common and elsewhere. However, these too have gradually disappeared, mainly through forestry activities.

One area of true heath I remember well, was at Earlswood Common, where seventy years ago we were taken by our parents, at the appropriate season, to gather "whinberries". During the 1939-45 war this heath was ploughed up to aid the national agricultural effort and later the area was excavated to extract "gravel" during the construction of the Llanwern Steelworks. The site is now covered in part by conifers but mostly by a dismal uniform green sward.

In the west of the county near Pontllanfraith is an interesting stretch of boggy heath which supports many species of scarce and rare plants and is the haunt of the Green Hairstreak and Marsh Fritillary butterflies and of locally scarce moths including the Lead-coloured Pug (*Eupithecia plumbeolata* Haw.), Chimney Sweeper (*Odezia atrata* Linn.), Beautiful Brocade (*Lacanobia contigua* D.& S.), Striped Wainscot (*Mythimna pudorina* D.& S.), Silver Hook (*Eustrotia uncula* Cl.) and the attractive Pyralid moth *Anania funebris* Ström.. Sadly within the last few years however much of this habitat has been lost to housing development.

Just as heathland is scarce in the county so also can Monmouthshire boast of little in the way of bogs. In the upland moors of the west and north, as on the Coity Mountain, boggy patches are found but true peat bogs are small and occur only in the headwaters of the Sirhowy and Ebbw rivers.

There are, however, several lowland peat-bogs but these are all small and have deteriorated markedly over the last sixty years or so. The largest of these is Trellech Bog which is typical Molinietum having a total area of some 17 ha. which includes an area of willow carr. It occupies the site of a glacial lake and has a rich flora including a number of species rare in this county. Unfortunately the bog is rapidly drying out and is being invaded by bracken and gorse and there are many young saplings of willow (*Salix atrocinerea* and *S. aurita*) and rowan. Some years ago it was declared a Local Nature Reserve and SSSI and is now referred to as "Cleddon Bog". It supports a number of locally scarce and rare species of moths, both macrolepidoptera and microlepidoptera. The former include the Welsh Wave (*Venusia cambrica* Curt.), Barred Chestnut (*Diarsia dahlii* Hb.), Neglected Rustic (*Xestia castanea* Esp.), Striped Wainscot (*Mythimna pudorina* D.& S.), the Suspected (*Parastichtis suspecta* Hb.) and the Marsh Oblique-barred (*Hypenodes turfosalis* Wocke), whilst among the microlepidoptera are to be found *Oecophora bractella* Linn., *Ancylis geminana* Don., *Ancylis subarcuana* Dougl. and *Rhopobota myrtillana* Humph. & Westw..

A smaller bog with a similar rich flora is found near by at Whitelye and a third lowland bog with a good wetland flora is to be seen at Llwyn-y-celyn near Shirenewton.

A small botanically-rich bog in the vicinity of Usk is the only site in the county where the Marsh Grass-veneer (*Crambus uliginosellus* Zell.) is to be found.

LEPIDOPTERA of the GWENT LEVELS and the SEVERN LITTORAL

When considering the moths and butterflies of this region it is clear that the insect fauna of Magor Marsh, the last remnant of the formerly extensive estuarine fen, should be dealt with first.

My first list of its lepidoptera recorded during numerous visits both by day and night in the years 1968 to 1971 and published in the *Gwent Naturalists' Trust Newsreel* of Spring 1972 comprised 154 species of macrolepidoptera including 13 butterflies. In addition there were 12 species of microlepidoptera.

Subsequently, by 1979, the list had increased to 190 "macros" and 81 "micros" which included 33 species of Pyralids and Plumes and 34 species of Tortricoid moths. As Magor Marsh lies on the northern edge of the Gwent Levels, it is obvious that many of these moths were either common ubiquitous species or vagrants from a variety of other non-marshy habitats in the general neighbourhood.

However some 34 typical marsh and fen species were recorded and it is of interest to note that the distribution of a number of these is mainly in south-eastern England or the fens of East Anglia while others are more widely distributed but of local occurrence.

Among the more notable Magor moths can be listed the Oblique Carpet (*Orthonama vittata* Borkh.), Small Seraphim (*Pterapherapteryx sexalata* Retz.), Round-winged Muslin (*Thumatha senex* Hb.), Dingy Footman (*Eilema griseola* Hb.), Water Ermine (*Spilosoma urticae* Esp.), Scarlet Tiger (*Callimorpha dominula* Linn.), Southern Wainscot (*Mythymna straminea* Treit.), Obscure Wainscot (*Mythymna obsoleta* Hb.), Crescent Striped (*Apamea oblonga* Haw.), Double Lobed (*Apamea ophiogramma* Esp.), The Crescent (*Celaena leucostigma* Hb.), Twin-spotted Wainscot (*Archanara geminipuncta* Haw.), and the Tortricoid moths *Piercea alismana* Rag. and *Hedya salicella* Linn.

Also associated with the drainage reens and the few remaining wet marshy areas of the Gwent Levels are a number of moths which are either absent or scarce elsewhere in the county. They include species such as the Powdered Quaker (*Orthosia gracilis* D.& S.), Bulrush Wainscot (*Nonagria typhae* Thunb.), The Brown-veined Wainscot (*Archanara dissoluta* Treit.), Large Wainscot (*Rhizedra lutosa* Hb.), Small Rufous (Coenobia rufa Haw.), Gold Spot (*Plusia festucae* Linn.) and the following Pyralid moths *Donacaula forficella* Thunb., The Water Veneer (*Acentria nivea* Ol.), Brown China-mark (*Nymphula nympheata* Linn.), Ringed China-mark (*Parapoynx stratiotata* Linn.) and the Small China-mark (*Cataclysta lemnata* Linn.).

There is little of note to be said about the butterflies of the Gwent Levels and the Severn littoral except that, historically, during the early 1870's, the Black-veined White (*Aporia crataegi* Linn.), now extinct in Britain, swarmed here from the vicinity of Llanwern westwards to Cardiff.

However nowadays the Common Blue, which in the county at large has greatly diminished in numbers over the last sixty years, is usually very plentiful on the sea-wall as are also the Small Copper and Small Heath. The Gatekeeper too and the Marbled White are also well-established along parts of the sea-wall and in "good" Clouded Yellow years this migrant is frequently seen while the Grayling, locally common in north-west Monmouthshire, occurs in small numbers.

As indicated earlier, the absence of sea cliffs, shingle beaches and sand dunes on the Severn Estuary shore deprive Monmouthshire of many of our coastal species of moths. However, among the salt-marsh and coastal species that do occur are the Tortricoid moth *Bactra robustana* Christ., the Saltern Ear (*Amphipoea fucosa paludis* Tutt) and Dog's Tooth (*Lacanobia suasa* D.& S.) while the Nutmeg (*Dicestra trifolii* Hufn.), generally scarce in Monmouthshire, is found most frequently along the Severn littoral.

The Dusky Sallow (*Eremobia ochroleuca* D.& S.), first recorded in this county in 1984, had by 1989 become well-establihed along some parts of the sea-wall. Another coastal species discovered here recently (1990) is the Bordered Sallow (*Pyrrhia umbra* Hufn.).

LEPIDOPTERA of the NORTHERN HILLS and MOORLAND

Twenty years ago little was known about the lepidoptera of the hills and moorland of northern Gwent other than from Rait-Smith's records which date from before the First World War. Unlike the more salubrious Wye Valley this depressed mining and industrial region failed to attract the visiting lepidopterist and entomologically the area was virtually unknown territory. However since the discovery of Eriopygodes imbecilla in 1972 and publication of the event in the entomological literature there has been a much greater awareness among lepidopterists of the district's possibilities even if the Silurian is still the main attraction.

Among the commoner moorland moths, as one would expect, are the day-flying Northern Eggar (*Lasiocampa quercus* callunae Palm.), Fox Moth (*Macrothylacia rubi* Linn.) and the Emperor Moth (*Saturnia pavonia* (Linn) while the night-flying moths include the Map-winged Swift (*Hepialus fusconebulosa* DeG.), Northern Spinach (*Eulithis populata* Linn.), July Highflyer (*Hydriomena furcata* Thunb.), Twin-spot Carpet (*Perizoma didymata* Linn.), Ling Pug (*Eupithecia goosensiata* Mab.), Glaucous Shears (*Papestra biren* Goeze) and the abundant True Lover's Knot (*Lycophotia porphyrea* D.& S.).

The scarcer moorland species include the Galium Carpet (*Epirrhoe galiata* D.& S.), Striped Twin-spot Carpet (*Nebula salicata latentaria* Curt.), Wood Tiger (*Parasemia plantaginis* Linn.), The Confused (*Apamea furva* D.& S.), and the Anomalous (*Stilbia anomala* Haw.). The Smoky Wave (*Scopula ternata* Schr.) is found and is not too uncommon while the Beautiful Yellow Underwing (*Anarta myrtilli* Linn.) occurs very sparingly.

Among the microlepidoptera the Tortricoid moth *Ancylis myrtillana* (Treit.) a moorland species common in northern Britain including North Wales but, less frequent in South Wales, is found on the Blorenge and and the Sugar Loaf. *Philedone gerningana* D.& S. a mountain heathland moth has also been recorded. The Pyralid moth *Eudonia murana* Curt. a local moorland and mountain moth is found in association with rocks and old stone walls.

Quarries in the area shelter several species such as the Northern Rustic (*Standfussiana lucernea* Linn.) and Light Feathered Rustic (*Agrotis cinerea* D.& S.). The Thyme Pug (*Eupithecia distinctaria* H.-S.) and the Pyralid moth *Pyrausta cingulata* Linn. have been noted in one quarry on the Carboniferous Limestone.

Several species of moths found in the hills of north Monmouthshire are here at the southern limit of their range in this country and include the Grey Mountain Carpet (*Entephria caesiata* D.& S.), Light Knot Grass (*Acronicta menyanthidis* Esp.) and the Scarce Silver Y (*Syngrapha interrogationis* Linn.).

Another moorland "northener" the Small Autumnal Moth (*Epirrita filigrammaria* H.-S.) was known to occur as far south as Leicestershire, mid-Wales and Cannock Chase. Michael Harper informed me that he had found it in the Black Mountains in Herefordshire. Therefore it seemed fairly reasonable to assume that it would occur in north Monmouthshire, so in 1986, Martin Anthoney and I set out to search for it. On Sept 15 several were located but during the ensuing week we found that in some localities it was literally "abundant" and currently this is probably the southernmost station for the species in this country.

Although twenty-one years have now elapsed since the discovery of *Eriopygodes imbecilla* Fabr. (the Silurian) in the hills of north-west Monmouthshire it has not yet been found in any other locality so it still remains one of the county's rarest insects.

MIGRANTS

Monmouthshire, due to its geographical situation on the northern shores of the Severn Estuary and Bristol Channel, is not a good place to observe migrant species from the Continent. While the commoner and more numerous butterfly and moth immigrants reach Monmouthshire as they spread through the country from the south the scarcer, numerically fewer, species tend to be "creamed off" by the southern and south-western English counties before reaching this second watery barrier so that few of them arrive here in this county.

The commoner migrant butterflies usually plentiful here include the Painted Lady, and Red Admiral but the Clouded Yellow often fails to turn up even in some of the years when it is fairly plentiful in southern England. Two rare immigrants, the Camberwell Beauty and the Queen of Spain Fritillary, have occurred in Monmouthshire.

The commoner migrant moths include the Dark Sword-grass (*Agrotis ipsilon* Hufn.), Pearly Underwing (*Peridroma saucia* Hb.), the often abundant Silver Y (*Autographa gamma* Linn.), the Small Mottled Willow (*Spodoptera exigua* Hb.) a regular visitor, the Gem (*Orthonama obstipata* Fabr.) and the Vestal (*Rhodometra sacraria* Linn.) a fairly frequent arrival.

A rare visitor from the Mediterranean region the Crimson Speckled (*Utetheisa pulchella* Linn.) has occurred once, in 1871. The White-speck (*Mythimna unipuncta* Haw.) another migrant from the same region was recorded in 1978.

Among the migrant Hawk-moths the Death's-head Hawk-moth (*Acherontia atropos* Linn.) and the Humming-bird Hawk-moth (*Macroglossum stellatarum* Linn.) are not infrequent visitors while the more uncommon migrant species viz. the Convolvulus Hawk-moth (*Agrius convolvuli* Linn.), the Bedstraw Hawk-moth (*Hyles gallii* Rott.) and the Striped Hawk-moth (*Hyles lineata livornica* Esp.) have been recorded.

Common migrant microlepidoptera include the Diamond-back Moth (*Plutella xylostella* Linn.) and the Pyralids *Udea ferrugalis* Hb. and Rush Veneer (*Nomophila noctuella* D.& S.). Several scarce or rare Pyrales have been recorded including *Palpita unionalis* Hb. six of which came to my trap in October 1977. *Uresiphita polygonalis* D.& S. turned up at Usk in 1969. This was the first Welsh record and one of only about a dozen that had been recorded in Britain. In 1978 *Euchromius ocellea* Haw. came to light, this being the 2nd Welsh and the 24th British record. Another Pyrale *Udea fulvalis* Hb. was recorded at Usk in 1971 and is believed to have become temporarily established in the district in the late 1960s.

VAGRANTS

Some insects are neither resident nor immigrant. These are the "wanderers", those which have flown in from other parts of the country. It is not always possible to be sure of their source of origin or even that they are not a scarce and unexpected resident species and such probable vagrants are listed below.

Two butterflies the Chalk Hill Blue (*Lysandra coridon* Poda) and the Large Tortoiseshell (*Nymphalis polychloros* Linn.) are certain vagrants here and the two examples of the former no doubt originated either in the Cotswolds (possibly Wotton-under-Edge) or in Somerset (Brean Down).

The vagrant moths include the Least Carpet (*Idaea vulpinaria atrosignaria* Lempke) recorded in 1991 this being the first Welsh record. The Royal Mantle (*Catarhoe cuculata* Hufn.), and White-banded Carpet (*Spargania luctuata* D.& S.) have also been seen. The Square-spot Dart (*Euxoa obelisca grisea* Tutt) represented by a single moth taken at Usk in 1974 probably came from the coastal cliffs of north Devon or south-west Wales. The Stout Dart (*Spaelotis ravida* D.& S.) has been turned up twice by Martin Anthoney in recent years and this species may have arrived from the

Midlands or the Thames Valley. On the other hand one wonders whether it has now become established here in Gwent.

The Great Brocade (*Eurois occulta* Linn.) has twice been recorded in the county and is probably a vagrant from the North. Robert Humphreys in 1969 took a fine specimen of the Double Line (*Mythimna turca* Linn.) at Usk Castle. This is the only Monmouthshire record and probably this specimen is a vagrant from Central Wales or Mid Glamorgan where it is known to occur, but again, one wonders if it is now a Monmouthshire resident.

Finally in 1983 my garden m.v.l. trap at Plas Newydd in Usk produced a fine specimen of the Cream-bordered Green Pea (*Earias clorana* Linn.), new to Monmouthshire and Wales and almost certainly a vagrant.

SPECIES NOW EXTINCT

Two species of butterflies found here during the last century have become extinct as they have throughout the British Isles viz. the Black-veined White (*Aporia crataegi* Linn.) and the Mazarine Blue (*Cyaniris semiargus* Rott.) while the Purple Emperor (*Apatura iris* Linn.) is probably now extinct in Monmouthshire but still occurs elsewhere in the country. The Brown Hairstreak (*Thecla betulae* Linn.) is another butterfly no longer found here.

Among the more notable moths which are known to have occurred but which have now disappeared are the Kentish Glory (*Endromis versicolora* Linn.) last recorded in the 1890s and the Conformist (*Lithophane furcifera suffusa* Tutt) not seen now for some thirty years or more.

LEPIDOPTERA RECORDING in MONMOUTHSHIRE

Resident entomologists have always been few in Monmouthshire and they still remain so.

Charles Conway (c. 1797-1860) owner of a tin-plate works and a keen naturalist lived at Pontrhydyryn and is best remembered as a botanist rather than an entomologist and the herbarium he formed in the 1830s is now in the Welsh National Herbarium. In 1833 he published a list of the butterflies found in the district which now embraces the new post-war town of Cwmbran in Monmouthshire's Eastern Valley. This was the first local list of the county's lepidoptera to be produced.

Edward Newman's *The Natural History of British Butterflies 1871* contains a number of Monmouthshire records supplied to him by George Lock one of his correspondents who, it would seem, lived in or near Newport. I have been unable to obtain any information about George Lock. Although his Monmouthshire records are of great interest unfortunately one or two of them are rather suspect especially that of the Black Hairstreak in St. Julian's Wood. Neither has it always been possible to recognise with certainty the localities he named but this is no fault of his but is due to the change in local place names.

W. Rait-Smith, a Kentish man and noted lepidopterist, was a mining surveyor who between 1905 and 1914 worked at Abertillery in north-west Monmouthshire. For a time he lived there and published many local moth and butterfly records in the *Ent. Rec. J. Var.*.

In the early 1930s the late G.F. Crowther arrived from Cheshire, living first at Bettws Newydd near Usk, and latterly in Monmouth until his death shortly after the last war. Some of his records appear in the literature but his son informed me that sadly, as far as he was aware, his father left no written records of his entomological activities.

My good friend the late Lieut.-Col. R.B. Humphreys (1914-1989), who as a boy had collected with Crowther, lived at Usk in his youth and throughout his life frequently returned here to visit his family and "run his moth-trap" at Usk Castle. He had a number of new Monmouthshire records to his credit including the Chocolate Tip and Double Line. He kindly permitted me to extract notes from his diary. Since the last war we often hunted moths and butterflies together from Cornwall to Cumbria and Northumberland but mostly here in Monmouthshire.

Colin Titcombe of Caerwent in the south-east of the county, a most observant and knowledgeable local naturalist, has made many interesting records of lepidoptera and other insects. A number of his records are included in this work.

A recent and welcome addition to our resident Gwent lepidopterists is Dr Martin Anthoney who some years ago moved here from Essex where he had made several notable discoveries of rare moths. Residing in Risca since 1976 he has greatly extended our knowledge of the macrolepidoptera in the west of the county and has added a number of species to the county list. He is a keen conservationist and non-collector and is currently vice-chairman of the Gwent Wildlife Trust.

Over the years, the county has had many visiting lepidopterists, both amateur and professional and they have far outnumbered the resident entomologists. Since the middle of the last century there have been many such visitors but until recent years, almost without exception, they have concentrated their efforts on the Wye Valley which is the most scenically attractive and debateably the most entomologically rewarding part of Monmouthshire. However for the lepidopterist, as this book will show, there is far more to Monmouthshire than the Wye Valley.

In the second half of the last century and in the earlier years of this many moth-hunters came to the Wye Valley and their records are to be found scattered through the contemporary zoological and entomological journals.

Perhaps the earliest such arrival was Parry who in 1839 recorded the Lappet Moth (*Gastropacha quercifolia*) from Monmouth town.. Palmer in 1890 gave records from Monmouth itself and also from Cwmcarvan in the nearby hills. In the same year Patten recorded insects on the banks of the Wye. In 1904 Thornewill reported that the Striped Hawk-moth (*Hyles lineata livornica*) had been captured in Monmouth.

Among the many visiting lepidopterists who came to the lower part of the Wye Valley were Piffard who in 1859 worked the Tintern area and Goss who noted the Scarlet Tiger (*Callimorpha dominula*) at Tintern in 1867 and again in 1890.

A.H. Jones gave Tintern records for 1876 and the following year Ince worked the Tintern district but also moved westwards into the neighbouring region referring in his records to "Monmouthshire in every part" and visiting Wentwood and the area between Chepstow and Usk where he recorded the High Brown Fritillary.

Nesbitt visited the Llandogo district in 1892 and 1893 while Bird worked the Tintern area in 1905, 1906, 1907 and again in 1912.

Barraud also visited the Wye Valley in 1906 and again in 1907 when he recorded the Conformist (*Lithophane furcifera suffusa*) on sallow at Bigsweir but on the Gloucestershire bank of the Wye.

Since the 1939-45 war many professional and notable amateur field lepidopterists have worked the lower Wye Valley and include such well-known names as R.A. Jackson, Blathwayt, J.M. Chalmers-Hunt, Charles de Worms, J.L. Messenger, Austin Richardson, G.M. Haggett, R.M. Mere, J.D. Bradley, D.S. Fletcher, R.W.J. Uffen, R.E.M. Pilcher, M.J. Leech, E.C. Pelham-Clinton, A. Maitland Emmet, M.W. Harper, J.R. Langmaid, D.J.L. Agassiz, Bernard Skinner and numerous others.

The discovery in this area in the 1960s of *Salebriopsis albicilla* and Fletcher's Pug (*Eupithecia egeneria*) together with the rediscovery of the Scarce Hook-tip (*Sabre harpagula*) which was thought to be extinct in Britain has led to a large increase in the numbers of lepidopterists visiting the area in recent years.

In conclusion it should be stressed that the foregoing survey concentrates on the entomologically richer and more interesting regions of the county. Even so, some of the less promising areas such as the agricultural districts in the north-east and the largely built-up industrial parts of the west can provide surprises. Examples are the presence of the High Brown Fritillary in the western valleys and the appearance of the Dark Green Fritillary and the Grayling (in large numbers) noted by Martin Anthoney on old slag heaps near Ebbw Vale.

The Systematic List and Records

For the sake of convenience the species list with its accompanying records is presented in several arbitrary sections. There are two main parts, viz. Part One, the Butterflies (Rhopalocera) and Part Two, the Moths (Heterocera) which is sub-divided into Section One, the Larger Moths (macrolepidoptera) and Section Two, the Smaller Moths (microlepidoptera). Within the limits of each section the individual species are numbered as in Bradley and Fletcher's *Recorder's Log Book of British Butterflies and Moths* 1979 and are listed in the currently accepted order of the National List as in their more recent *Indexed List of British Butterflies and Moths* 1986.

Records of species are so arranged that the older ones, gleaned from the literature, precede the more recent ones. The observers, to whom the recent records are attributable, are identified by their initials as listed on a previous page. Records not otherwise attributed are the author's and these represent only a very small proportion of the many thousands accumulated over the years. Usually the date of the first available sighting is given, followed by several others noted at intervals over the ensuing twenty-five years or so. In the case of nocturnal species the date of appearance given is that of the beginning of the night on which insects were seen in the field or caught at light. Wherever possible, records of an individual species are noted from a number of different stations within the county and all sites named are shown in the appendix which gives their National Grid References. In the case of some rare or scarce species all known records are included and where it is a new county record this is usually indicated thus, "1st v-c. rec.". Where records are believed to be new to Wales this is also noted.

Since his arrival from Essex in 1976 Dr. M.E. Anthoney, a keen lepidopterist and conservationist, has proved a very assiduous recorder of macrolepidoptera especially in the western half of Gwent. He kindly placed his numerous records at my disposal and it has usually been found expedient to group together near the end of the accounts of individual species those records attributed to him.

All information regarding habitat, distribution and status, duration of flight period, earliest dates (Ed.), latest dates (Ld.) etc. is based solely on Monmouthshire observations. Similarly, larval food-plants have only been named when they have actually been noted as such within this county.

Following the accounts of individual species are listed the 10 Km. Squares in which the insect is known to have occurred in Momouthshire and these squares are delineated in the foregoing county map of Gwent.

Altogether 1,164 species of Lepidoptera are listed comprising 588 species of macrolepidoptera and 576 microlepidoptera.

Systematic List

Part One
Rhopalocera (Butterflies)

HESPERIIDAE

HESPERIINAE

1526 ***Thymelicus sylvestris*** Poda   Small Skipper

Widespread and common, often abundant, flying in July and August and frequenting waysides, rough grassy places and woodland rides and clearings.

Ed. 9/7/68 (Wentwood), 3/7/87 (Pontllanfraith). Ld. 31/8/58 (Pont-y-saeson), 27/8/79 (Hendre Wds).

Pontnewydd, "common" (Conway 1833). Monmouth (Palmer 1890). Wye Valley 1906 (Bird 1906a). Llanock Wood, Crumlin 1906 and 1911 (Rait-Smith 1906, 1912). Abertillery dist. 1906 and 1912 (Rait-Smith 1906, 1913).

Pont-y-saeson 28/7/53 (abdt.) and subsequently noted most years until 28/7/81. Usk 19/7/64, 11/8/81 etc. Tredean Wds 31/7/65, 28/7/76. Wentwood 9/7/68, 19/7/68 (abdt.). Runston 23/7/68. St. Pierre's Great Wds 23/7/68 (abdt.), 12/7/89. Prescoed 12/7/69. The Glyn, Itton 13/7/70. Hendre Wds 2/8/70, 28/7/81. Trelleck Bog 11/8/70. Craig-y-dorth 1/8/70 (abdt.). Redding's Inclosure 13/7/71 (plfl.). Hadnock Quarries 20/7/71. Slade Wds 21/7/71. Great Barnet's Wds, Mounton 27/7/76. Llwyn-y-celyn Bog 27/7/76. Coed-y-Fferm, Llangybi 28/7/76. Coed-y-paen (The Forest) 29/7/76. Ysguborwen, Llantrisant 10/8/77. Llangeview Hill 28/7/79. Uskmouth 15/7/88. Fedw Fach, Trelleck 13/7/89. Magor Pill 18/7/89. (MEA) : Pontllanfraith 3/7/87, Dixton Bank 21/7/87, Fleur-de-lis 15/8/87.

Sq. **SO** 20,30,40,50,41,51. **ST** 38,48,19,29,39,49,59.

1531 ***Ochlodes venata*** Brem. & Grey   Large Skipper
   ssp. ***faunus*** Turati

Widespread and common, frequenting rough grassy banks and woodland margins, rides and clearings, flying from late May until the end of July. Ed. 31/5/42 (Llanfrechfa), 29/5/82 (Hendre Wds). Ld. 31/7/65 (Pont-y-saeson), 4/8/81 (Hendre Wds).

Pontnewydd, "common" (Conway 1833) Monmouth (Palmer 1890). Wye Valley 1906 (Bird 1906a). Abertillery dist. 1906 and 1912 (Rait-Smith 1906, 1913).

Llanfrechfa 31/5/42 (GANH). Pont-y-saeson 5/6/56 and noted annually until 1980, often in abundance. Usk 19/6/59. Tredean Wds 20/6/65. Wentwood, June 1968 (JMD). Bica Common 20/6/68. Trelleck Bog 21/6/68. Coed-y-paen 19/7/68. St Pierre's Great Wds 23/7/68. Runston 23/7/68. Llangeview Hill 28/6/70. Cicelyford 6/7/70. Hendre Wds 2/8/70. Slade Wds 6/7/71. Redding's Inclosure 13/7/71 (abdt.). Piercefield Park 30/6/73. Mynyddislwyn 21/6/74. Magor Marsh 10/6/75. Prescoed 4/8/75. Cwm Tyleri 29/6/76. Llanllowell 5/7/87. Pontllanfraith 20/6/87. Dixton Bank 18/6/88. Fforest Coal Pit 24/6/88. Fedw Fach (Trelleck) 14/6/88. Goetre

(Wern Fawr) 14/6/88. Gaer Hill (Newport) 16/6/88. Uskmouth 15/7/88. (MEA) : Abercarn 14/7/85, Ochrwyth 20/6/87, Cwmfelinfach 20/6/87, Fleur-de-lis 6/7/87.

Sq. **SO** 20,30,40,50,41,51,22. **ST** 28,38,48,19,29,39,49,59.

PYRGINAE

1532 *Erynnis tages* Linn.   Dingy Skipper
ssp. *tages* Linn.

Flies during May and June in woodland rides and clearings, rough fields and on grassy embankments. Still not uncommon in Monmouthshire but mainly in the east of the county and in declining numbers.

Pontnewydd, "common" (Conway 1833). Monmouth (Palmer 1890). Llandogo (Nesbitt 1893). Wye Valley (Barraud 1906). Wye Valley 1906 (Bird 1906a). Abertillery dist. 1906, 1912 (Rait-Smith 1906, 1913).

Croesyceiliog (The Plantations) 1927 and 1928 (GANH). Tredean Wds, May and June 1951 (frqt.), 3/5/68 (1). Pont-y-saeson, 15/5/56 (plfl.), and noted most years until 1981, 1/6 (1). Caerwent (Brockwells) 7/6/64 (CT). Wentwood, June 1968 (JMD), 9/6/76 (CT). Cicelyford 9/6/69, 2/6/70 (GANH). Dinham 6/5/74 (CT). Hendre Wds 6/6/78. Fedw Fach (Trelleck) 14/5/88 (4), 27/5/88 (20), 12/6/89.

Sq. **SO** 20,40,50,41,51. **ST** 48,29,39,49,59.

1534 *Pyrgus malvae* Linn.   Grizzled Skipper

This species flies during May and June in the rides and clearings of open woods and in rough grassy hollows. In Monmouthshire apparently it has never been common and is certainly less frequent now than it was some thirty years ago. It was unrecorded between 1971 and 1987 when it was found on unimproved limestone pasture within a M.o.D establishment at a site not subject to modern agricultural methods and where a large rabbit population helps to keep the herbage short. The form ab. *taras* Bergsträsser has been recorded in the county.

Pontnewydd, "one only" (Conway 1833). Monmouth (Palmer 1890). Llandogo (Nesbitt 1893). Wye Valley (Barraud 1906). Tintern 1906 (Bird 1906a). Wye Valley inc. ab. *taras* (Bird 1906a). Llanhilleth 1906 (Rait-Smith 1906). Pontllanfraith, one, 1909 (Rait-Smith 1912).

Croesyceiliog (The Plantations) 1927 and 1928 (GANH). Tredean Wds, (frqt.), June 1951 (GANH). Pont-y-saeson 1953; 1956 (plfl.), 15/5, 5/6, 13/6, 27/6; 1958 8/6; 1963 10/6. Bica Common 1968 20/6 (GANH). Slade Wds 7/5/71 (CT pers. comm.). Dinham 22/5/87 (3) (MEA) – the first county record for sixteen years; 15/6/88 (CT).

Sq. **SO** 20,50,51. **ST** 48,19,29,39,49,59.

PIERIDAE
DISMORPHIINAE

1541 ***Leptidea sinapis*** Linn.   Wood White
ssp. ***sinapis*** Linn.

This species, frequenting woodland borders, rides and clearings in May and June, has apparently always been scarce and local in Monmouthshire where it is at the western limits of its present range in this country to the north of the Severn Estuary. However, it occurs commonly near Monmouth about half a Km. inside the Herefordshire border. I have known it at this site for thirty years or so but it was not until 1981 that I saw my first Wood White at a locality actually within Monmouthshire.

Monmouthshire, "Scarce and only taken in St Julian's Wood" (George Lock in Newman 1871). Monmouth dist. "local but plentiful in some woods in June" (S.G. Charles c. 1937).

Trelleck Bog 1965 (PNH). Hendre Wds 1981 7/7 (1) (GANH), 1982 29/5 (2) (JDB, GANH). Dixton Bank 1987 (PS)

Sq. **SO** 50,41,51. **ST** 38.

COLIADINAE

1543 ***Colias hyale*** Linn.   Pale Clouded Yellow
or 1544 C. ***australis*** Ver.   Berger's Clouded Yellow

Usk (Plas Newydd garden) 30/7/83, a single specimen of one of these scarce continental migrants. It eluded capture so its determination is in doubt. However it was definitely not *C. colias* f. *helice*.

Sq. **SO** 30.

1545 ***Colias croceus*** Geoffr.   Clouded Yellow

A sporadic immigrant from the Mediterranean area with numbers varying greatly from year to year. In Monmouthshire, its numbers are small, or it is absent, even in years when plentiful in the southern English counties. However, it was extremely abundant in the 1947 invasion and fairly common in 1983. The butterfly is usually seen from July to September or October Only once have I noted it earlier, when in the Monmouth district on 16/6/79, I saw one in the late afternoon sunshine headed north in direct and rapid flight. The pale female variant f. *helice* Hübn. has been recorded.

Pontnewydd, "a pair taken some years ago by a neighbour" (Conway 1833). Near Tintern (Piffard 1859). Wye Valley (Nesbitt 1892). Bettws Newydd (Crowther 1933). Monmouth dist., "rather erratic in its appearance" (S.G. Charles c. 1937).

1941: Llanfrechfa 21/8 (GANH). 1947: abundant in Sept. and Oct. eg. 5/10/47 "swarmed" at Llantarnam and Llanfrechfa inc. f. *helice* (GANH). 1955: Usk dist., several 8/8 – 22/8; Llantrisant 8/8; Llangwm 12/10. 1964:

Usk 13/7 (1). 1967: Wentwood 8/9 (1), Usk 21/9 (1). 1970: Llangybi 28/9 (1). 1979: Wonastow 16/6 (1). 1983 a "good year" for this species in Gwent: Usk, frequent 30/7 – 18/8; Llangybi 4/8; Glascoed 10/8 (2); Llangovan 8/8 (1); Slade Wds, 13/8 (plfl.); Gwernesney 17/8 (1); Llanllowell 24/8. 1987: Fleur-de-lis 6/7 (1) (MEA). 1988: Llansoy 2/8 (1). 1992: Ebbw Vale 31/7 (6) (MEA). Llansoy (♀ 1) 7/8, Monmouth 24/8 (♀ 1) (GANH).

Sq. **SO** 10,30,40,50,41,51. **ST** 48,19,39,49.

1546   *Gonepteryx rhamni* Linn.   The Brimstone
          ssp. *rhamni* Linn.

A common and widespread resident frequenting waysides, woodland rides and open woods where buckthorns (*Rhamnus and Frangula*) grow. This single-brooded butterfly flies from July to September and, following hibernation, reappears in early spring. Formerly very plentiful and often abundant in Gwent its numbers have markedly declined in recent years so much so that in 1989 I did not see a single specimen. Ed. 11/3/57 (Usk), 1/3/61 (Llancayo). Ld. 16/11/53 (Usk).

Pontnewydd, "abundant" (Conway 1833); 7/2/1833, "several" (Conway, idem). Monmouth (Palmer 1890). Wye Valley (Barraud 1906). Wye Valley 1906 (Bird 1906a). Abertillery dist. 1906 and 1912 (Rait-Smith 1906, 1913). Usk (Castle gardens) 14/4/32 (RBH). Bettws Newydd (Crowther 1935).

Croesyceiliog (The Plantations) 1927, 25/4/28 (GANH). Llantarnam 10/4/31 (frqt.), 14/4/68. Llanfrechfa 21/8/41. Usk: 13/4/52, 16/11/53 (1), 10/4/54 (plfl.); on a sunny grassy bank in my Usk garden (Plas Newydd) it was noted every year, from 1957 to 1984 and was often plentiful, but over the years its numbers gradually diminished. Pont-y-saeson 27/7/52, 10/8/68 (plfl.), 6/9/69, 2/6/70 (abdt.), 28/7/81 (1), 10/5/88 (a few). The Cwm (Shirenewton) 29/7/52 (plfl.). Bettws Newydd 23/3/53 (7). Llantrisant 24/4/55, 15/4/77. Llangwm 14/10/62. Llandegfedd 16/4/64. Glascoed 16/3/67. Wentwood 20/8/68. St Pierre's Great Wds 11/8/68. Cicelyford 9/6/69. Piercefield Park 8/4/69. Devauden 13/4/71. Hendre Wds 21/4/72. Coed-y-paen (The Forest) 5/9/73. Bishton 19/5/75. Prescoed 4/8/75 (plfl.). Raglan 17/4/76. Chepstow Park Wood 27/7/76, in profusion on hemp agrimony flowers. Llandogo 26/5/78. Llanishen 10/9/82. Slade Wds 17/5/88 (1). Goetre (Wern Fawr) 11/6/88 (12). (MEA) : Ynysyfro 2/5/76, Rhiwderin 22/5/76, Risca 24/9/83, Cwm Merddog 14/6/86, Ochrwyth 7/8/86, Cwmfelinfach 14/4/87.

Sq. **SO** 10,20,30,40,50,41,51. **ST** 28,38,48,19,29,39,49,59.

PIERINAE

1548   *Aporia crataegi* Linn.   Black-veined White
This species, now extinct in the British Isles, disappeared from Monmouthshire about 1893.

Pontnewydd, "in general plentiful in this neighbourhood" (Conway 1833). Near Tintern (Piffard 1859). "In 1867 found in large numbers, about mid-summer, in hayfields in Monmouthshire" (South 1906). Monmouthshire: "common, its range extending from about a mile below Cardiff to a place called Llanwern, a distance of about fifteen miles". (George Lock in Newman 1871). "Recorded from Catbrook, and in the utmost profusion at Tintern" (Goss 1887). "Occurred plentifully some dozen or so years ago, at or near Llantarnam, about four miles from Newport" (Birkenhead 1891). Newport dist. 1893, "several caterpillars and four butterflies were noted on May 22" (South 1906); this was the last county record.

Sq. SO 50. ST 28,38,29,39,59.

1549 *Pieris brassicae* Linn.   Large White

This butterfly is widespread, virtually ubiquitous, and generally very common, often abundant, in meadows, open woodlands and especially in gardens, allotments and arable land where the larvae are a pest on cabbage (*Brassica*) species, often completely defoliating the food-plant. Numbers vary widely from year to year and in some years are augmented by immigrants but because of our geographical situation this is not very obvious in Gwent. The species is double-brooded and flies from April until late in the Autumn. Ed. 4/3/28, (Croesyceiliog), 31/3/56 (Usk), 19/3/90 (Llansoy). Ld. 29/10/61 (Tintern); 19/10/67 (Usk dist.); 18/10/79 svl. (Usk). Huge numbers noted at Usk on 3/8/55 and 1/8/82.

Pontnewydd, "very common" (Conway 1833). Monmouth (Palmer 1890). East Monmouthshire, larva on *Lunaria biennis* (honesty) (Bird 1905). Wye Valley (Barraud, 1906). Wye Valley 1906 (Bird 1906a). Abertillery dist. 1906 and 1912. (Rait-Smith 1906, 1913).

Croesyceiliog (The Plantations), 4/3/28 (1), 25/4/28 (svl.) (GANH). Llantarnam 1930. Llanfrechfa 21/8/41. Llangeview 21/4/53. Usk 28/7/55 (plfl.). Cwm Tyleri 15/9/79, (on heather bloom at 1700 ft.). etc., etc. Usk, twice recorded in garden m.v.l. trap, 12/5/71 and 9/10/79. Noted very commonly at flowers of hemp agrimony, buddleia, dahlia and lavender.

Recorded in all squares.

1550 *Pieris rapae* Linn.   Small White

Very common and widespread, frequenting rough ground, open woodland, gardens and cultivated ground especially where cabbage (*Brassica* spp.) and other Cruciferae grow. Flying from late March until October in two or three generations its numbers are often boosted by immigrants. In late Summer and Autumn it abounds on such flowers as hemp agrimony (*Eupatorium cannabinum*) and, in gardens, on buddleia, lavender, Michaelmas-daisies etc. At Usk I noted it on ivy blossom on 16/9/82. Once, after a heavy mid-day shower at Priorees Mill, near Usk (19/7/82), I watched a group of seven Small Whites settled close together and sipping moisture from a small area of damp earth which had recently been burned bare by a garden bonfire. It was very reminiscent of streamside congregations of butterflies including Pierids to be seen in the Tropics sipping moisture from patches of damp sand and a habit also noted in this country with *P. napi.* Ed. 25/3/31 (Llantarnam), 26/3/57 and 26/3/82 (Usk). Ld. 25/10/55 and 23/10/78 (Usk).

Pontnewydd, "very common" (Conway, 1833). Monmouth (Palmer 1890). Wye Valley (Barraud 1906, Bird 1906a). Abertillery dist. (Rait-Smith 1906).

Recorded from numerous localities throughout the county eg.: Llantarnam 25/3/31, Llanfrechfa 21/5/41, The Cwm (Shirenewton) 29/7/52, Tal-y-coed 18/5/54, Llandegfedd 16/4/64, Coed-y-paen 18/8/68, Graig Syfyrddin 20/7/71, Llanfaenor 20/7/71, Hendre Wds 22/7/72, Henllys 16/8/73, Goldcliff 20/8/73 (abdt.), Mynyddislwyn 25/8/73 (abdt.), Cwm Coed-y-cerrig 4/9/73, Bishton 11/8/75, Usk 28/8/77 (abdt.) etc. etc.

Recorded in all squares.

1551   *Pieris napi* Linn.   Green-veined White
       ssp. *sabellicae* Steph.

This butterfly, frequenting open woodland, hedgerows, damp meadows and marshy ground, is common and widespread in Gwent, but as with other common species it is now less plentiful than thirty or forty years ago. There are two broods. The first flies from April to June but individuals have even been noted in March while the second flies in late July and August and it has sometimes been seen in September. Ed. at Usk, 12/4/68, 28/3/73, and 3/4/80. Ld. at Usk, 23/9/64, 20/9/65, and 23/9/75.

Pontnewydd, "very common" (Conway 1833). Monmouthshire (Ince 1887). Wye Valley (Barraud 1906). Wye Valley 1906 (Bird 1906a). Abertillery dist., 1912, "the commonest of the three whites" (Rait-Smith 1906, 1913). Bettws Newydd (Crowther 1933).

Llanfrechfa 21/8/41. The Cwm (Shirenewton) 29/7/52. Pont-y-saeson 8/8/53 (abdt.), 26/8/68 (d°). Usk 3/8/55 (abdt.); 28/3/73 (1) (SYH). Tredean Wds 20/7/65, 28/7/76. Wentwood 29/5/68. St Pierre's Great Wds 11/8/68 (plfl.). Magor Marsh, 23/8/68, in huge numbers on flowers of purple loosestrife (*Lythrum salicaria*) with many in cop. Trelleck Bog 1/8/70 Redding's Inclosure 9/7/71. Llantarnam 17/4/73. Llanvetherine 24/5/73. Henllys 16/8/73. Goldcliff 20/8/73. Mynyddislwyn 25/8/73. Llandogo 19/5/78. Slade Woods 16/5/82. (MEA) : Cwm Merddog 31/5/85. etc.

Recorded in all squares.

1553   *Anthocharis cardamines* Linn.   Orange-tip
       ssp. *britannica* Ver.

Flying from the second half of April until mid-June this species is common and widespread in Monmouthshire, frequenting gardens, lanes, hedge-banks, damp meadows and woodland rides and margins. Though still common, it is far less numerous than thirty years ago. Numbers declined very noticeably in the early sixties at the time when, in this county, herbicidal sprays were used indiscriminately on grass verges and even on the banks of minor country lanes. When this deplorable practice was discontinued, numbers increased but the current use of mechanical trimmers certainly contributes to the continuing comparative paucity in its numbers. Earliest recorded dates in the Usk district are 10/4/54, 5/4/57 and a female on 16/3/61; latest date, Fforest Coal Pit 24/6/88 (♀).

Pontnewydd, "very common" (Conway 1833). Monmouth (Palmer 1990). Llandogo dist. (Nesbitt 1892). Wye Valley (Barraud 1906). Wye Valley 1906 (Bird 1906a). Abertillery dist. (Rait-Smith 1906, 1913, 1915).

Croesyceiliog 28/4/28 (GANH). Usk dist. 25/4/52, 10/5/54 (abdt.). Llangeview (Allt-y-bela lane) May 1952 (plfl.). Trostrey Common 17/5/56. Usk (Plas Newydd garden) 29/4/58, 17/5/70 (abdt.), 11/5/79 (SYH), 27/4/82 (plfl.) etc. Llangybi 15/4/65. Monkswood 15/4/65. Pont-y-saeson 13/6/65 (plfl.), 5/6/78. Prescoed 1/6/68, 5/6/78. Newchurch West 8/5/69, 19/5/77. Trelleck Bog 2/6/70. Cwm-mawr 19/5/71. Llanvetherine 24/5/73. Llanfrechfa 26/5/73 (plfl.). Llantilio Crossenny 28/4/73. Wentwood 11/6/74. Whitson 19/5/75. The Glyn (Itton) 20/5/75 (plfl.). Llanddewi Fach 23/5/75. Henllys 29/5/75. Gaer-fawr 19/5/77. Aberffrwd 20/5/77. Hendre Woods 24/5/77 (plfl.).

Llandogo 19/5/78. Magor Marsh 27/5/78. Pontllanfraith 17/5/88. Slade Wds 17/5/88. Fforest Coal Pit 24/6/88. (MEA) : Cwm Merddog 31/5/85, Risca 2/6/85, Ochrwyth 8/6/86.

Recorded in all squares.

LYCAENIDAE
THECLINAE

1555 *Callophrys rubi* Linn.   Green Hairstreak
This species, inhabiting woodland margins and rides, hedgerows, damp heathy places and moorland, flies in May and June and has been recorded until late July. Widespread in Monmouthshire, it is of sporadic and infrequent occurrence in the eastern half of the county while in the hilly western districts it appears more frequently. Earliest recorded dates 17/5/56 (Trostrey Common), 16/5/88 (Blaenavon Moor); latest dates 8/7/79 and 28/7/81, both at Hendre Wds.

Monmouth (Palmer 1890). Twmbarlwm Mountain (Knight, *Entomologist* **26**:199). Llandogo, male, 1906 (Bird 1906a). Pentwyn and Pen-y-fan (Rait-Smith 1906). Abertillery dist. "abundant" 1912 and 1914 (Rait-Smith 1913, 1915).

Trostrey Common 17/5/56 (GANH); this hedgerow colony died out several years later. Pont-y-saeson c. 1965. Trelleck Bog 2/6/70 (1).

Llangeview (Allt-y-bela lane) 25/5/73 (1). Hendre Wds, 8/7/79 one on broom, 28/7/81 (1). Mountain Air (near Pontypool) 17/5/80, plentiful on gorse bushes and one on the tumulus on top of Twmbarlwm Mountain (1300 ft.) 24/5/80 (G.A. Hill pers. comm.). Gray Hill 29/5/84 (10), Blaenavon Moor 16/5/88 (CT). Pontllanfraith 17/5/88 (GANH). (MEA) : Cwm Tyleri 14/6/86, Clydach 17/6/89.

Sq. **SO** 10,20,30,40,50,21,41,51. **ST** 19,29,49,59.

1556 *Thecla betulae* Linn.   Brown Hairstreak
One old record only for Monmouthshire viz. Pontnewydd, "rare" (Conway 1833). No later records.
Sq. **ST** 29 or 39.

1557 *Quercusia quercus* Linn.   Purple Hairstreak
This species, flying during July and August, is locally plentiful in the eastern half of Monmouthshire in woods containing oaks. When first recorded in the county in 1833, Conway found it abundant at Pontnewydd while during the last forty years I have noted it in many eastern localities, often in abundance, but there do not appear to be any recent records from the western half of the county. Ed. 13/7/71 (Redding's Inclosure), 15/7/88 (Llansoy). Ld. 22/8/72 (St Pierre's Great Wds).

Pontnewydd, "abounds" (Conway 1833). Monmouthshire (Patten 1890). Monmouth (Palmer 1890). East Monmouthshire (Bird 1905). Wye Valley 1906 (Bird 1906a).

Tredean Wds 29/7/52, numerous about the crown of a 30 ft. wych elm on the edge of this mixed deciduous wood. For several years the butterflies were seen to congregate around this particular tree until it was eventually felled. Tintern (Angidy Valley) 26/7/55 ♀ (1). Craig-yr-iar (Llangeview) 4/8/55 (2), 12/8/55 (1). Pont-y-saeson 11/8/64 and 31/7/65, abundant in plantation of young oaks, 23/7/66 (a few), 1/8/70. St Pierre's Great Wds 23/7/68 (abdt.), 22/8/72 (1), 31/7/87, one flying around a common lime. Bica Common 20/8/66, swarming about the top of an oak, one taken by an attendant Spotted Flycatcher. Hendre Wds: 2/8/70, 10/8/74, 9/8/77 (abdt.), 31/7/79 (swarmed around the crown of a sweet chestnut in full flower), 21/7/83 (abdt.). Redding's Inclosure 13/7/71 (1), 21/7/83 (a few). Near Llanfaenor 20/7/71, (plfl.). Slade Wds 21/7/70 "abundant", (JMD). Chepstow Park Wood 25/7/76. Coed-y-paen (The Forest) 29/7/76. Usk (Plas Newydd garden) 10/8/76. Wentwood 12/8/76 (CT). Rhadyr (Usk) numerous on the foliage of an ash tree 15/7/82. Llansoy 15/7/88 (1).

Sq. **SO** 30,40,50,41,51. **ST** 48,39,49,59.

1558 *Satyrium w-album* Knoch   White-letter Hairstreak
Occurs locally in the eastern half of Monmouthshire in woods, thickets and

hedgerows where wych elm (*Ulmus glabra*), the larval food-plant grows. The butterfly, flying in July and August, is usually noticed when visiting flowers, especially those of hemp agrimony and bramble. In 1953 there was a well-established colony of this Hairstreak on wych elms at Llangeview (Allt-y-bela lane) which I kept under observation for twenty two years until 1974 when the trees succumbed to Dutch elm disease. During the nineteen fifties and sixties the butterfly flourished at a number of other sites and I knew strong colonies in the Angidy Valley near Tintern, at The Cwm (Shirenewton), St Pierre's Great Woods and the Hendre Woods. Dutch elm disease had become rampant in Monmouthshire by 1972 and many wych elms were lost. As a result this Hairstreak became very scarce and it was feared the species would perish. The last I saw were single specimens at Pont-y-saeson on 4/6/77 and the Hendre Woods on 24/7/79. However, in 1987 I found a small active colony associated with a small group of three or four young hybrid elms planted beside a new trunk road in the Usk Valley little more than a mile from the site of the Llangeview colony referred to above, so hopefully the species is recovering from the disaster of the seventies. I know of no records or sighting of the White-letter Hairstreak in this county to the west of the River Usk.

Near Tintern (Piffard 1859). Monmouthshire (Patten 1890). Monmouth, one, 1890 (Palmer 1890). Llandogo (Nesbitt 1892). East Monmouthshire (Bird 1905). Wye Valley 1906 (Bird 1906a). Tintern, 1912, larva on wych elm (Bird 1913). Usk, Castle gardens, on buddleia 18/7/33 (RBH). Usk, Castle gardens, 1934, four on hemp agrimony July 17 to July 26 (RBH). Monmouth dist., "very plentiful in some years" (S.G. Charles c. 1937).

1953: Llangeview (Allt-y-bela lane), abundant around wych elms in July and August, descending to bramble and hemp agrimony flowers. The Cwm near Shirenewton, plentiful on hemp agrimony and bramble flowers. 1954: 26/7, Llangeview, (plfl.). Tintern, July and August, (plfl.). 1955: Llangeview abdt. 28/7 and 31/7. 1958: Pont-y-saeson, Aug 2nd. 1962. Cae Kenfy, Abergavenny (Brig. Daniel). 1964: Pont-y-saeson 4/8 (1), on bramble flowers. Llangeview 13/8, colony numerically very strong and flourishing. 1965: Pont-y-saeson 31/7 (1). 1968: St Pierre's Great Wds 11/8. 1971: Llangeview 5/8, several flying around wych elms. 1972: St Pierre's Great Wds 22/8 (2). 1974: Llangeview colony 19/8 (3) on hemp agrimony. This was the last record from this site where all the elms succumbed to Dutch elm disease. 1976: Hendre Wds 20/7, one. Brockwells, Caerwent 17/7 and 21/7 (CT pers. comm.). 1977: Pont-y-saeson 4/8, one on hogweed flowers. Hendre Wds 9/8, one on hemp agrimony. 1979: Hendre Wds, 24/7 one (GANH). St Pierre's Great Wds, "imagines in clearing" (JJW). 1980-81: "Adults basking on hazel leaves" Black Cliff (Tintern), (JJW). 1987: Usk dist new colony on young elms 5/7 – 13/7 (six seen at a time); 1993: this colony still active 17/7 (GANH). 1992: Penallt 19/7 (1); 26/7 one on mint flowers (SJT in litt.).

Sq. **SO** 30,40,50,31,41,51. **ST** 38,48,39,49,59.

1559 *Satyrium pruni* Linn.   Black Hairstreak

George Lock in Newman 1871 writes: Monmouthshire, "I have taken a single specimen in St. Julian's Wood." [This wood, is situated on a hillside on the east bank of the R. Usk between Caerleon and Newport and what remains of it is now within the Newport Borough boundaries]. I have always considered this record to be erroneous and J.M. Chalmers-Hunt 1966, referring to the same record wrote "is very doubtful in my view and refers almost certainly, I suspect, to S. *w-album*".

LYCAENINAE

1561 *Lycaena phlaeas* Linn.   Small Copper
 ssp. *eleus* Fabr.

Common and widespread in Monmouthshire, frequenting waste ground, rough fields, embankments, waysides and woodland rides and clearings. Two or three broods a year occur and it flies from May until late October in hot summers such as 1989 when it was more plentiful than for many years. Ed. 6/6/28 (Croesyceiliog), 30/5/71 (Slade Wds), 27/5/78 (Pont-y-saeson). Ld. 25/10/55 (Usk). It has been recorded at numerous sites throughout Gwent and a few of these are given below.

Pontnewydd, "very common" (Conway 1833). Monmouthshire (Ince 1887). Monmouth (Palmer 1890). Llandogo (Nesbitt 1893). East Monmouthshire, abs. (Bird 1905). Wye Valley (Barraud 1906). Tintern 1912 (Bird 1913). Abertillery dist. and ab. (Rait-Smith 1906, 1912, 1913, 1915).

Croesyceiliog 6/6/28 (GANH). Llanfrechfa 21/8/41. Pont-y-saeson 8/8/53, 26/8/68, 28/8/82 etc. Wentwood 20/8/53 (plfl.). Usk: 13/6/54, 3/10/55 (two on Michaelmas-daisies), 25/10/55, 28/7/82 etc. Trelleck Common 14/6/67. St Pierre's Great Wds 23/7/68. Magor Marsh 23/8/68. Hendre Wds 2/8/70. Bica Common 11/8/70 (plentiful on ling blossom). Undy (sea-wall) 30/8/70. Slade Wds 30/5/71. Mynyddislwyn 25/8/73. Redding's Inclosure 15/6/71. Craig-y-dorth 19/8/73. St Brides Wentlooge 26/8/73. Cwm Coed-y-cerrig 4/9/73. Newchurch West 10/10/75, one on *Cotoneaster* berries. Cwm Tylori 29/6/76. Cicelyford 25/7/76. Chepstow Park Wood 27/7/76. Coed-y-paen (The Forest) 29/7/76. Magor Pill (sea wall) 1/8/89. Fedw Fach (Trelleck) 11/8/89.

(MEA) : Risca 24/9/83, Cwm Merddog 31/5/85.

Recorded in all squares.

1571 *Plebejus argus* Linn.   Silver-studded Blue
 ssp. *argus* Linn.

George Lock, in Newman 1871, gave "common at Castle-y-Bwch".

Castell-y-Bwch lies just to the south of Cwmbran New Town and I have known the area since boyhood. It has always seemed to me a most unlikely site for this insect and I find this record very suspect (GANH pers. obs.).

1572 *Aricia agestis* D. & S.   Brown Argus

This species which is very local and scarce in Monmouthshire is now restricted to the Carboniferous Limestone areas in the south-east where it is seen only sparingly and sporadically. It frequents woodland rides and clearings and rough fields. It is bivoltine, the first brood being seen in June and the second brood insects in late July and August.

Pontnewydd, one, 1833 (Conway 1833).

Pont-y-saeson 1965, 13/6 one (GANH). This apparently freshly-emerged specimen, was sitting in the late afternoon sunshine on the flower of a spotted orchis growing in a woodland ride; St Pierre's Great Wds 1968, 23/7 (1), 11/8 (1); Great Barnets Wood (near Mounton) 1976, 27/7 (1) (GANH). The Nedern (Caldicot) 1976, 26/7 (pair in cop.); 28/7 (twelve or more with some in cop.) (DJU pers. comm.). Caerwent (Brockwells) 1976, 31/7 (2) (CT). Slade Wds 1983, 13/8 (1) (GANH).

Sq. ST 48,39,59.

1574 *Polyommatus icarus* Rott.   Common Blue
       ssp. *icarus* Rott.

Common and widespread, inhabiting rough fields and hillsides, grassy embankments, waysides, woodland rides etc. and flying in two generations from May to September, with numbers varying greatly from year to year. Now much less common than fifty years ago but is still recorded from many sites throughout the county. It is now only to be seen regularly in large numbers along the sea-walls of the Severn Estuary. Ed. 31/5/60 (Llangwm), 21/5/71 (Pont-y-saeson); Ld. 17/9/68 (Usk), 7/9/69 (Magor Marsh).

Pontnewydd, a few, 1833 (Conway 1833). Monmouthshire (Ince 1887). Monmouth (Palmer 1890). East Monmouthshire, abs. (Bird 1905). Wye Valley 1906 (Bird 1906a); d° (Barraud 1906). Abertillery dist. (Rait-Smith 1906,12,13,15). Bettws Newydd (Crowther 1933, 35).

Croesyceiliog 1923, (abdt.); Llanfrechfa 21/8/41; The Cwm (Shirenewton) 29/7/52 (GANH). Wentwood 1953. Pont-y-saeson 14/4/53, 22/8/82. Usk (Plas Newydd garden) 3/6/60. Nant-y-derry 16/8/65 (abdt.). Llansoy 21/8/67. Magor Marsh 23/8/68. St Pierre's Great Wds 11/8/68. Hendre Wds 2/8/70. Bica Common 11/8/70, plentiful on flowers of ling. Undy (sea-wall) 30/8/70 (abdt.). Slade Wds 30/5/71. Runston 22/8/72. Prescoed 14/6/73. Henllys 16/8/73. Mynyddislwyn 25/8/73. St Brides Wentlooge 26/8/73. Cwm Tyleri 4/9/73. Llwyn-y-celyn Bog 27/7/76. Tredean Wds 28/7/76. Llangeview Hill 24/7/79. Usk (Flagpole Hill), "abundant" 25/8/79 (RBH). Dixton Bank 11/6/88. Magor Pill (sea-wall), 1/8/89 (abdt.). Fedw Fach (Trelleck) 11/8/89. (MEA) : Risca 3/8/84, Ochrwyth 13/6/86, Dinham 22/5/87, Fleur-de-lis 9/7/87.

Recorded in all squares.

1575   *Lysandra coridon* Poda   Chalk Hill Blue

There are only two Monmouthshire records for this butterfly viz. Abertillery, one male on Aug. 26th 1905 (Rait-Smith 1906) and Tintern, one male on Aug. 11th 1906 (Bird 1906). The larval food plant of this species is, almost exclusively, the horse-shoe vetch (*Hippocrepis comosa*) which is not found in Gwent. The male butterfly, however, is known to wander so these records are undoubtedly of vagrants from beyond the Severn, either from the Cotswolds where many colonies existed eg. Wotton-under-Edge or from Somerset where there were also strong colonies as at Brean Down.

Sq. **SO** 20,50.

1578   *Cyaniris semiargus* Rott.   Mazarine Blue

As a resident species the Mazarine Blue became extinct in Britain before the end of the last century and had always been regarded as scarce. There are two old Monmouthshire records for this readily recogniseable butterfly. In 1833 Conway wrote of it "Pontnewydd, very scarce and local and captured in but one meadow". George Lock (Newman, 1971) wrote "I have taken one specimen at St Julian's". In 1891 G.A. Birkenhead wrote "This insect has occurred in the Rhymney Valley annually, the last report being of two pupae being secured from last year's brood". The R. Rhymney was the boundary between Monmouthshire and Glamorgan so this record may well apply to the latter county.

Sq. **ST** 38,39.

1580   *Celastrina argiolus* Linn.   Holly Blue
         ssp. *britanna* Ver.

Widespread and common in Gwent, inhabiting gardens, shrubberies, hedgerows and open woods. It is bivoltine and has been noted from mid-April to late August. For nearly thirty years it was often plentiful in my garden at Usk where in "good years", as 1958 and 1970, six or seven at a time were to be seen flying around the hollies and Portugal laurels. Its numbers varied greatly from year to year and although recorded most years, I saw none in the county in 1973, 1974 and 1975, yet in the subsequent four years until 1979 it was plentiful almost everywhere, as it also was in 1991. Ed. 11/4/31 (Llantarnam), 11/4/61 and 11/4/82 (Usk). Ld. 27/8/64 (Llansoy), 30/8/70 and 2/9/71 (Usk).

   Pontnewydd (Conway 1833). Monmouth (Palmer 1890). Twm Barllwm Mountain (Knight, *Entomologist* **26**:199). Llandogo (Nesbitt, 1893). Wye Valley 1906 (Bird 1906a). Pentwyn, one (Rait-Smith 1906). Llanhilleth, one 1911, (Rait-Smith 1912). Abertillery dist. 1914 (Rait-Smith 1915).

   Llantarnam 11/4/31 (GANH). Usk 23/5/55. Goetre 8/5/56. Gaer-fawr 1/6/56. Pont-y-saeson 10/5/57. Usk (Plas Newydd garden) eg. 1958 29/4 (svl.), 4/5 (plfl.), etc. and most years until 1983. Llansoy 27/8/60, 8/7/89.

Llanbadoc (Cefn Ila) 17/6/65 (JWY). The Cwm (Shirenewton) 12/5/70. Llandenny 12/5/70. Jingle Street 16/5/70. Trostrey Hill 20/5/70, 3/8/70 (plfl.). Wentwood 1/6/70. Prescoed 31/7/70. Trelleck Bog 1/8/70. Penallt 1/8/70. Craig-y-dorth 1/8/70. Hendre Wds 2/8/70 (frqt.), 23/8/77. Cwm-mawr 19/5/71. Blorenge 19/5/71. Slade Wds 30/5/71, 17/5/88 (the only record for 1988). Hadnock Quarries 27/7/71. Llantrisant 30/4/76. Newchurch West 26/7/76. Chepstow Park Wood 27/7/76. Great Barnets Wood 27/7/76. Coed-y-paen (The Forest) 29/7/76. Redding's Inclosure 2/8/77. Runston 9/8/77. Raglan 31/8/77. Wonastow 16/6/79. St Pierre's Great Wds 12/7/89 (svl.). Magor Marsh 5/7/89. Magor Pill 18/7/89. (MEA) : Cwm Merddog 31/5/85, Risca 2/6/85.

Recorded in all squares.

NYMPHALIDAE

1584 *Ladoga camilla* Linn.   White Admiral

Formerly, this striking woodland butterfly, in flight during July and August, was mainly restricted to the eastern and southern counties of England south of a line running roughly from The Wash to the Severn Estuary, but during the nineteen thirties especially, it extended its range to the north and west. Monmouthshire lay at the western limits of its range and it was unrecorded in the county until 1952 when I saw my first White Admiral in the Angidy valley near Tintern. I kept the colony under observation untiil 1962. By this time it had spread a further two miles up this little valley. Then, however, it completely disappeared from these woods as a result, I suspect, of the extremely severe and prolonged winter of 1962-63.

I did not see the butterfly again until 1970 when several appeared in woods near Monmouth. This colony has expanded and steadily increased in numbers and still flourishes. In good years the butterfly is abundant and has also spread to several other woods in the neighbourhood.

In 1977 I saw a single White Admiral in St Pierre's Great Woods in the south-east of the county. Ten years later several were to be seen there and in 1988 it was numerous and was also spreading to neighbouring woods in the south. The species is now (1989) found in a number of woods on the eastern side of the Wye Valley and by 1985 it had also re-appeared in its old haunts along the Angidy Brook. Thus it would seem that the White Admiral is now firmly established in eastern Monmouthshire and still continues to extend its range westwards.

Angidy Valley (near Tintern), 1952, 22/7 (1) (GANH), (1st v-c. rec.); 1954, 10/8 (4); 1955, 22/7 – 27/7 (frqt.); 1956, 11/7; 1958, 19/7 (plfl.). Pont-y-saeson (Panta Arch), 1958 19/7 (5), 2/8 (1); 1962 29/7 (1), 1/8 (1) (the last record at this site until 1985 q.v.); 1992, 30/6 (plfl.). Hendre Wds, 1970, 2/8 (svl.) and annually until 1987 and often abdt. eg. 22/7/78, 21/7/83. St Pierre's Great Wds, 1977 2/8 (1); 1987, 31/7 (6); 1988, 5/7 (plfl.); 1989, 12/7 (3).

Treowen (Dingestow), 1982, July (frqt.) (GDT pers. comm.). Pen-y-clawdd Wds (near Dingestow), 1985, 86, 87, 88 (HB pers. comm.). Great Barnets Wood, Mounton, 1987 (svl.) (TGE, in litt.). Pont-y-saeson (Panta Arch), 1985, 86, 87 (CE, pers. comm.). Hael Wds, 1987 5/8 (1); 1992 30/6 (abdt.). Slade Wds, 1989 18/7 (1) (GANH).

Sq. **SO** 40,50,41. **ST** 48,59.

1585 *Apatura iris* Linn. Purple Emperor
Regrettably this fine woodland butterfly has disappeared from Monmouthshire. There have been no definite records since the early nineteen thirties.

W. Langley (in Newman, 1871) in addition to giving "Gloucestershire. Forest of Dean" also gave "Monmouthshire. In the Forest of Dean". This would indicate Redding's Inclosure (near Monmouth) the only part of the Forest of Dean within Monmouthshire. There are a few other old records viz. Monmouth, "several" (Palmer 1890); Tintern, " ♂ July 1905" (Bird 1905). Tintern, " ♂ and ♀ 1906" (Bird 1906). S.G. Charles writing c. 1937 of the Purple Emperor in the Monmouth district stated "not seen lately", though it had previously been seen or caught by him. Conway had written (1833) "I have been not a little surprised that *Apatura iris* has never been found in this locality (ie. Pontnewydd) which is almost entirely overrun with oak." Conway was a keen naturalist and this observation suggests that he was probably aware of its occurrence in the east of the county.

Sq. **SO** 50,51.

1590 *Vanessa atalanta* Linn. Red Admiral
An immigrant species from the Continent varying greatly in numbers from year to year and rarely seen in Gwent before late July or early August. It is commonest in the autumn, when in September and October the butterflies congregate on flowers or on ripe or rotting fruit. The garden flowers most frequently visited are buddleia, Michaelmas-daisies, *Sedum spectabile*, and dahlias; while wild flowers such as thistles, hemp agrimony, ivy and *Polygonum cuspidatum* are favoured. Ed. 15/6/76 (Usk), Llansoy 24/4/88 and 13/6/89. Ld. 23/10/64 (Usk) and 25/10 69 (Llantarnam), though Conway in 1832 recorded Nov 25th. The butterfly is seen most years, but was very scarce in 1953 and 1954 when only one or two were seen. None were seen in 1956 while in the years 1962, 1983 and 1986 it was again very scarce. However in 1969, 89, 91 and 92 it was widespread and abundant.

Pontnewydd, "generally abounds, but scarce in 1832 and only two or three specimens taken, but one on the wing as late as 25th Nov." (Conway 1833). Tintern, Oct 14 1876 (A.H. Jones 1876). Monmouth (Palmer 1890). Wye Valley (Bird 1906, 1907). Abertillery dist. (Rait-Smith 1906, 1913).

Bettws Newydd (Crowther 1933, 1935).

I have recorded it in many Gwent localities eg.: Llantarnam 1930, 25/10/69. Tredean Wds 22/8/52, 28/7/76. The Cwm (Shirenewton) 29/7/52. Undy 27/9/52, 14/10/76. Bettws Newydd 28/8/52 (plfl.). Llangwm 26/8/54. Usk 27/8/54, one on garden rose. Llandenny 14/10/54 one on fallen fruit. Llangeview Hill 4/8/55. Usk (Plas Newydd garden): 9/10/57 on Michaelmas-daisies; 13/10/62 (1) (the only 1962 rec.); 26/9/66, five on *Sedum spectabile*; 9/9/82, on red valerian. Coed-y-paen 29/6/64 (1). Pont-y-saeson 11/8/64 several on hemp agrimony, 20/9/64 (1). St Pierre's Great Wds 11/8/68. Llangybi 7/10/68. Kingcoed 1/10/69. Trostrey Common 28/9/69, (7) on rotting pears. Wentwood 27/8/70, on buddleia. Hendre Wds 2/8/71. Henllys 16/8/73. Cwm Tyleri 4/9/73. Cwm Coed-y-cerrig 7/9/73. Prescoed 7/9/73, a number on dahlia flowers. Piercefield Park 5/10/74. Magor Marsh 3/7/76. Trelleck Bog 19/9/76. Nash 22/9/81. Usk (old railway line), a number on ivy blossom and flowers of *Polygonum cuspidatum*. Ysguborwen (Llantrisant) 3/10/82. Slade Wds 17/7/87. Llansoy 24/4/88, 13/6/89. Fedw Fach (near Trelleck) 11/8/89 (10+) on hemp agrimony. Magor Pill (sea-wall) 1/8/89. (MEA) : Pontllanfraith 15/8/87.

Recorded in all squares.

1591 *Cynthia cardui* Linn. Painted Lady

This immigrant butterfly from the Mediterranean basin is widespread but from year to year its date of appearance and numbers vary greatly. Rarely plentiful in Monmouthshire, it is often absent or scarce. For 14 of the 38 years in the period from 1952 to 1989 I have no v-c. records for this insect (viz. 1953, 54, 56, 57, 60, 61, 63, 67, 71, 74, 75, 78, 81 and 1984) and it was very scarce in some other years. Generally it is first seen from mid May onwards eg. 28/5/66 (Usk), 10/5/77 (Llangwm), 14/5/88 (Llansoy) but occasionally it appears much earlier as 1/4/28 (Croesyceiliog) (GANH), 1/4/52 (Llanbadoc) and 24/4/55 (Llantrisant). Those insects seen late in the season are either fresh migrants or result from eggs laid by earlier arrivals. The species, however, does not survive the winter in this country. Ld. 30/9/66 (Usk), 1/10/69 (Kingcoed).

Pontnewydd, "several a few years back and in 1832 a few taken not many miles away" (Conway 1833). Monmouth (Palmer 1890). Wye Valley (Monmouthshire), "very abundant this spring" (Nesbitt 1892). Wye Valley (Barraud 1906); (Bird 1906). Abertillery dist. (Rait-Smith 1906,1913). Bettws Newydd (Crowther 1933).

Recorded in many localities eg.: Croesyceiliog (The Plantations), 1/4/28 and 16/6/28. Croesyceiliog 19/8/28. Llanbadoc 1/4/52. Bettws Newydd 22/8/52 (on buddleia in Crowther's old garden). Undy (sea-wall) 27/9/52, 30/8/70. Llantrisant 24/4/55. Pont-y-saeson 1/8/58. Llangovan 28/7/62 (only record this year). Coed-y-paen 28/8/64. Usk (Plas Newydd) 3/8/64, 14/9/82 (on red valerian). Goetre 18/9/68. Newchurch West 16/8/69. Kingcoed

1/10/69. Wentwood 11/8/70 (several on hemp agrimony). Trelleck Bog 5/9/72 (only rec. this year). Chepstow Park Wood 27/7/76. Hendre Wds 27/8/79. Raglan 10/8/80. Peterstone Wentlooge 7/6/80 (CT). Llanishen 10/9/82. Risca 24/9/83 (MEA). Slade Wds 13/8/83. Llansoy 14/5/88. Pontllanfraith 30/6/88. St Pierre's Great Wds 5/7/88. Fedw Fach (Trelleck) 11/8/89 (svl.). 1991, generally plentiful.

Llandegfedd 26/7/87, larvae on common nettle, reared (MEJ pers. comm.).

Recorded in all squares.

1593   *Aglais urticae* Linn.   Small Tortoiseshell

This common and virtually ubiquitous butterfly is bivoltine, the larvae, whose food-plant is the common nettle, being found in June and again later in August. Butterflies of the second brood go into hibernation and reappear in early spring, flying freely on sunny days, usually by the second week of March. Prior to hibernation they often congregate in large numbers to sip nectar at their favourite flowers. Occasionally at night they are attracted to m.v.l. Ed. 5/3/61 (Usk), 5/3/66 (Coed-y-paen), 5/3/76 (Llandenny). Ld. 20/11/63 (2) (Llangybi), 2/11/72 (Llanbadoc), 28/10/69 (Usk). Though still common in Monmouthshire this species is now much less abundant than formerly.

Pontnewydd, "generally abounds" (Conway 1833); Pontnewydd 1833, 9/2 one (Conway, idem). Monmouthshire, "everywhere" (Ince 1887). Monmouth (Palmer 1890). Wye Valley (Bird 1906). Abertillery (Rait-Smith 1906). Bettws Newydd (Crowther 1933).

Recorded in many localities on numerous occasions e.g. Croesyceiliog 19/4/28, Llantarnam 25/3/31, Llanfrechfa 21/8/41 (GANH). Usk dist. 9/10/57 (2) on Michaelmas-daisies. Tintern 4/8/64, abdt. on hemp agrimony. Monkswood 14/9/64, (28) on a small clump of *Sedum spectabile*. Newchurch West 17/8/69, abdt. on buddleia. Wentwood (Foresters' Oaks) 9/9/71, (70+) on one buddleia. Goldcliff 20/8/73, abdt. on ragwort and fleabane flowers on sea-wall. Dlorenge Mountain 22/8/73, abdt. on ling (*Calluna vulgaris*). Mynyddislwyn 25/8/73 (abdt.). Prescoed 4/9/73, plfl. on dahlias. Usk (Plas Newydd garden) 9/9/82 on red valerian. Usk 16/9/82, on ivy blossom and flowers of *Polygonum cuspidatum*. Usk 19/9/69, four in garden m.v.l. trap.

Larvae on common nettle (*Urtica dioica*) Magor Marsh (7/8/73) and Llantrisant (27/6/87).

Recorded in all squares.

1594   *Nymphalis polychloros* Linn.   Large Tortoiseshell

There are a few records of sporadic Monmouthshire sightings viz. Monmouth, one, Aug. 23 1890 (Palmer 1890); Wye Valley, one, 1906 (Bird 1906). 1976,

Usk (Plas Newydd garden), one, 16/4 (GANH). 1979, Wentwood, one, 4/9/79 (MEA).

Sq. **SO** 30,50,51. **ST** 49.

1596 *Nymphalis antiopa* Linn. Camberwell Beauty
One record only: Usk Castle, one on buddleia August (c. 1931). (RBH pers. comm.).

Sq. **SO** 30.
The distribution map in *M.B.G.B.I* Vol 7 illustrating the 1976 Camberwell Beauty "invasion" indicates that some five to ten were recorded in Gwent (1976 to mid-1977) but I have no details and saw none myself.

1597 *Inachis io* Linn. The Peacock
This univoltine resident butterfly is virtually ubiquitous and still fairly common in Monmouthshire but during my lifetime its numbers have greatly diminished. Following hibernation it appears, flying freely on sunny days, usually in March and April or even earlier. Pairing occurs and the larvae which feed on the common nettle (*Urtica dioica*) are to be seen in June and July. The resulting imagos, prior to hibernation, often congregate in large numbers during the late summer and autumn to sip nectar from flowers such as thistles, hemp agrimony, buddleia, Michaelmas-daisies etc. Earliest recorded dates include 3/3/61 (Usk dist.), 24/2/75 (Newchurch West), 12/3/82 at Wyesham (SYH). Latest dates at Usk include 12/10/61 (svl.), 29/10/69 and 6/10/71.

Reviewing my Gwent records for this species covering the last sixty years it is apparent that its numbers have declined dramatically, especially in the last two decades. There is little doubt that the use of herbicides is a big factor especially when combined with the now prevalent practice of "tidying up" the odd weedy corners that sustain the nettle its larval food-plant.

Pontnewydd, "generally abounds" (Conway 1833). Monmouthshire (Ince 1897). Monmouth (Palmer 1890). Wye Valley (Bird 1906). Abertillery dist. (Rait-Smith 1906, 1913). Bettws Newydd (Crowther 1933, 1935).

Croesyceiliog 19/4/28, Llantarnam 1930, Llanfrechfa 21/8/41 (GANH) The Cwm (Shirenewton) 29/7/52. 1953, extremely abundant inc. very many hundreds in late summer on thistles at Gwehelog and abundant at Pont-y-saeson 8/8. 1955, abundant on banks of R. Usk at Llantrisant 29/4.

Pont-y-saeson 11/8/64 very plentiful with 40 on one small clump of hemp agrimony. Coed-y-paen (The Forest) 18/8/68, and Wentwood 11/8/70 (abdt.). Occasionally as in 1954 it was very scarce. Since 1970 it has been noted in numerous localities but though widespread its numbers have been much reduced especially in the late eighties. Magor Marsh, 12/6/79, larvae on common nettle.

Recorded in all squares.

1598   *Polygonia c-album* Linn.   The Comma

Common and widespread in Gwent, frequenting hedgerows and woodland rides and clearings. In late summer and autumn prior to hibernation, the Comma commonly visits flowers such as hemp agrimony, buddleia etc., ripe and rotting fruit and also unsavoury organic matter. Latest recorded dates include 9/10/57 on Michaelmas-daisies, 7/10/68 on ivy bloom and 21/10/79. It reappears on sunny spring days after hibernation. Earliest recorded dates include 11/4/31, 5/3/61 and 18/3/73.

Records indicate that in the early years of the 19th century the species was generally common in Monmouthshire. When its range in southern Britain contracted in the latter part of the century and it had disappeared from many of the southern English counties where it had formerly flourished it continued to be reported regularly from the Wye Valley. By the first two decades of this century it had become virtually restricted to counties adjacent to the English-Welsh border including Herefordshire and Monmouthshire, although at a much reduced level. However in the 1920s and subsequently it again extended its range through the southern counties and here in Gwent, especially from the 1950s onwards, its numbers increased and it is now plentiful. I well recall my excitement at seeing my first Monmouthshire specimens in 1930 and 1931 when it was still scarce here.

Pontnewydd, "generally abounds" but in 1832 "was scarce" and no more than two or three specimens taken (Conway 1833). Near Tintern (Piffard 1859). Tintern 1867 (Goss). Monmouthshire: "Rather scarce in Heullis' and St Julian's Woods" (George Lock in Newman 1871). Tintern Oct. 14 1876 (A.H. Jones). Tintern Abbey etc. (Ince 1887). Monmouthshire (Patten 1890). Monmouth (Palmer 1890). Wye Valley (Nesbitt 1892). Llandogo (Nesbitt 1892). East Monmouthshire, larvae on *Ulmus montana* (wych elm), *Urtica dioica* (common nettle), *Humulus lupulus* (hop) (Bird 1905). Wye Valley 1906 (Bird 1906a). Bettws Newydd (Crowther 1933). Wyesham 1933 "along old railway line 10/7" (RBH). Monmouth dist., "The banks of the Wye abound in Comma" (S.G. Charles c. 1937).

Llantarnam, 1930, two on buddleia and 1931, several on rockery flowers 11/4 (GANH). Subsequently noted in numerous localities throughout the county including the following The Cwm (Shirenewton) 1952 (frqt.) Llangeview 2/8/52 Tredean Wds 22/8/52. Pont-y-saeson 29/8/52. Tal-y-coed 18/4/54. Usk 9/10/57, 5/3/61, 23/9/67 on *Sedum spectabile*. Llansoy 21/8/67, 15/7/89. St Pierre's Great Wds 11/8/68 (plfl.), 12/7/89 (abdt.). Coed-y-paen (The Forest) 18/8/68. Magor Marsh 23/8/68. Trostrey Common 28/9/69 (7) on rotting pears and one on horse "droppings". Wentwood 11/8/70 on hemp agrimony flowers. Hendre Wds 22/7/72, one on moist cow dung; 7/9/75, hibernating specimen beaten from foliage of stool alder; 9/10/79 two on ripe blackberries; 18/7/81, two on fox excrement. Redding's Inclosure 21/7/83.

1992, 18/6 Llansoy, larva feeding on wych elm, bred. (MEA) : Pontllanfraith 15/8/87.

Recorded in all squares.

1599 *Araschnia levana* Linn. European Map

This species was added to the British list as the result of the capture of a single, possibly immigrant, specimen in Surrey in May 1982.

However this butterfly, occurring widely in Europe including north-eastern France, appears to have been artificially introduced in 1912, but by whom is uncertain, to a locality in the Forest of Dean, Monmouthshire and also as a second small colony near Symond's Yat in Herefordshire.

E.B. Ford (1945) considered it had been a successful introduction of a foreign species as it not only survived but also increased in numbers for several years. It was then intentionally exterminated by a collector said by Frohawk to have been the well-known entomologist A.B. Farn who apparently disagreed with the introduction of any foreign species.

It was not seen with certainty after 1914 and Austin Richardson in his *Supplement to Donovan's Catalogue* of the *Macrolepidoptera* of *Gloucestershire* 1945 stated "Nothing further reported of its existence in the Forest for the last quarter of a century".

1600 *Boloria selene* D. & S. Small Pearl-bordered Fritillary
ssp. *selene* D. & S.

Frequents moist open woods, woodland rides and clearings and damp heathy places flying during June and July and sometimes noted in August. Much less common in Monmouthshire now than fifty years ago but still seen most years in scattered localities, usually in small numbers.

Ed. 31/5/42 (Llanfrechfa), 2/6/70 (Cicelyford). Ld. 14/8/53 (Pont-y-saeson). Ab. *flavus-pallidus* Spuler has been recorded.

Pontnewydd, "in a bog at a considerable elevation on the mountain side" [i.e. Mynydd Maen GANH.] (Conway 1833). Monmouthshire, "common" (George Lock 1871). Monmouth (Palmer 1890). Twmbarlwm Mountain (Knight). Llandogo, one, 1906 (Bird 1906a). Abertillery dist. (Rait-Smith 1906, 1912, 1913). Pontllanfraith 1914 (Rait-Smith 1915). Monmouth dist. (Charles c. 1937).

Croesyceiliog (The Plantations), plentiful 1927 and 1928; Llanfrechfa (Candwr Brook) 31/5/42 (GANH). Tredean Wds 1951 (plfl.). Pont-y-saeson 14/8/53 and often abundant e.g. 1956 5/6 and 27/6, 8/6/58, 29/6/65. Numbers at this site gradually declined until in 1979 a few were noted and one only on 8/7/80. The habitat was lost mainly through forestry activities including coniferisation and the use of herbicides. I took a male specimen of the pale yellowish – white aberration ab. *flavus-pallidus* here on 27/6/56 (GANH). Trelleck Bog 14/6/67, 2/7/87. Wentwood 24/6/69. Redding's Inclosure 15/6/71. Cwm-mawr 20/6/71. Slade Wds 15/6/73.

Mynyddislwyn 21/6/74. Hendre Wds 9/7/68, 16/6/79 (abdt.). Cwm Tyleri 29/6/76, 25/7/87. Pontllanfraith 4/7/87. Fforest Coal Pit 15/6/88 (abdt.). Fedw Fach (Trelleck) 14/6/88. Penterry (Tintern) June 1989 (plfl.) (CE). (MEA) : Abercarn 14/7/85, Cwmfelinfach 20/6/87, Carno Reservoir 4/7/87, Fleur-de-lis 6/7/87, Ebbw Vale 22/7/87, Clydach 17/6/89, The British 1989.

Sq. **SO** 10,20,30,40,50,11,21,41,51,22. **ST** 48,19,29,39,49,59.

1601 *Boloria euphrosyne* Linn.   Pearl-bordered Fritillary

A woodland butterfly, fomerly common, and flying from mid-May to mid-June in open deciduous woods, woodland rides and clearings. I remember it as quite common in the county but over the last three decades it has virtually disappeared and is now rarely seen in Monmouthshire.

Pontnewydd, "in great abundance some seasons" (Conway 1833). Monmouth (Palmer 1890). Llandogo (Nesbitt 1893). Wye Valley (Barraud 1906). Wye Valley 1906 (Bird 1906a). Wentwood and Twmbarlwm Mountain (Knight, *Entomologist* **26**:199). Abertillery dist. (Rait-Smith 1906, 1913, 1915). Llanock Woods, Crumlin, 1911 (Rait-Smith 1912); [= Llanerch Woods GANH].

Croesyceiliog (The Plantations) 1927 and 1928, (plfl.) (GANH). [Redbrook, Glos. (Knockall's Inclosure, Forest of Dean) near county boundary, 15/5/56, (plfl.)]. Pont-y-saeson 15/5/56 (abdt.), seen most years until 1981 e.g. 10/6/63, 20/5/71 (abdt.), and 6/6/81 but not subsequently. Cilfeigan Park, Llanbadoc 16/5/56 (plfl.). 1966, St Pierre's Great Wds (PC-W). Wentwood (Foresters' Oaks) June 1968 (JMD). Bica Common 20/6/68. Slade Wds 30/5/71 (plfl.) and last recorded here on 3/6/80. Hendre Wds 6/6/78 (plfl.) but not seen here after 29/5/82. (MEA) : Cwm Merddog 14/6/86 (1).

Sq. **SO** 10,20,30,50,41,51. **ST** 48,29,39,49,59.

1603 *Argynnis lathonia* Linn.   Queen of Spain Fritillary

An occasional immigrant from the Continent noted several times in Monmouthshire.

Rhiwderin, "one captured by a schoolboy on railway embankment during the 1930s" (per E.W. Ecutt 1958, pers. comm.). Monmouth dist. "only seen on two occasions" (S.G. Charles c. 1937).

Sq. **SO** 51. **ST** 28.

1606 *Argynnis adippe* D. & S.   High Brown Fritillary
       ssp. *vulgoadippe* Ver.

Old records suggest that this woodland butterfly was formerly widespread and locally common in Monmouthshire. It is now very scarce, as it has been for the last four decades. Until a few were seen in 1989 it was thought to have died out in Gwent. This species usually frequents large open deciduous woods flying from mid-June to early August. In the Angidy Valley my earliest recorded date was 14/6/67 and the latest date 5/8/62.

Pontnewydd dist. "in this neighbourhood – only in one little mountain glen" (Conway 1833). Monmouthshire, "rather scarce", Heullis' Wood (George Lock in Newman 1871). Between Usk and Chepstow (Ince 1887). Monmouth (Palmer 1890). Banks of Wye (Patten 1890). Llandogo (Bird 1905). Llandogo, ♀ 1906 (Bird 1906a). Abertillery dist. common at Crumlin, Llanhilleth and Mynyddislwyn, singletons elsewhere (Rait-Smith 1906, 1912). Bettws Newydd (Crowther 1935). Wyesham (old railway line), 29/7/32 (RBH). Monmouth dist. (Charles c. 1937).

Angidy Valley, (Panta Arch): 1958, 2/8 (1); 1961, 24/6 (2); 1962, 29/7 (1), 5/8 (svl.); 1965, 18/7 (1). It then disappeared from this locality. 1962, Llanvair (Llanishen) 29/7 (1). 1968, Wentwood 19/6 (1). 1980, Raglan 23/7 (1). 1984, Slade Wds 30/6 (1). 1989, Llantrisant 10/7 (1) (GANH); Brynawel (Sirhowy Valley) on buddleia 11/7 (1), 12/7 (1) (AH). Brynawel, 1990 29/6 (3); 1991 28/7 (2); 1992 28/6 (1), 4/7 (1) (MEA).

Sq. **SO** 20,30,40,50,51. **ST** 48,19,29,39,49,59.

1607 *Argynnis aglaja* Linn. Dark Green Fritillary
ssp. *aglaja* Linn.

Flying from mid-June to early August and frequenting lightly wooded hillsides, woodland clearings and moorland this butterfly is now very scarce in Gwent. Ed. 14/6/57 (Pont-y-saeson). Ld. 5/8/78 (Hendre Wds).

Pontnewydd dist. "in this neighbourhood confined to the mountainside and only in one little mountain glen" (Conway 1833). Monmouthshire, rather scarce, Heullis' Wood (Lock 1871). Between Usk and Chepstow (Ince 1887). Abertillery, Aug 6 1905; Crumlin, female, Aug. 9 1906 (Rait-Smith 1906, 1912). Bettws Newydd (Crowther 1933, 1935). Monmouth dist. (Charles c. 1937).

Angidy Valley (Panta Arch) 1956 (svl.) 26/6 and 27/6; 1957 14/6, 16/6, 18/6. Hendre Wds, 1978 July 9 (2),13 (svl.),14,15,22, Aug. 5; 1979 8/7 ♀ (1), ♂ (svl.); 1981, 7/7 (1) (GANH). Chepstow Park Wood 25/7/76 (TGE); 23/7/83 (CT). Ebbw Vale 22/7/87 (svl.) (MEA).

Sq. **SO** 10,20,30,41,51. **ST** 19,29,49,59.

1608 *Argynnis paphia* Linn. Silver-washed Fritillary

Flying in July and August, this butterfly frequents lanes, flowery river banks and woodland margins and rides and was formerly very plentiful in the eastern half of the county but sadly over the last two or three decades it has become relatively scarce in many of its old haunts and is now really numerous in only a few favoured localities. Ed. 4/7/76 (Slade Wds), 30/6/92 (Pont-y-saeson). Ld. 29/8/78 (Hendre Wds).

Pontnewydd dist. "in this neighbourhood almost entirely confined to the mountainside and very abundant." (Conway 1833). Tintern Abbey (Ince 1887). Banks of the Wye (Patten 1890). Monmouth (Palmer 1890). Wye Valley (Nesbitt 1892a). Wye Valley 1906 (Bird 1907). Llanock Wood, [= Llanerch Wood] Crumlin 1906, 1911 (Rait-Smith 1906, 1912). Bettws Newydd (Crowther 1933, 1935). Wyesham, old railway line, 10/7/33 (RBH). Monmouth dist. "The banks of the River Wye abound in the Silver-washed Fritillary" (Charles c. 1937).

Penallt 18/8/43. Tintern 1952 (plfl.); 10/8/54 (abdt.); 26/7/55 and 4/8/64 (d°). The Cwm (Shirenewton), 1952, 22/7 and 29/7 (abdt.); 7/8/53 (d°); 5/8/71. Pont-y-saeson, abundant most years from 1953 to 1968 then numbers gradually diminished but still fairly plentiful on 4/8/77 a few noted on 22/8/78 and an early specimen on 30/6/92. Bettws Newydd 18/7/56, (abdt.). Usk (garden) 3/8/64, 25/7/76 (on Buddleia). Magor Marsh 23/8/68 (1). St Pierre's Great Wds 1966 (PC-W), 11/8/68 (plfl.), 31/7/87. Prescoed 15/7/70. Hendre Wds 2/8/70, (abdt.); 25/7/78 (d°, 14 on the flowers of one small bramble bush); continues to be plentiful here most years. Llancayo 3/8/70. Redding's Inclosure 13/7/71. Newchurch West 23/8/73. Trelleck Bog 25/7/76. Chepstow Park Wood 27/7/76 (in profusion). The Nedern (Caldicot) 28/7/76 (DJU). Slade Wds 21/7/79, 18/7/89. Hael Wds 17/7/89 (3). Fedw Fach (Trelleck) 11/8/89. (MEA) : Risca 31/8/86.

Sq. **SO** 30,40,50,41,51. **ST** 48,29,39,49,59.

1610   *Eurodryas aurinia* Rott.   Marsh Fritillary
       ssp. *aurinia* Rott.

Until fifty years or so ago colonies of this species were widespread in Monmouthshire and the butterfly was often abundant, flying from late in May until the end of June. The colonies occupied damp meadows, heaths and the open glades of mixed deciduous woods where its larval food plant the devil's-bit scabious (*Scabiosa succisa*) grew.

Through a variety of causes there has been a marked decline in the species in the last three or four decades. Over the years I knew four colonies in the eastern half of the county all of which, sadly, have long since disappeared. Sixty-two years ago at Croesyceiliog it was well-established in a boggy lightly wooded area but succumbed to sporadic housing development prior to the 1939-45 war. Two colonies formerly existed in the Angidy Valley in the Tintern district. One colony at Old Furnace, well known to some of our country's leading lepidopterists, was obliterated about 1970 through over-grazing by ponies. Not far distant, on a wooded hillside, flourished a very strong colony where one could see many scores of the butterfly in flight at any one time. I had kept this colony under observation for some fifteen years until its destruction, also about 1970. In this case it succumbed to forestry activities including spraying of the undergrowth with herbicides and replanting with conifers. Coniferisation also destroyed a colony in Tredean Woods near Wolvesnewton.

However in the last few years the species has been found at four sites in Monmouthshire. The colony at Pontllanfraith "found" in 1985 is almost certainly the one referred to by Rait-Smith (1906 – 1915). In 1987 the butterfly was noted at yet another site in the far west, but since the political readjustments of the county boundaries in 1974, this latter locality is no longer in Gwent but nevertheless is still within Watsonian vice-county 35. Also in 1987, the Marsh Fritillary was noted from two other sites, one in the centre of the county and the other in the north. This latter colony has proved to be a very strong and flourishing one.

Pontnewydd, "– so abounds in this neighbourhood that almost any quantity might be captured during the season" (Conway 1833). Monmouthshire, very common near Heullis' and St Julian's Woods (George Lock in Newman 1871). Cwmcarvan (Palmer 1890). Pontllanfraith (Rait-Smith 1906, 1912, 1913, 1915); and another colony near Abertillery found in 1911 (Rait-Smith 1912). Bettws Newydd (Crowther 1935). Monmouth dist. (S.G Charles c. 1937).

Croesyceiliog (The Plantations), abundant 1927 and 1928 (GANH). Tredean Woods, Wolvesnewton 1951 plfl. (GANH).

Angidy Valley (Panta Arch), near Pont-y-saeson, 5/6/56 (GANH), abundant and recorded annually until 1969. Larval webs were numerous on devil's-bit scabious on 11/4/67 and on 9/5/67 a few larvae were still feeding on honeysuckle leaves while a chrysalis was found attached low down on the trunk of a red oak sapling. The colony was still very strong on 5/6/67 and many scores were to be seen in flight at a time. In 1968 a few, mostly small dark specimens, were noted on 8/6 and the last imagines were seen here on 9/6/69.

Old Furnace (Tintern) 9/6/69 (plfl.) (CGMdW, GANH). This well known colony was destroyed through overgrazing by ponies about a year later. Pontllanfraith, 1985 (PG); 16/6/87 (4) (SK); 30/6/88 (3) (GANH).

Goetre 10/6/87 (1) (TH). Bedwellty, June 1987, one (TGE), (v-c. 35). North Monmouthshire June 1987 (5) (TGE); 15/6/88 (abdt.), a strong and very flourishing colony (GANH).

Sq. **SO** 20,30,40,50,51,22. **ST** 38,19,39,59.

Currently, in Sq. **SO** 30,22. and **ST** 19.

SATYRIDAE

1614 *Pararge aegeria* Linn.   Speckled Wood
    ssp. **tircis** Butl.

Widespread and very common, flying from mid-April to mid-October in two or three generations. It frequents woodland rides and margins, gardens and shady lanes. Ed. 16/4/61 (Usk). Ld. 16/10/64 (Llanbadoc).

Pontnewydd, "generally very numerous" (Conway 1833). Monmouthshire (Ince 1887). Monmouth (Palmer 1890). East Monmouthshire, larva (Bird 1905). Wye Valley 1906 (Bird 1906a).

Llanfrechfa 21/8/41, 26/5/73 (GANH). The Cwm (Shirenewton) 29/7/52. Pont-y-saeson 29/8/52, 19/7/67 (abdt.). Coed-y-paen (The Forest) 18/8/68 (abdt.). Tintern 10/8/54. Gwehelog 21/4/55. Usk (Plas Newydd garden), 23/4/60; 4/10/67 on Michaelmas-daisies; 22/9/69 on *Sedum spectabile*; 16/9/82 on Ivy blossom. Llanbadoc (Cefn Ila) 16/10/64 (JWY). Llantrisant 13/5/70. Hael Wds 25/5/79. Hendre Wds 2/8/70, 22/7/72 (abdt.). Wentwood 11/8/70. Undy (sea-wall) 30/8/70. St Pierre's Great Wds 29/9/70. Redding's Inclosure 9/5/71. Slade Wds 30/5/71. Trelleck Bog 5/9/72. Henllys 18/9/72. Craig-y-dorth 19/8/73. Blorenge 20/8/73. Mynyddislwyn 25/8/73. St Brides Wentlooge 26/8/73. Prescoed 7/9/73. Newchurch West 18/9/74. Magor Marsh 14/6/75. Bishton 11/8/75. Chepstow Park Wood 27/7/76. Penallt 1/6/79. Trostrey Common 2/10/79. Also many other localities. (MEA) : Llanarth 2/5/76, Caerleon 23/5/76, Risca 31/5/85, Cwmfelinfach 9/5/87, Ochrwyth 20/6/87, Fleur-de-lis 6/7/87.

Recorded in all squares.

1615   *Lasiommata megera* Linn.   The Wall

Bivoltine, flying in May and June, and as a second brood in August and September this species frequents gardens, dry south-facing banks and sunny woodland rides and lane-sides. Though widespread and fairly common in Monmouthshire, it is decidedly less plentiful than thirty and forty years ago. Ed. 9/5/61 (Llangibby Park), 26/5/70 (Pont-y-saeson), 20/5/76 (Gwehelog). Ld. 7/9/75 (Hendre Wds), 4/9/78 (Magor Marsh), 8/9/79 (Usk Plas Newydd garden).

Pontnewydd, "generally very numerous" (Conway 1833). Monmouthshire (Ince 1887). Monmouth (Palmer 1890). Wye Valley (Barraud 1906). Wye Valley 1906 (Bird 1906a). Abertillery, one, 1906; Crumlin, one, 1906 (Rait-Smith 1906).

Llanfrechfa (Candwr lane) 21/8/41, 31/5/42 (GANH). Llansoy 4/8/52. Pont-y-saeson 8/8/53, 22/8/82. Bettws Newydd 13/8/53. Newchurch West 7/6/54. Usk (Plas Newydd garden) 31/5/58. Llanbadoc (Cilfeigan Park) 24/8/65. Prescoed 31/5/68. Wentwood (Foresters' Oaks) 10/6/68. St Pierre's Great Wds 11/8/68. Magor Marsh 23/8/68. Craig-y-dorth 1/8/70. Hendre Wds 2/8/70, 10/8/76 (plfl.). Undy (sea-wall) 30/8/70. Slade Woods 15/6/71. Henllys 16/8/73. Llanvapley 18/8/73. Mynyddislwyn 25/8/73. Usk (Park Wood) 10/6/75. Chepstow Park Wood 27/7/76. Tredean Wds 28/7/76. Coed-y-paen (The Forest) 29/7/76. Raglan 24/8/80. Magor Pill (sea-wall) 1/8/89 (plfl.). (MEA) : Ochrwyth 13/6/86, Risca 1/8/86.

Recorded in all squares.

1620 *Melanargia galathea* Linn.   Marbled White
    ssp. *serena* Ver.

Colonies of this species frequent unimproved grassland and rough grassy places such as field margins, embankments, grass verges and woodland rides especially in limestone areas. The butterfly is univoltine and flies from late June to early August. Old records suggest that it was formerly common in Monmouthshire, mainly in the south-east and in the Wye Valley, associated with the Carboniferous Limestone of those districts. However, during the earlier years of this century and between the wars it had become rare in Gwent. As recently as 1953 I would drive to the Cotswolds to see this insect in any numbers. In 1949 Rex Jackson noted a colony near Chepstow and this I believe was the Runston colony whence in favourable years vagrants would appear in neighbouring localities as in Wentwood and St Pierre's Great Woods in 1968. Over the ensuing two decades the Marbled White has become common and widespread in the eastern half of the county and now, in 1989, may be seen in many localities. Without doubt the construction of the M4 motorway and the A499 trunk road aided its recovery in Gwent by providing easy uninterrupted routes for its spread along their grassy embankments, often enhanced in the south of the county by the exposure, in places, of the Carboniferous Limestone. It is now well-established and numerous at many new sites along and adjacent to these roads.

Pontnewydd, "generally very numerous" (Conway 1833). Monmouthshire, common near Heullis' Wood (George Lock in Newman 1871). Wentwood near Usk (Ince 1887). Monmouth (Palmer 1890). Near Chepstow, 1949 (R.A. Jackson, per JPS-B in litt. 1966). Caerwent Quarry, one, 25/7/65 (CT in litt.).

Pont-y-saeson 28/7/62 (1) and 29/7 (1). Wentwood (Foresters' Oaks) 17/7/68 (JMD), 24/7/68 (1) (GANH), 1969, 1971. St Pierre's Great Wds 23/7/68, 25/7/78. Runston 23/7/68 (abdt.), 12/7/74 (d°), 21/7/79 (GANH). Nedern (Caldicot) 15/7/69, Portskewett 28/6/70, Newport (Coldra) 28/7/72, Dinham 6/7/73 (CT). Llanwenarth 13/7/71 (1) (SS-B pers. comm.). Usk (Plas Newydd garden) 21/7/74 (1), 28/7/76 (1). Coity Mountain (1700 ft.), one 6/7/76. Chepstow Park Wood 27/7/76. Hendre Wds 13/7/80. Slade Wds 5/7/84 (plfl.). Ysguborwen (Llantrisant) 3/7/84 (abdt.). Llansoy 8/7/84. Llangeview (Maerdy cutting) 7/7/85, 21/6/92 (abdt.). Abernant (Bulmore) 16/7/85. Llanllowell (Pentwyn) 17/7/85 (abdt.), 5/7/87 (d°). Trelleck Bog 6/8/87. Uskmouth 15/7/88 (abdt.). Fedw Fach (Trelleck) 13/7/89. (MEA) : Dixton Bank 21/7/87, Brynawel (Sirhowy Valley) 19/7/89.

Sq. **SO** 20,30,40,50,21,41,51. **ST** 38,48,29,39,49,59.

1621 *Hipparchia semele* Linn.   The Grayling
    ssp. *semele* Linn.

This univoltine species, flying from mid-July to mid-September, frequents stony hillsides, dry stony pathways and woodland rides, embankments and old quarries. While of very scarce and sporadic appearance in the east of the county it is more frequent in the hilly western districts where in a few colonies it has proved to be abundant. Ed. 6/7/73 (Dinham), 6/7/76 (Cwm Tyleri). Ld. 6/9/73 (Prescoed), 15/9/79 (Cwm Tyleri).

Tintern, a male, Aug. 8 (Bird 1906). Trinant 1906 and a colony found at Mynyddislwyn on Aug. 6 (Rait-Smith 1906). Trinant, a few only, 1911 (Rait-Smith 1912). Usk Castle, one, 25/7/34 (RBH). Monmouth dist. "local but plentiful on hilly, rocky ground" (Charles c. 1937).

Llanfrechfa (Candwr Lane) 21/8/41 (GANH). Wentwood (near Five Paths) 6/8/54 (GANH), dº 14/8/70 (JMD). 1966, St Pierre's Great Wds (PC-W). 1973, Cwm Tyleri 4/9 (1), Newchurch West 6/9 (1), Prescoed 6/9 (1). 1976, Cwm Tyleri 6/7 (3), 10/8 (2). 1979, Slade Wds 21/7 (1), Cwm Tyleri 15/9. 1989, Magor Pill 18/7, one on sea-wall.

(CT in litt.):- 1961, Caldicot (Dewstow Quarry) 23/7 (1), (Ballan Moor) 8/8 (1); 1963, Caerwent (Trewen) 7/7 (svl.), Brockwells 12/8 (1); 1972, Newport (Coldra) 28/7; 1973, Dinham 6/7, Abercarn 5/8/73; 1974, Brockwells 2/8, Trewen 3/8; 1976, Trefil 24/7, Mynyddislwyn 15/8, Machen 15/8; 1977 Mynydd Farteg Fawr 2/8 (svl.); 1980, Newport Docks 20/8. (1).

(MEA pers. comm.):- 1987, Ebbw Vale 22/7 and 10/8 "swarming" on old slag heaps at the designated site of Garden Festival Wales 1992, Cwm Merddog 22/7, Fleur-de-lis 15/8; 1988, Ochrwyth 15/8; 1989, Brynawel (Sirhowy Valley) 19/7, Blaenavon-Brynmawr road 19/7 (abdt.), British 19/7 (abdt.).

Sq. **SO** 10,20,30,50,11,51. **ST** 19,29,39,49,28,38,48.

1625 *Pyronia tithonus* Linn.   The Gatekeeper (Hedge Brown)
    ssp. ***britanniae*** Ver.

This butterfly flying in July and August has occasionally been noted in June and even in May. It frequents open woods, thickets, woodland rides, embankments and laneside hedges and verges and is particularly attracted to bramble flowers. Ed. 8/6/58 and 5/6/78 (Pont-y-saeson), 21/5/88 (Fedw Fach). Ld. 31/8/58 (Pont-y-saeson), 29/8/78 (Hendre Wds). Old records indicate that it was widespread and plentiful in Monmouthshire during the last century and it continued to be so, at least in the eastern half of the county, until the 1960s. However, it then became decidedly scarce along many lanes and roadside hedges where it was formerly plentiful although still remaining common in some woods. This I believe was mainly due here to the indiscriminate spraying of roadside verges and hedgebanks with herbicides. When this practice was largely discontinued some years later the species gradually recovered in these situations but, as a wayside butterfly, it still remains less common than formerly.

Pontnewydd, "generally very numerous" (Conway 1833). Monmouthshire, "reported to be common" (Newman 1971). Monmouthshire, "in every part" (Ince 1887). Monmouth, "not common" (Palmer 1890).

Llandogo, male, 1906 (Bird 1906). Usk, "Llangeview cross-roads and Allt-y-bela lane, plentiful", Aug. 1930 (RBH.).

Llanfrechfa (Candwr lane) 21/8/41. The Cwm (Shirenewton) 29/7/52. Pont-y-saeson 28/7/53, 8/6 – 31/8/58 (plfl.), 5/6/78 etc. Llangeview (Allt-y-bela lane) 2/8/53 (abdt.), 19/8/73 one, on hemp agrimony. Llandegfedd 11/8/65 (1), the only specimen seen this year. St Pierre's Great Wds, 1968, abundant on 23/7 and 11/8. Hendre Wds, plentiful on 2/8/70, 5/8/78 and 9/8/80. Wentwood 11/8/70. Prescoed 22/7/71. Runston 22/8/72. Undy (sea-wall) 7/8/73. Henllys 16/8/73. Tredunnoc 17/8/74. Magor Marsh 24/8/75. Chepstow Park Wood 27/7/76. Tredean Wds 28/7/76. Usk (Plas Newydd gard- en) 9/8/82. Fedw Fach (Trelleck) 1988, 21/5 (2) and 27/5 (svl.). Gaer Hill (Newport) 16/6/88. Dixton Bank 18/6/88. Slade Wds 18/7/89. Magor Pill (sea-wall) 1/8/89 (abdt.). (MEA) : Cwm Merddog 14/6/86, Pontllanfraith 15/8/87.

Recorded in most squares

1626 *Maniola jurtina* Linn.   Meadow Brown
ssp. *insularis* Thomson

This univoltine species, flying from mid-June to early September, is common and widespread on grassland including hay meadows, woodland clearings and rides, grassy hillsides and embankments. It visits such flowers as knapweed, hemp agrimony and bramble and in gardens is often to be seen on Buddleia. Charles Conway in 1833 gave it as "generally very numerous" and the same can still be said of it today but, even so, it has dramatically declined in numbers over the last fifty years, probably as a result of altered farming practices including the change from hay to silage production. Ed. 13/6/61 (Pont-y-saeson), 15/6/71 (Slade Wds), 10/6/75 and 5/6/82. Ld. 4/9/73 (Cwm Tyleri), 26/9/72 (St Pierre's Great Wds), 8/9/77 (Usk, Plas Newydd garden).

Pontnewydd, "generally very numerous" (Conway 1833). Monmouthshire (Ince 1887). Monmouth (Palmer 1890). Wye Valley (Bird 1906). Abertillery dist. (Rait-Smith 1906, 1912, 1913).

Croesyceiliog, 1923 and 1928, abundant in hayfields (GANH). Llanfrechfa 21/8/41 (GANH). Pont-y-saeson, abdt. on 8/8/53, 6/7/56 and 20/7/65. Llangybi 16/7/65 abdt. in hayfields. Usk (Plas Newydd garden) 28/6/64. Craig-y-iar (Llangeview) 5/7/68 (abdt.). Wentwood 9/7/68 (abdt.). St. Pierre's Great Wds 23/7/68 (abdt.). Llanfaenor and Graig Syfyrddin 20/7/71. Piercefield Park 1/7/73 (abdt.). Llanvapley 18/8/73. St Brides Wentlooge 26/8/73. Cwm Tyleri 4/9/73. Trelleck Bog 25/7/76. Llantrisant 26/7/76 (on buddleia). Hendre Wds 25/7/78 (abdt.). Usk (Plas Newydd garden) 24/8/80. Pontllanfraith 4/7/87. Uskmouth 15/7/88.

Recorded in all squares.

1627   *Coenonympha pamphilus* Linn.   Small Heath
       ssp. *pamphilus* Linn.

Locally common on moorland, heaths, embankments, rough hilly pastures, grass verges and dry woodland rides and flying from late May to mid-September. Ed. 31/5/68 (Prescoed), 20/5/71 (Pont-y-saeson). Ld. 5/9/76 (Magor Marsh), 15/9/79 (Cwm Tyleri).

Pontnewydd, "generally very numerous" (Conway 1833). Monmouthshire (Ince 1870). Monmouth (Palmer 1890). Llandogo (Nesbitt 1893). Wye Valley 1906 (Bird 1906). Abertillery dist. (Rait-Smith 1906, 1912, 1913, 1915).

Croesyceiliog (The Plantations) 16/6/28 (GANH). Llanfrechfa 21/8/41. Pont-y-saeson 8/8/53, 6/6/81. Wentwood 1954 (plfl.). Trelleck Common 14/6/67. Prescoed 31/5/68. Bica Common 20/6/68. Craig-yr-iar (Llangeview) 28/6/70. Undy (sea-wall) 30/8/70 (abdt.). Blorenge (1500 ft.) 4/9/73 (abdt.). Cwm Tyleri 29/7/72. St Pierre's Great Wds 22/8/72. Piercefield Park 1/7/73. Mynyddislwyn 21/6/74. Trelleck Bog 1/7/75. Magor Marsh 12/8/75. Hendre Wds 6/6/78 (plfl.). Pontllanfraith 4/7/87. (MEA) : Ochrwyth 13/6/86, Cwm Merddog 14/8/86.

Recorded in all squares.

1629   *Aphantopus hyperantus* Linn.   The Ringlet

This univoltine species, on the wing from late June to mid-August, and locally common in the eastern half of the county, occurs in colonies frequenting lanes, hedgerows and woodland rides and margins. The insects often visit bramble flowers to sip nectar. The form ab. *arete* Müller is not uncommon. Ed. 20/6/74 (Prescoed), 18/6/88 (Dixton Bank). Ld. 19/8/73 (Craig-y-dorth), 23/8/77 (Hendre Wds).

Pontnewydd, "generally very numerous" (Conway 1833). Monmouthshire, "in every part" (Ince 1887). Monmouth (Palmer 1890). Wye Valley (Bird 1906). Wyesham, "old railway line", inc. one ab. *arete*, 10/7/33 (RBH).

The Cwm (Shirenewton) 29/7/52 (plfl.). Tintern 10/8/54. Craig-yr-iar (Llangeview) 4/8/55. Pont-y-saeson 10/7/56 (abdt.) inc. many ab. *arete*, 16/7/64 (abdt.), 28/7/89. Penallt 20/7/65 (abdt.). Tredean Wds 20/7/65 (abdt.). Wentwood 18/7/68 (plfl.). Coed-y-paen (Porthlong Barn) 19/7/68 (plfl.). St Pierre's Great Wds 23/7/68 (abdt.), 12/7/89 (abdt.). Hendre Wds 2/8/70, 20/7/76 (plfl.). Slade Wds 6/7/71. Redding's Inclosure 13/7/71 (abdt.). Llanfaenor 20/7/71. Prescoed 20/6/74. Runston 6/8/74. Trelleck Bog 25/7/76. Wyndcliff 31/7/91, one at m.v.l.

Sq. **SO** 40,50,41,51. **ST** 48,39,49,59.

Systematic List

Part Two
Heterocera (moths)

Section 1: The Larger Moths (macrolepidoptera)

HEPIALIDAE

14 *Hepialus humuli* Linn. Ghost Moth
    ssp. *humuli* Linn.

Widespread and common from mid-June to early August in rough fields, waste ground, roadside verges and other grassy places. Both sexes fly at dusk and both, especially the female, will come to light.

Wye Valley 1906 (Bird 1907). Abertillery dist. (Rait-Smith 1912, 1915).

Llantarnam 1931 plentiful in garden (GANH). Usk, occurred most years from 1955 to 1983 and recorded frequently in garden m.v.l. trap from 1966 eg. 1970, ♂ (1) and ♀♀ (5) (26/6 – 1/8). Ed. 6/6/82. Ld. 10/8/68. Magor 10/7/70, males abundant on roadside verge. Shirenewton 12/6/71. Magor Marsh 30/6/75, 2/7/89 ♀♀ (3) to m.v.l. Brockwells (Caerwent) 11/7/76 (CT). Hendre Wds 11/7/86. Llansoy 6/7/87. (MEA) : Mescoed Mawr 3/7/76, Ynysyfro 28/6/76, St Anne's Vale 7/7/77, Wyndcliff 21/6/77, Black Vein 12/6/81, Risca 14/7/81, Newport Docks 11/7/84, Cwm Tyleri 10/7/85, Dinham 7/7/87.

Sq. **SO** 20,30,40,21,41. **ST** 28,38,48,29,39,49,59.

15 *Hepialus sylvina* Linn. Orange Swift

In Monmouthshire this moth occurs sparingly in gardens, open woods and moorland during July and August and comes to light.

Wye Valley 1906 (Bird 1907). Abertillery dist. 1911 (Rait-Smith 1912). Usk Castle, one 16/8/34 (RBH).

Wentwood (Foresters' Oaks) 1969, frequent at m.v.l. inc. 30/7, and 31/7 (JMD). Usk, garden m.v.l. trap, singletons 21/8/75 and 20/8/77. Cwm Tyleri 8/7/75 (1), 29/7/75 (1). Llansoy 17/8/86, 26/8/88. (MEA) : Pontllanfraith 16/8/87, Cwm Coed-y-cerrig 17/8/87, Magor Pill 6/8/89

Sq. **SO** 20,30,40,22. **ST** 48,19,49.

16 *Hepialus hecta* Linn. Gold Swift

Found in open woodland, heathy places etc. in June and July but local and usually at low density in Gwent. Flies at dusk and comes to light.

Tintern 1906 (Bird 1907). Abertillery dist., "swarmed in every wood" (Rait-Smith 1912, 1913).

Usk (Park Wood) 1967 to m.v.l. Usk, garden m.v.l. trap, 11/7/74 (1), 16/6/76 (1). Wentwood, 1/7/69, flying in abundance at dusk. Hendre Woods 5/7/70. Hael Wds, to m.v.l., 12/7/71, 4/7/87, 6/7/88. Pont-y-saeson, one in spider's web, 11/7/72. Trelleck Bog 1986, 18/7 (1), 24/7 (1). Magor Marsh

26/6/87. (MEA) : Mescoed Mawr 1/7/76, Ynysyfro 4/7/76, Black Vein 3/7/81, Blorenge 7/7/84, Cwmfelinfach 6/7/87.

Sq. **SO** 20,30,50,21,41. **ST** 28,48,19,29,49,59.

17 **_Hepialus lupulinus_** Linn.   Common Swift
This moth, flying from late May through June and July, is widespread and common, frequenting agricultural land, gardens, open woods and moorland. It readily comes to light.
   Wye Valley 1906 (Bird 1907). Abertillery, "one in town" (Rait-Smith 1913).
   Usk, garden m.v.l. trap, recorded most years from 1966 to 1983 eg. 3/6/68, 1977 (8) 29/5 – 24/6. Ed. 23/5/80. Ld. 7/7/71. Prescoed 8/6/69. Rhyd-y-maen 6/6/70. Cwm-mawr 2/7/71. Cwm Tyleri 30/7/72. Llansoy 4/6/85. (MEA) : Llantarnam 10/6/76, Wentwood 4/7/77, Mescoed Mawr 11/6/78, Risca 31/5/80, Slade Wds 8/6/82, Trelleck Common 11/6/82, Wyndcliff 1/7/85, Cwm Merddog 14/7/86, Ochrwyth 27/6/86.

Sq. **SO** 10,20,30,40,50. **ST** 28,48,29,39,49,59.

18 **_Hepialus fusconebulosa_** De G.   Map-winged Swift
   ssp. **_fusconebulosa_** De G.
Flies from the end of May until mid-July and comes to light. In the eastern half of Gwent it occurs locally and at low density in open woodland but it is common and widespread in the hills and moorland in the west and north of the county where the form ab. *gallicus* Led. is frequent. Abertillery dist. 1911, "fairly common, one ab. *gallicus*" (Rait-Smith 1912).
   Wentwood (Foresters' Oaks) to m.v.l. 18/6/69 (2), 31/5/71 (JMD). Wentwood, one at dusk 1/7/69 (GANH). Usk, garden m.v.l. trap, 19/6/73 (1), 31/5/82 one ab. *gallicus*. Cwm Tyleri 30/7/72, abundant, inc. ab. *gallicus*, also 1975, 1976 (abdt. on 8/7 inc. ab. *gallicus*), 1977, 28/6/86. Trelleck Bog 29/6/88 (2). (MEA) : St Anne's Vale 7/7/77 (ab. *gallicus*), Mescoed Mawr 11/6/78, Black Vein 12/6/81, Blorenge 7/7/84, Cwm Merddog 7/7/85, Ochrwyth 27/6/86, Pontllanfraith 2/7/87, Trefil 3/7/87, Clydach 18/7/89.

Sq. **SO** 10,20,30,50,11,21. **ST** 28,19,29,49.

COSSIDAE
ZEUZERINAE

161 **_Zeuzera pyrina_** Linn.   Leopard Moth
Widespread and fairly common in gardens, woods etc. from late June until early September. Both sexes come to light.
   Usk, garden m.v.l. trap, recorded most years from 1967 to 1983 eg. 9/7/67 (1), 1976 (4) 21/6 – 2/7, 1977 (5) 20/7 – 2/8, 1983 (8) 11/7 (3) – 18/7. Usk Castle m.v.l. trap (RBH): single moths on 22/7/68, 13/7/69, 17/7/73.

Magor Marsh 6/7/71; 1976 July 2,3,7. Abergavenny (♀) 12/7/73. Slade Wds 2/8/82 (♀), 17/7/87. Tal-y-coed 9/9/83. Hendre Wds 11/7/86. (MEA) : Wyndcliff 17/7/83, 21/7/87; Risca 26/7/84; Cwmfelinfach 15/7/86, 6/7/87; Dinham 7/7/87.

Sq. **SO** 30,40,31,41. **ST** 48,19,29,49,59.

COSSINAE

162 *Cossus cossus* Linn.   Goat Moth
Abergavenny (Chapman 1871).
    Newbridge-on-Usk 9/2/71, the trunk of an old solitary ash standing in a marshy field found to be riddled with larval borings (GANH).

Sq. **SO** 31. **ST** 39.

ZYGAENIDAE
PROCRIDINAE

163 *Adscita statices* Linn.   The Forester
Very scarce and local with few recent records.
    Llandogo dist. 16/5/92 (Nesbitt 1892). Llandogo dist. 22/4/93 (Nesbitt 1893). Abertillery dist., "in two spots only" (Rait-Smith 1912, 1913).
    Monmouth dist. (S.G. Charles c. 1937).
    Dinham, 1982 5/6, (1), 1984 7/6 (15), 1985 3/7 (a few), 1988 15/6 (svl.) (CT); 8/6/89, one on flower of marsh thistle. Gaer Hill (Newport), 1984 22/6 (2) (CJR, in litt.). Raglan dist. (Croes Llwyd), June 1989 (DL).

Sq. **SO** 20,30,50. **ST** 28,49.

ZYGAENINAE

169 *Zygaena filipendulae* Linn.   Six-spot Burnet
    ssp. *stephensi* Dup.
    = *anglicola* Trem.
Colonies of this day-flying moth are commmon in Monmouthshire in flowery meadows and on sunny woodland rides, embankments, hillsides and wayside verges. The moths fly from late June to August. The yellow form, ab. *flava* Robson, has been recorded.
    Wye Valley 1906 (Bird 1907). Crumlin; Pontllanfraith (Rait-Smith 1912, 1913).
    Croesyceiliog, 1923, plentiful in hay-meadows (GANH). Wentwood 3/7/68, colony at Foresters' Oaks. Craig-yr-iar (Llangeview) 5/7/68. St Pierre's Great Wds, 24/6/69. Piercefield Park 30/6/73, strong colony within Chepstow Racecourse. Runston, 6/8/74, plentiful on roadside verges and banks. (At all five of these sites the species flew in company with *Z. lonicerae*). Magor Marsh 3/7/76. Usk (Conigar), 8/7/76. Dixton Bank

18/6/88. (MEA) : Gaer Hill (Newport) 17/7/86, Fleur-de-lis 9/7/87. Allt-yr-yn, ab. *flava* 2/9/90 (1) (CJR pers. comm.).

Sq. **SO** 30,40,51. **ST** 28,48,19,29,39,49,59.

170 *Zygaena trifolii* Esp.   Five-spot Burnet
    f. *palustrella* Ver.

Only one colony of this moth was known in Monmouthshire but sadly it has been exterminated.

Tintern 31/5/68 (JMC-H. 1969). This strong colony flourished in a hillside meadow at Old Furnace in the Angidy Valley and the moth was flying in numbers when I visited the locality on 9/6/69 with Charles de Worms, but soon afterwards, the colony was obliterated through overgrazing by ponies as was a colony of the Marsh Fritillary which flew in the same meadow. Unfortunately this was the only known Monmouthshire station for this Burnet and none has been discovered since. (GANH 1993).

Sq. **SO** 50.

171 *Zygaena lonicerae* Schev.   Narrow-bordered Five-spot Burnet
    ssp. *latomarginata* Tutt
    = *transferens* Ver.

Widespread and common in Monmouthshire, frequenting sunny banks and hillsides, woodland rides, wayside verges etc., this species flies in late June and July often in company with *Z. filipendulae*.

Wentwood (Foresters' Oaks) 3/7/68, strong colony on roadside verges. Craig-yr-iar (Llangeview), 5/7/68, 30/6/78. St Pierre's Great Wds 24/6/69, in woodland clearing. Piercefield Park 30/6/73, abundant within Chepstow racecourse. Usk, 8/7/73, abundant on roadside verge. Runston 12/7/74. Pont-y-saeson 3/7/79. Hendre Woods 8/7/79. Pontllanfraith 4/7/87. Dixton Bank, 18/6/88 (abdt.). Trelleck (Fedw Fach) 13/7/89. Llangeview (Maerdy cutting) 21/6/92 (abdt.). (MEA) : Fleur-de-lis 6/7/87.

Sq. **SO** 30,40,41,51. **ST** 19,49,59.

LIMACODIDAE

173 *Apoda limacodes* Hufn.   The Festoon
    = *avellana* auct.

One Monmouthshire record only viz. Usk, garden m.v.l. trap, 4/6/67 male (1) (GANH).

Sq. **SO** 30.

SESIIDAE

Of the fifteen British species of Sesiidae or Clearwings only two have been noted in Monmouthshire. The neighbouring counties of Brecon and Hereford can also claim only one and three species respectively, while on the other hand, eleven species have been reported from both Gloucestershire and Glamorgan, the other two counties adjacent to Gwent.

SESIINAE

371 *Sesia bembeciformis* Hb.   Lunar Hornet Moth
Pantygelli (Near Abergavenny) Aug. 1968 "one on currant-bush two days running", (WEK pers. comm.).

Sq. **SO** 21.

PARANTHRENINAE

373 *Synanthedon tipuliformis* Cl.   Currant Clearwing
One record only, viz. Tintern, one, 1911 (Bird 1912).

Sq. **ST** 59.

LASIOCAMPIDAE

1631 *Poecilocampa populi* Linn.   December Moth
Widespread and fairly common in gardens, hedgerows and deciduous woodland, flying from late October to mid-December and both sexes coming readily to light.

Wye Valley 1906 (Bird 1907). Tintern 1912 (Bird 1913). Abergavenny (Chapman teste Tutt *Br. Lep.* **2**: 480). Cwmyoy 1965 (SHR).

Usk, garden m.v.l. trap, recorded annually from 1966 to 1982 and sometimes abundant eg. 30/11/67 (54 inc. 10 ♀♀), 5/12/71 (50 inc. 13 ♀♀). Ed. 23/10/71. Ld. 14/12/69. Wentwood 1968. Caerwent (Brockwells) 1/12/68 (CT). Rhyd-y-maen 2/11/70. Redding's Inclosure 2/11/71. Prescoed. (MEA) : Risca 13/11/86, Cwmfelinfach 1986.

Sq. **SO** 30,40,31,51,32. **ST** 48,19,29,39,49,59.

1632 *Trichiura crataegi* Linn.   Pale Eggar
Local and scarce in the eastern half of Monmouthshire, frequenting gardens, hedgerows and woodland from mid-August to mid-September.

Usk, one to m.v.l. 5/9/66 (1st v-c. rec.) (RBH). Usk, garden m.v.l. trap, 31/8/68 and sporadically in small numbers eg. 1970 (4) 27/8 – 1/9, 1979 (3) 7/9 – 16/9. Ed. 26/8/73. Ld. 16/9/79. Usk (Park Wood) 3/9/69 (2). Rhyd-y-maen 1/9/73 (3), 2/9/73 (3). Hendre Wds 4/9/73. Pont-y-saeson 18/9/79 (2). Tal-y-coed 9/9/83. Llansoy 1986 (10) 30/8 – 21/9, 1987 1/9 (3), 5/9 (1).

Sq. **SO** 30,40,31,41. **ST** 59.

1633   *Eriogaster lanestris* Linn.   Small Eggar
Rare in Monmouthshire with few records.

   Abergavenny (Chapman, teste Tutt, *Br. Lep.* **2**: 520). Monmouth dist. (S.G. Charles c. 1937).

   Usk, garden m.v.l. trap, 7/4/76, one freshly-emerged male (GANH). Magor Pill 16/7/83, larval webs on hawthorn (DU, CT).

Sq. **SO** 30,31,51. **ST** 48.

1634   *Malacosoma neustria* Linn.   The Lackey
Locally common in eastern Gwent but scarce in the western half of the county, flying during July and August in open woods, hedgerows etc. and coming readily to light.

   Abergavenny (Chapman, teste Tutt, *Br. Lep.* **2**: 566).

   Usk, garden m.v.l. trap, recorded every year from 1966 to 1983 and sometimes frequent eg. 1970 (25) 3/7 – 13/8. Ed. 23/6/76. Ld. 13/8/70. Wentwood 12/7/68. Usk (Park Wood) 23/7/68. Magor Marsh 11/7/69, 22/7/69 (plfl.). Wyndcliff 13/7/69. Rhyd-y-maen 18/7/70. Cwm Tyleri 30/7/72. Undy (Collister Pill) 13/7/73. Hendre Wds 17/7/79. Slade Wds 23/7/79. Llansoy 28/7/84. (MEA) : Dinham 7/7/87, Uskmouth 16/8/88.

   Usk 28/5/73, larvae feeding on wych elm (*Ulmus glabra*), bred. Magor Marsh 12/6/79, larvae feeding on foliage of meadowsweet (*Filipendula ulmaria*).

Sq. **SO** 20,30,40,50,31,41. **ST** 38,48,49,59.

1637   *Lasiocampa quercus* Linn.   Oak Eggar and Northern Eggar
This moth, flying from late June to August, is local and scarce in the eastern half of Monmouthshire where it frequents open woodland and here the sub-species *L. quercus quercus* Linn. has been noted. The moth is much more plentiful in the hills and moorland of the western and northern parts of the county where it is represented by the sub-species *L. quercus callunae* Palm., the males being commonly seen flying by day and at night both sexes, but the females more readily than the males, come sparingly to light.

   Monmouthshire, larvae (Patten 1890). Abergavenny (Chapman, teste Tutt, *Br. Lep.* **3**: 110). Llandogo (Nesbitt, teste Tutt, loc. cit.). Abertillery dist. 1911 and 1912 (Rait Smith 1912, 1913).

   Caerwent 29/8/62 ♀, Crick 1/6/73 larvae, Dinham 26/7/73 ♀, Gray Hill 15/8/73 ♀ (CT in litt.). Magor Marsh 22/7/69 ♀ to m.v.l. Slade Wds 21/7/79 (by day), 23/7/83 ♀ to m.v.l. (GANH). Risca 12/7/83 (MEA).

Sq. **SO** 20,30,50,21,31. **ST** 48,29,59.

   L. quercus Linn.   Oak Eggar
   ssp. *quercus* Linn.
Pont-y-saeson 9/5/67, full-fed larva in a hibernaculum formed of last year's

dried leaves still attached to the twigs of a small shrubby oak some eighteen inches above the ground and surrounded by the previous year's dead herbage. A typical *L. quercus quercus* Linn. emerged on 17/7/67 (GANH).

Sq. **ST** 59.

    L. quercus Linn.   Northern Eggar
    ssp. ***callunae*** Palm.

1968 Blorenge, "I found two cocoons; one moth emerged in 1969 of the callunae form". (J. Newton, *Macrolepidoptera in Gloucestershire* 1984).

1971 Blorenge 24/6, ♂♂ (2) flying by day (GANH). Cwm Tyleri 1973 to m.v.l. 19/6 (♀ 1), 1/7 (♀♀ 2), 9/7 (♀ 1); 1974 5/8 (♂ 1) to m.v.l. and one full-grown larva (GANH); 1977, 9/7 by day (♂♂ 8) (CGMdW, JLM, GANH); 10/7 by day ♂♂ (2), ♀ (1). (MEA) : Trefil 4/7/87, Nant Gwyddon 25/6/87.

Sq. **SO** 20,11,21. **ST** 29.

1638   *Macrothylacia rubi* Linn.   Fox Moth

In the east of the county this moth occurs locally and at low density in open woods and the few remaining fragments of semi-heath but in the hills and moorland of the west and north of Gwent it is abundant. In May and June the males are seen flying by day, and at night, the females come to light while in the late summer and autumn the full-fed larvae are commonly seen sunning themselves on the herbage.

Abertillery dist. (Rait-Smith 1912, 13, 15).

Wentwood, 17/6/68 ♀ to m.v.l. (JMD), 23/8/74 larva (CT). Sugar Loaf 22/9/68, larvae (13) on bilberry; Newchurch West 23/9/68, larvae (2) on sunny bank; Twmbarlwm mountain 24/8/71, larvae (GANH). Usk 3/6/70, ♀ (1) to garden m.v.l. trap. Trelleck Bog 5/9/72, larvae; 25/5/87 ♀♀ (2) to m.v.l. Gray Hill 24/9/74, larvae (CT). Cwm Tyleri to m.v.l. ♀ (1) on 27/6/76 and on 12/7/77. (MEA) : Risca, to act.l. 12/7/83 ♂ (1), 20/6/84; flying by day Blorenge 7/7/84 and Cwm Tyleri 10/7/85; to m.v.l. Ochrwyth 27/6/86, Trelleck Bog 25/5/87, Pontllanfraith 22/6/87, Abercarn 16/6/88.

Sq. **SO** 20,30,50,21. **ST** 28,19,29,49.

1640   *Philudoria potatoria* Linn.   The Drinker

Common and widespread in moist open woodland, marshes and damp grassy places generally, flying from late June to mid-August.

Abergavenny (Chapman, teste Tutt, *Br Lep.* **3**: 182).

Croesyceiliog 1928, larvae plentiful in the "Plantations", a boggy deciduous open wood; July 1929, ♀ visiting bramble flowers in hedgerow at dusk (GANH). Usk, garden m.v.l. trap, recorded most years from 1966 to 1983 and sometimes frequent. Ed. 21/6/76. Ld. 21/8/67. Wyndcliff 10/7/67. Prescoed 5/7/68 (plfl.). Wentwood 8/7/68. Usk (Park Wood) 23/7/68. Magor

Marsh 23/8/68, 13/7/83 (abdt.). Hendre Wds 5/7/70. Rhyd-y-maen 18/7/70. Tintern 1/8/70. Redding's Inclosure 20/7/71, 24/7/83 (plfl.). Cwm Tyleri 28/7/72. St Pierre's Great Wds 13/7/73. Piercefield Park 18/7/73. Slade Wds 11/8/81. Llansoy 12/7/86. Ysguborwen (Llantrisant) 17/7/86. Trelleck Bog 19/7/86. Sugar Loaf (St Mary's Vale) 8/7/87. (MEA) : Mescoed Mawr 13/8/80, Risca 9/7/81, Ochrwyth 12/7/86, Cwm Merddog 8/8/86, Cwmfelinfach 12/8/86, Dinham 7/7/87, Pontllanfraith 10/7/87, Dixton Bank 18/6/88, Uskmouth 16/8/88.

Sq. **SO** 10,20,30,40,50,21,31,41,51. **ST** 28,38,48,19,29,39,49,59.

1642 *Gastropacha quercifolia* Linn.   The Lappet
Local and scarce in Monmouthshire flying in July and early August. Both sexes come to light. Ed. 30/6/75 (Magor Marsh). Ld. 3/8/74 (Magor Marsh).

Abergavenny (Chapman, teste Tutt, *Br. Lep.* **3**: 223).

Magor Marsh, 10/7/70 ♀ (1) to m.v.l. (GANH), 3/8/74 ♂ (1) to m.v.l. (MJL). 1975 single males on 30/6, 1/7 and 7/7. 1976 3/7 ♂ (1) and ♀ (1). 1978 9/7 ♀ (1), 11/7 ♂ (1). Usk, garden m.v.l. trap, 7/7/76 ♂ (1), 1979 (1), 1982 (2), 1983 (1) – (the only records over the years 1966 to 1983). Caerwent, larva 12/5/83 (CT). Llansoy 1986 ♀♀ (4) 11/7 – 15/7 (GANH). Risca 1983 (MEA). Twyn-y-Sheriff, 1989 (DL).

Sq. **SO** 30,40,31. **ST** 48,29.

SATURNIIDAE

1643 *Saturnia pavonia* Linn.   Emperor Moth
Flying in April and May, this species occurs sporadically and at low density in the eastern half of Monmouthshire, frequenting open woodland and the remnants of former heathland, but is plentiful on the moorland and hills in the north and west of the county. The males may be seen in flight on sunny days and at night the females come fairly readily to light.

Abertillery dist. 1911, 12, 14 (Rait-Smith 1912, 13, 15). Sugar Loaf 1931, larvae on heather 5/9, reared (RBH).

Llantrisant 20/4/61 ♀ and Newchurch West 30/4/68 ♀ (GANH). Gray Hill, 25/8/69 full-fed larva (JMD). In 1970 the moth was unusually frequent in central Monmouthshire when numerous males were seen flying in Wentwood on 12/5 (GANH) and (5) females came to m.v.l. 6/5 – 14/5 (JMD). A number of gravid females were picked up eg. Earlswood dist. (2), Tredunnoc (1), Usk churchyard (1), and two females came to my garden m.v.l. trap on 26/5 and 27/5. 1971 Wentwood, males frequent 20/4. 1972, Prescoed ♀ (1) to light 16/4. (MEA) : Cwm Merddog 31/5/85, Cwm Tyleri 14/6/86, Cwmfelinfach 8/5/88.

Sq. **SO** 10,20,30,21. **ST** 19,39,49.

ENDROMIDAE

1644 *Endromis versicolora* Linn.   Kentish Glory
A few old records only, viz. Monmouth, female, taken in the town, 30/4 1839 (Parry, *Ent. Wkly Intell.* **2**: 43). "On June 5 1894, Mr Clark exhibited three females bred from Monmouthshire ova". (*Ent. Rec.* **5**: 207).

Sq. SO 51.

DREPANIDAE

1645 *Falcaria lacertinaria* Linn.   Scalloped Hook-tip
This species, flying from the end of May until late August, frequents areas of heath and open woodland where birches grow. Widespread but at low density in Monmouthshire. Ed. 26/5/87 (Abercarn). Ld. 21/8/79 (Slade Wds).

Wye valley 1906, larvae (Bird 1907). Pont-y-saeson 13/6/67 (2) to m.v.l. Usk (Park Wood) 23/7/68. Wentwood (Foresters' Oaks) 8/7/68, 29/6/73 (svl.) (GANH), 1969 (3) to m.v.l. trap 25/7 – 7/8 (JMD). Tintern 1/8/70. Trelleck Bog 16/6/73, 28/5/87. Hael Wds 30/5/78, 31/7/82. Slade Wds 28/5/78, 21/8/79, 11/8/81. Usk garden m.v.l. trap, 21/6/79, 12/7/82. Kilpale 30/6/87. Hendre Wds 9/7/79. Redding's Inclosure 2/8/90. (MEA) : Risca 10/8/81, Mescoed Mawr 31/5/85, Abercarn 26/5/87, Cwmfelinfach 10/6/87, Pontllanfraith 16/8/87, Nant Gwyddon 7/6/88.

Sq. SO 30,50,41,51. ST 48,19,29,49,59.

1646 *Drepana binaria* Hufn.   Oak Hook-tip
This moth, found in oakwoods, is common and widespread in Gwent. Double-brooded, it flies in May and June and again from late July to mid-September. Ed. 15/5/82 (Usk). Ld. 16/9/79 (Hendre Wds).

Bettws Newydd (Crowther 1933, 1935).

Usk, garden m.v.l. trap, seen every year from 1966 to 1983 and often frequent eg. 1970 (19), 30/5 – 26/8; 1982 (16) 15/5 – 9/8. Usk (Park Wood) 6/6/66, 29/7/82. Prescoed 8/6/69. Magor Marsh 2/8/70. Cwm-mawr 31/5/71. Slade Wds 1/6/71. Redding's Inclosure 27/8/71. Hael Wds 21/5/73, 31/8/86. Wentwood 12/8/76 (CT). Hendre Wds 23/8/77, Wyndcliff 3/8/81. Llansoy 28/7/84. (MEA) : Risca 19/6/79, Ochrwyth 27/6/86, Abercarn 26/5/87, Pontllanfraith 16/8/87.

Sq. SO 20,30,40,50,41,51. ST 28,48,19,29,39,49,59.

1647 *Drepana cultraria* Fabr.   Barred Hook-tip
This woodland moth flying among beeches, apart from one or two favoured sites in the Wye Valley, is local and scarce and mainly confined to the eastern half of the county. Double-brooded, it flies from May to August.

Usk, garden m.v.l. trap, recorded sporadically in small numbers eg. 1966 (1) 19/8, 1969 (2), 1978 (13) 28/5 – 26/8. Ed. 28/4/67. Ld. 26/8/78.

Usk (Park Wood) 28/4/67 (2) to m.v.l. Wyndcliff 12/6/67, 8/8/84 (3). Wentwood 25/7/69 (JMD). Cwm-mawr 31/5/71. St Pierre's Great Wds 6/8/74. Hael Wds, noted most years, eg. 6/8/74; 30/8/77 abundant at m.v.l. (ten at a time on sheet); 31/8/86. Slade Woods 28/5/78. Magor Marsh 13/7/83. Redding's Inclosure 24/7/83. Clydach 18/7/89.

Sq. **SO** 20,30,50,21,51. **ST** 48,49,59.

1648 *Drepana falcataria* Linn.   Pebble Hook-tip
       ssp. *falcataria* Linn.

Widespread and fairly common in open woodland, flying in late May and June and, as a second brood, in August.

Llandogo dist. "to light" 26/5/92 (Nesbitt 1892). Wye Valley 1906, larvae (Bird 1907). Llanock Wood [= Llanerch Wood. GANH], Crumlin 1911 (Rait-Smith 1912). Bettws Newydd (Crowther 1933, 1935). Bettws Newydd "to light" 31/5/35 (RBH).

Usk, garden m.v.l. trap, a few most years from 1966 to 1983 eg. 1970 (7), 26/6 (1) and 2/8 – 12/8 (6); 1977 (8), 26/5 (1) and 9/8 – 23/8 (7). Ed. 26/5/77. Ld. 31/8/79. Usk 24/9/68, three larvae on birch, reared. Usk (Park Wood) 5/6/67. Pont-y-saeson 13/6/67. Prescoed 31/5/68. Wentwood, to m.v.l. 8/7/68, 27/8/68 (6). Cwm-mawr 31/5/71. Slade Wds 1/6/71 (plfl.). Hendre Wds 26/5/70, 11/7/86. Hael Wds 3/7/71. Tintern 1/8/70. Redding's Inclosure 7/6/71, 23/6/81. St Pierre's Great Wds 13/7/73. Trelleck Bog 16/7/79. Wyndcliff 23/7/82. Llanfoist 30/6/83 (CT). (MEA) : Risca 12/8/81, Cwm Merddog 7/7/85, Cwmfelinfach 15/7/86, Dixton Bank 18/6/88.

Sq. **SO** 10,20,30,50,21,41,51. **ST** 48,19,29,39,49,59.

1650 *Sabra harpagula* Esp.   Scarce Hook-tip

This very local species, flying in June and July, was formerly known from Leigh Woods near Bristol where it was first found in 1837. It was last reported from there in 1938 and was thought to be extinct but in 1961 it was rediscovered in the lower Wye Valley by Messrs D.S. Fletcher and J.D. Bradley. It is associated with the small-leaved lime (*Tilia cordata*) growing in the woods on both sides of the Wye below Tintern (Monmouthshire and Gloucestershire) but is not now found elsewhere in this country. It is recorded most years and is sometimes plentiful at m.v.l. eg. 1967, to m.v.l., 6/6 (1), 23/6 (svl.), 10/7 (abdt.); 1986 25/6 (1). 1991, feral last instar larva noted on 10/9 by BS (in litt.).

Sq. **ST** 59.

1651 *Cilix glaucata* Scop.   Chinese Character

Widespread and fairly common, frequenting open woodland, rough bushy places, hedgerows etc. It is double-brooded, the first brood flying in the latter half of May and the second brood in late July and August. This moth comes readily to light.

Abertillery dist. one 1911 (Rait-Smith 1912), May 13 1912 (Rait-Smith 1913).

Usk, garden m.v.l. trap, noted every year from 1966 to 1983 eg. 1968 (6) 15/4 – 13/6 and (12) 12/8 – 30/8. 1979, (11) 15/5 – 29/6 and (3) 23/7 – 3/9. Ed. 15/4/68. Ld. 8/9/82. Magor Marsh 23/8/68. Prescoed 8/6/69. Usk (Park Wood) 9/6/69. Undy (Collister Pill) 30/8/70. Trelleck Bog 22/5/73. Slade Wds 28/5/78. Hendre Wds 19/5/80. Hael Wds 29/5/82. Wyndcliff 10/8/83. Llansoy 26/7/84, 3/5/90 (7). Tal-y-coed 9/9/83. Wentwood. (MEA) : Risca 3/6/80, Mescoed Mawr 14/8/81, Pontllanfraith 16/8/87, Ochrwyth 17/6/88, Clydach 18/7/89.

Sq. **SO** 20,30,40,50,21,31,41. **ST** 28,48,19,29,39,49,59.

THYATIRIDAE

1652 *Thyatira batis* Linn.   Peach Blossom
A widespread and common woodland moth, flying from mid-May until the end of July.

Llandogo dist. 13/5 1892; Llandogo, "to light" 25/5/1892 (Nesbitt 1892). Wye Valley 1906 (Bird 1907). Abertillery dist. 1911 (Rait-Smith 1912). Bettws Newydd (Crowther 1933). Usk (Cockshoots Wood) 11/7/33 (RBH).

Usk, to light, 1958 (GANH). Usk, garden m.v.l. trap, most years from 1966 to 1983. Ed. 11/5/67. Ld. 2/8/81. Wyndcliff 12/6/67. Pont-y-saeson 13/6/67. Usk (Park Wood) 15/6/67. Prescoed 5/7/68. Trelleck Common 9/7/68. Wentwood 2/6/70. Hendre Wds 5/7/70. Magor Marsh 10/7/70. Cwm-mawr 31/5/71. Slade Wds 1/6/71. Redding's Inclosure 7/6/71. Hael Wds 3/7/71. Cwm Coed-y-cerrig 27/5/73. Trelleck Bog 16/6/73. St Pierre's Great Wds 13/7/73. Llansoy 28/6/87.

(MEA) : Mescoed Mawr 1/7/76, Ynysyfro 28/6/76, Tintern 30/6/78, Black Vein 3/7/81, Risca 5/7/82, Abercarn 16/6/85, Pant-yr cos 6/7/85, Ynysddu 12/7/85, Cwm Merddog 14/7/86, Ochrwyth 27/6/86, Cwmfelinfach 12/8/86, Pontllanfraith 10/7/87, Lower Machen (Park Wood) 14/5/88, Nant Gwyddon 10/6/88, Dixton Bank 18/6/88.

Sq. **SO** 10,20,30,40,50,41,51,22,32. **ST** 28,48,19,29,39,49,59.

1653 *Habrosyne pyritoides* Hufn.   Buff Arches
A moth of open woodland and bushy places flying from mid-June to early August. Though widespread and often plentiful, its numbers in this county appear to have declined markedly over the last two decades.

Tintern 1887 (Goss 1887). Llandogo dist. (Nesbitt 1892). East Monmouth-shire, "ovipositing on bramble" (Bird 1905). Wye Valley 1906 (Bird 1907). Abertillery dist. 1911 (Rait-Smith 1912). Usk 1932, "one taken"; Usk (Cockshoots Wood), one at sugar July 1933 (RBH). Bettws Newydd

(Crowther 1935).

Usk, garden m.v.l. trap, recorded annually 1966 to 83 and often plentiful eg. 1978, (65) 27/6 – 4/8; 1979, (110) 22/6 – 12/8. Ed. 29/5.69 (a very early moth), 12/6/81. Ld. 17/8/79. Pont-y-saeson 13/6/67. Wyndcliff 14/6/67 (REMP). Prescoed 5/7/68. Wentwood 8/7/68. Trelleck Common 9/7/68. Usk (Park Wood) 23/7/68. Hendre Wds 5/7/70. Hael Wds 5/7/70. Magor Marsh 10/7/70. Tintern 1/8/70. Redding's Inclosure 20/7/71. Cwm Tyleri 30/7/72. St Pierre's Great Woods 13/7/73. Piercefield Park 18/7/73. Newport (Coldra) 27/6/75. Trelleck Bog 16/7/79. Slade Wds 23/7/79. Llansoy 26/7/84. Kilpale 30/6/87. (MEA) : Mescoed Mawr 3/7/76, Risca 12/7/83, Newport Docks 11/7/84, Abercarn 16/6/85, Ynysddu 12/7/85, Ochrwyth 12/7/86, Cwmfelinfach 12/8/86, Pontllanfraith 2/7/87, Sugar Loaf (St Mary's Vale) 8/7/87, Dixton Bank 18/6/88, Clydach 18/7/89.

Sq. **SO** 20,30,40,50,21,41,51. **ST** 28,38,48,19,29,39,49,59.

1654   *Tethea ocularis* Linn.   Figure of Eighty
　　　ssp. *octogesimea* Hb.
Local and at low density in Monmouthshire, flying from late May to mid-July in damp woods and river valleys where poplars grow.

Usk Castle 1966, 9/6 (1) to m.v.l. (RBH) (1st v-c. rec.); also 1967 (svl.) and 1976 21/6 (3), 22/6 (1) (RBH). Usk (Plas Newydd), garden m.v.l. trap, 1966 (2), 1967 (2), 1968 (1) and a few most years until 1983 eg. 1975 (7) 11/6 – 10/7, 1979 (9) 12/6 – 14/7 (GANH). Ed. 29/5/82. Ld. 14/7/79. Usk (Park Wood) 9/6/69 (1) to m.v.l. (CGMdW, GANH). Magor Marsh 19/6/79 to m.v.l. ♀♀ (2), ♂ (1). Hael Wds 29/5/82 (2), 4/7/87 (1). Slade Wds 4/7/84 (1). Hendre Wds 1/7/86 (1). (MEA) : Wyndcliff 8/7/85, Ochrwyth 27/6/86, 17/6/88.

Sq. **SO** 30,50,41. **ST** 28,48,59.

1655   **Tethea or** D. & S.   Poplar Lutestring
Local and very scarce in Monmouthshire, this moth flies in damp woodland during June and July.

Usk Castle, one to m.v.l. 8/6/66 (RBH) (1st. v-c. rec.). Usk (Park Wood), 5/6/67 (1) to m.v.l. (GANH). Usk, garden m.v.l. trap, 1969 (3) 8/6 – 16/7. Wentwood 28/7/71 (1) (JMD). Hael Wds 10/7/73 (1), 14/6/86 (1) (GANH). Wyndcliff 6/7/86 (MEA).

Sq. **SO** 30,50. **ST** 49,59.

1656   *Tetheella fluctuosa* Hb.   Satin Lutestring
In Monmouthshire this moth, frequenting woodland rides and open woods containing birches, is of local occurrence but, where found, is often abundant especially in the east of the county including the Wye Valley. It flies from

early June to mid-August and comes readily to light. Earliest recorded dates 7/6/66 (Wyndcliff and Pont-y-saeson), 5/6/85 (Llansoy). Latest dates 10/8/74 (Hael Wds), 12/8/86 (Cwmfelinfach).

Tintern, "a few at light in 1947 and one early one in 1948" (Blathwayt 1967).

Wyndcliff 7/6/66. Pont-y-saeson 7/6/66, 13/7/81. Wentwood 8/7/68, to m.v.l., abundant (20 or more on the sheet at a time). Trelleck Common 9/7/68. Hael Wds 5/7/70 and plentiful most years, 31/7/71 (abdt.). Tintern 1/8/70. Redding's Inclosure 7/6/71. Trelleck Bog 16/6/73, 24/6/87. Usk, garden m.v.l. trap, 25/6/73 (1), the only record in the eighteen years from 1966 to 1983. St Pierre's Great Wds 13/7/73. Slade Wds 6/8/83. Llansoy 5/6/85. (MEA) : Mescoed Mawr 1/7/76, 8/7/84 (common), Risca 24/7/84, Abercarn 16/6/85, Ynysddu 12/7/85, Cwmfelinfach 12/8/86, Cwmcarn 11/7/87, Nant Gwyddon 10/6/88, Dixton Bank 18/6/88.

Sq. SO 30,40,50,51. ST 48,19,29,49,59.

1657 *Ochropacha duplaris* Linn.   Common Lutestring

Flying in deciduous woodland from the end of May to early August this species is widespread in Monmouthshire but rarely numerous. Earliest recorded date 24/5/76 (Cwm Coed-y-cerrig). Latest recorded date 6/8/74 (Hael Woods).

Bettws Newydd (Crowther 1935). Usk (Park Wood) 6/6/66, 9/6/69 to m.v.l. Usk, garden m.v.l. trap, noted sporadically from 1968 to 1982 eg. 1968 (3) 11/6 – 22/7, 1975 (3) 23/6 – 9/7, 1982 (1) 30/5. Wyndcliff 13/7/69. Hael Wds 3/7/71, 31/7/82. Cwm Coed-y-cerrig 5/6/73 to m.v.l., 5/6/78 (7) to act.l. Trelleck Bog 16/6/73. Wentwood 29/6/73. Slade Wds 23/7/79. Hendre Wds 4/6/79.   (MEA) : Mescoed Mawr 3/7/76, Risca 13/7/82, Ynysddu 12/7/85, Cwmfelinfach 12/8/86, Sugar Loaf (St Mary's Vale) 8/7/87, Cwmcarn 11/7/87, Abercarn 16/6/88.

Sq. SO 30,50,21,41,22. ST 48,19,29,49,59.

1658 *Cymatophorima diluta* D. & S.   Oak Lutestring
      ssp. *hartwiegi* Reisser

Flying from mid-August through September, this moth occurs locally in some oakwoods in the east of the county. It is often abundant where found and comes readily to light. Earliest recorded date 11/8/82 (Wyndcliff). Latest recorded date 30/9/79 (Pont-y-saeson).

Llandogo (Nesbitt 1892). Wye Valley 1906 (Bird 1907).

Redding's Inclosure 27/8/71 to m.v.l. (abdt). Pont-y-saeson 7/9/73 (2), 18/9/79 (abdt), 30/9/79. Wyndcliff 11/8/82. Hael Wds 16/8/76 (1), 18/9/82, 19/8/89 (plfl.). Llansoy 31/8/87 (1).

Sq. SO 40,50,51. ST 59.

1659   *Achlya flavicornis* Linn.   Yellow Horned
         ssp. *galbanus* Tutt

This moth, frequenting woods containing birches and flying during March and April, is local and at low density in Monmouthshire.

Wye Valley 1906, larva (Bird 1907). Abertillery dist. (Rait-Smith 1912, 1913). Cwmyoy, March 1965 (SHR). Usk, garden m.v.l. trap, 4/3/67, 28/3/68 and sporadically in small numbers until 1983. Ed. 29/2/80. Ld. 20/4/76. Wentwood 11/4/70, 1971 (12) 13/3 – 19/4 (JMD); 6/4/80 (CT). Redding's Inclosure 20/3/72. Hendre Wds 15/4/79 (3). Hael Wds 19/3/83 (7) to m.v.l. (MEA) : Wyndcliff 9/3/78 (4), Risca 5/4/87, Cwmfelinfach 4/8/88.

Sq. **SO** 20,30,50,41,51,32. **ST** 48,19,29,49,59.

1660   *Polyploca ridens* Fabr.   Frosted Green

Flying from early April to mid-May, this moth is locally common and often abundant in oakwoods in the eastern half of the county. Earliest recorded date 30/3/81 (Hael Wds). Latest date 21/5/79 (Hael Wds).

Usk, garden m.v.l. trap, a few noted most years from 1967 to 1982 eg. 2/5/68, 12/4/80. Usk (Park Wood) 28/4/67, 22/4/68 (abdt.). Llangybi (Park Wood) 23/4/68. Wentwood 30/4/70 (JMD). Redding's Inclosure 4/4/72, 21/4/81. Hael Wds 16/4/73 (plfl.). Pont-y-saeson 16/4/73. Wyndcliff 26/4/76. Hendre Wds 15/4/79, 27/4/79 (abdt.), 14/4/80 (abdt.). Llansoy 3/5/85.

Sq. **SO** 30,40,50,41,51. **ST** 39,49,59.

GEOMETRIDAE
ARCHIEARINAE

1661   *Archiearis parthenias* Linn.   Orange Underwing

Flies by day in March and April on heaths and in open woodland containing birches. Scarce and local in Monmouthshire.

Tintern and Llandogo 1906 (Bird 1907). Wye Valley 29/3/1907 (Barraud 1907b). Abertillery dist. (Rait-Smith 1912, 1913, 1915).

Pont-y-saeson 26/3/68. Prescoed. Wentwood (Foresters' Oaks) 7/4/69 (JMD), 20/4/71 (GANH). Hadnock Quarries 30/3/71. Hendre Wds 11/4/71 plfl., (44 counted), 3/4/79 (6), 22/3/84 (1). (MEA) : Cwmfelinfach 14/4/87 (2), 7/4/88; Black Vein 7/4/88.

Sq. **SO** 20,50,41,51. **ST** 19,29,39,49,59.

1662   *Archiearis notha* Hb.   Light Orange Underwing

Barrett 1900 gives "Monmouthshire" **6**: 336). Wye Valley, near Bigsweir, 29/3/1907 (Barraud 1907b). No recent records.

Sq. **SO** 50.

OENOCHROMINAE

1663 *Alsophila aescularia* D. & S.   March Moth

A widespread and common woodland moth. The female is flightless, but the male flying in March and April, comes readily to light.

Wye Valley 1906 (Bird 1907). Tintern 1912 (Bird 1913). Abertillery dist. (Rait-Smith 1912, 13, 15). Caerwent 15/3/63 (CT). Cwmyoy 1965 (SHR).

Usk, garden m.v.l.trap, recorded annually 1966 – 84 eg. 1969 (28) 15/3 – 3/5; Ed. 25/2/76; Ld. 12/5/70. Llangybi (Cae Cnap) 23/4/68. Usk (Park Wood) 3/5/69. Hadnock Quarries 30/3/71. Cwm-mawr 6/4/71. Redding's Inclosure 7/3/72. Pont-y-saeson 20/3/72. Hendre Wds 25/3/72. Prescoed 16/4/72. St Pierre's Great Wds 23/3/73. Hael Wds 26/3/73, (abdt.). Slade Wds 26/3/83. Llansoy 13/3/87. (MEA) : Mescoed Mawr 27/2/77, Llantarnam 18/3/77, Wyndcliff 9/3/78, Risca 3/4/82, Cwmcarn 23/3/87, Cwmfelinfach 11/4/87, Ochrwyth 13/4/87.

Sq. **SO** 20,30,40,50,21,41,51,32. **ST** 28,48,19,29,39,49,59.

GEOMETRINAE

1665 *Pseudoterpna pruinata* Hufn.   Grass Emerald
ssp. *atropunctaria* Walk.

This moth, not uncommon in heathy places and woodland clearings, flies from mid-June to early August. By day it is easily disturbed from herbage and at night occasionally comes to light.

Llanock Wood [= Llanerch Wood, GANH.] 1911 and 1912 (Rait-Smith 1912, 1913).

Usk, garden m.v.l. trap, recorded sporadically over the years eg. 13/7/67, 4/8/70, 1976 (4) 15/6 – 5/7. Ed. 15/6/76. Ld. 10/8/73. Usk (Park Wood) 23/7/68. Trelleck Bog 12/7/70 (MWH in litt.), 17/7/79. Redding's Inclosure 20/7/71. Hael Wds 10/7/73, 22/6/76. (MEA) : Risca 8/7/71, Wyndcliff 6/7/84, Cwmfelinfach 15/7/86, Fleur-de-lis 6/7/87.

Sq. **SO** 20,30,50,51. **ST** 19,29,49,59.

1666 *Geometra papilionaria* Linn.   Large Emerald

Common and widespread in Monmouthshire, flying in open woodland and heathy places from late June to August and coming readily to light.

Wye Valley, "larvae on birch, May 1906, reared" (Bird 1907).

Usk, garden m.v.l. trap, 21/7/66 and a few most years until 1982. Ed. 22/6/76. Ld. 5/8/74. Glascoed 10/8/65. Wyndcliff 10/7/67, Wentwood 22/7/68. Usk (Park Wood) 23/7/68. Hendre Wds 5/7/70, 11/7/86. Tintern 1/8/70. Redding's Inclosure 20/7/71, 24/7/83. Hael Wds 12/7/74. Pont-y-saeson 10/8/74. Magor Marsh 3/7/76. Cwm Tyleri 8/7/76. Llansoy 3/8/77. Trelleck Bog 16/7/79. Slade Wds 23/7/79. (MEA) : Risca 13/7/83, Mescoed Mawr 8/7/84, Cwmfelinfach 15/7/86, Sor Brook 17/7/86, Cwmcarn 11/7/87,

Nant Gwyddon 15/8/88.

Sq. **SO** 20,30,40,50,41,51. **ST** 48,19,29,39,49,59.

1667 *Comibaena bajularia* D. & S.   Blotched Emerald
  = *pustulata* Hufn.

Occurs locally in oakwoods in the eastern half of Monmouthshire. It flies from late May to mid-July and comes to light.

Tintern dist. (Piffard 1859). Wye Valley (Nesbitt 1892).

Usk, garden m.v.l. trap, 23/6/66 and most years until 1983 eg. 1971, (5) 24/5 – 10/7; 1983, (17) 2/7 – 12/7. Ed. 24/5/71. Ld. 21/7/72. Usk (Park Wood) 30/6/68. Wyndcliff 13/7/69. Hendre Wds 5/7/70, 11/7/86. Hael Wds 3/7/71. Pont-y-saeson 18/7/72. Slade Wds 8/7/83. Llansoy 6/7/87.

Sq. **SO** 30,40,50,41. **ST** 48,59.

1669 *Hemithea aestivaria* Hb.   Common Emerald

Widespread and common, flying from late in June to mid-August, sometimes later, in open woodland and bushy places and coming to light.

Usk, garden m.v.l. trap, 30/6/67 and every year until 1982, eg. 1969 (11) 10/7 – 22/7. Ed. 23/6/76. Ld. 12/8/68. Wentwood 8/7/68. Pont-y-saeson 14/7/68. Wyndcliff 13/7/69. Magor Marsh 22/7/69. Hendre Wds 5/7/70. Hael Wds 5/7/70. Redding's Inclosure 20/7/71, 24/7/83. Hadnock Quarries 27/7/71. St Pierre's Great Wds 22/8/72. Usk (Park Wood) 23/8/72. Slade Wds 29/8/72. Llansoy 14/7/86. (MEA) : Ynysyfro 28/6/76, Mescoed Mawr 3/7/76, Newport Docks 11/7/84, Ynysddu 12/7/85, Risca 13/7/85, Cwmfelinfach 15/7/86, Pontllanfraith 10/7/87, Dixton Bank 18/6/88.

Sq. **SO** 30,40,50,41,51. **ST** 28,38,48,19,29,49,59.

1673 *Hemistola chrysoprasaria* Esp.   Small Emerald
  = *immaculata* Thunb.

This moth occurs locally, mainly in the east and south of the county, in woods, hedgerows and bushy places where traveller's joy (*Clematis vitalba*) flourishes. It flies from late June to early August and comes to light.

Tintern dist. (Piffard 1859).

Usk, garden m.v.l. trap, 10/7/77 and recorded every year until 1983 eg. 1969 (9) 10/7 – 26/7, 1974 (6) 7/7 – 5/8. Ed. 26/6/76. Ld. 5/8/74. Trelleck Common 14/6/67. Wyndcliff 10/7/67. Hendre Wds 5/7/70. Redding's Inclosure 20/7/71. Wentwood 4/7/75 (CT). Magor Marsh 25/6/76. Slade Wds 21/7/83. (MEA) : Upper Llanover 18/7/83, Risca 6/7/84, 27/6/88.

Sq. **SO** 20,30,50,41,51. **ST** 48,29,49,59.

1674 *Jodis lactearia* Linn.   Little Emerald

A small woodland moth common and widespread in Gwent flying from late May to mid-July and readily disturbed from undergrowth by day. It comes sparingly to light.

Llandogo (Nesbitt 1893). Wye Valley (Barraud 1906).

Usk, garden m.v.l. trap, recorded occasionally eg. 26/6/69, 15/7/73. Pont-y-saeson 7/6/66. Trelleck Common 14/6/67. Wentwood 21/6/69. Wyndcliff 22/6/71. Cwm Coed-y-cerrig 27/5/73. Slade Wds 15/6/73. Hael Wds 10/7/73. St Pierre's Great Wds 1/6/75. Hendre Wds 6/6/78. Trelleck Bog 8/7/78. Llansoy 30/6/86. (MEA) : Allt-yr-yn 20/6/76, Llantarnam 24/6/76, Mescoed Mawr 25/6/76, Ynysyfro 4/7/76, Risca 7/7/86.

Sq. **SO** 30,40,50,41,22. **ST** 28,38,48,29,39,49,59.

STERRHINAE

1676 *Cyclophora annulata* Schulze   The Mocha
Scarce in Monmouthshire, occurring locally in woods in the east of the county. It is double-brooded and has been noted from late May through June, July and August. It comes to light.

Tintern dist. (Piffard 1859). Wye Valley (Barraud 1906). Llandogo dist. "to light" 26/5/92 (Nesbitt 1892).

Wyndcliff 16/6/67 (GANH); 1/7/85 (MEA); 1986 13/6, 25/6, 6/7 (GANH, MEA); 22/5/91. Wentwood 27/5/70 (JMD), 17/8/83 (MEA). Usk (Plas Newydd), garden m.v.l. trap, 26/8/76 (one) (GANH).

Sq. **SO** 30,50. **ST** 49,59.

1677 *Cyclophora albipunctata* Hufn.   Birch Mocha
Llandogo dist. (Nesbitt 1892). No recent records.

Sq. **SO** 50.

1679 *Cyclophora porata* Linn.   False Mocha
Scarce and local in the eastern half of the county. It comes to light and has been noted from May to September.

Usk, garden m.v.l. trap, 1968 13/9; 1969 27/5, 27/7; 1971 (3) 4/6 – 31/7; 1973 18/8; 1976 19/8. Hendre Wds 26/5/70. Rhyd-y-maen 6/6/70.

Sq. **SO** 30,40,41.

1680 *Cyclophora punctaria* Linn.   Maiden's Blush
This moth, found locally in oakwoods in the eastern half of the county, is double-brooded, flying from late May to early July and again in August and early September. It comes to light.

Wye Valley (Barraud 1906). Tintern, one, 1906 (Bird 1907).

Usk garden m.v.l. trap, 1968 25/8, 3/9 and most years until 1983 eg. 1970 (4) 31/7 – 9/8; 1977 (9) 7/7 – 20/8. Ed. 25/5/80. Ld. 3/9/68.

Hendre Wds 26/5/70. Pont-y-saeson 6/6/72. Hael Wds 21/5/73, 3/8/90. Rhyd-y-maen 1/9/73. Redding's Inclosure 23/6/81. Slade Wds 2/8/82. Trelleck Bog 25/5/87. Wentwood. (MEA) : Wyndcliff 6/7/84.

Sq. **SO** 30,40,50,41,51. **ST** 48,49,59.

1681   *Cyclophora linearia* Hb.   Clay Triple-lines
This species is widespread and fairly common in Gwent in woods containing beeches. It flies from late May to mid-July and is occasionally noted in August and September and comes to light.

Llandogo dist. (Nesbitt 1892). Wye Valley, larvae on beech, August 1906 (Bird 1907). Redding's Inclosure 31/5/68 (JMC-H 1969).

Wyndcliff 7/6/66, 11/8/82. Wentwood 12/7/68. Usk, garden m.v.l. trap, recorded sporadiclly eg. 23/7/68; 1969 (4) 26/6 – 19/7; 11/8/82. Hael Wds 25/5/70. Redding's Inclosure 15/6/71. Pont-y-saeson 18/7/72. Piercefield Park 18/7/73. St Pierre's Great Wds 2/6/75. Slade Wds 21/7/83. Llansoy 4/6/85. Hendre Wds 11/7/86. Clydach 18/7/89. (MEA) : Llantarnam 10/6/76, Alt-yr-yn 20/6/76, Yynysyfro 22/6/76, Mescoed Mawr 25/6/76, Tintern 30/6/78, Risca 13/7/83, Ochrwyth 27/6/86 Cwm Merddog 14/7/86, Cwmfelinfach 5/9/86, Abercarn 27/6/87, Cwmcarn 11/7/87, Nant Gwyddon 31/5/88.

Sq. **SO** 10,30,40,50,21,41,51. **ST** 28,48,19,29,49,59.

1682   *Timandra griseata* Peters.   Blood-vein
Widespread and locally common in gardens, waste ground, damp open woods and marshy ground, this moth flies in two broods, the first in June and July and the second in August and the first half of September.

Usk, garden m.v.l. trap, noted annually 1976 – 83 eg. 1971, 1st brood (10) 4/6 – 8/8, 2nd brood (8) 11/8 – 17/9. Redding's Inclosure 7/6/71. Magor Marsh, frequent, 19/7/71 etc. Hael Wds 26/6/79. Llansoy 9/7/85. Hendre Wds 11/7/86. Dinham 4/6/87. Kilpale 30/6/87. (MEA) : Llantarnam 10/6/76, Allt-yr-yn 20/6/76, Ynysyfro 22/6/76, Wyndcliff 3/7/77, Tintern 30/6/78, Risca 13/7/83, Cwmfelinfach 5/9/86, Magor Pill 19/7/89.

Sq. **SO** 30,40,50,41,51. **ST** 28,48,19,29,39,49,59.

1690   *Scopula imitaria* Hb.   Small Blood-vein
Recorded fairly frequently in central Monmouthshire (Usk), but appears to be scarce elsewhere in the county. It flies in July and August and comes to light. It has occasionally been noted in June and September.

Abertillery dist. 1911 and 1912 (Rait-Smith 1912, 1913).

Usk, garden m.v.l. trap, 1967 7/7 and 13/7, then every year until 1983 eg. 1971, (7) 8/7 – 27/7; 1976, (7) 18/6 – 26/6; 1979 (5) 6/7 – 15/8. Magor Marsh 1974 (MJL in litt.)., 8/7/77 (MEA). Wyndcliff 7/7/83 and 1984 (MEA). Slade Wds 6/8/84, Llansoy 15/7/86 (GANH).

Sq. **SO** 20,30,40. **ST** 48,59.

1693   *Scopula floslactata* Haw.   Cream Wave
         ssp. *floslactata* Haw.
Common and widespread, flying in deciduous woodland from mid-May to

mid-July this moth is readily disturbed from foliage by day and at night comes to light.

Wye Valley (Barraud 1906). Abertillery dist. 1911 and 1912 (Rait-Smith 1912, 1913). Deri-fach 2/6/68 (JMC-H 1969).

Pont-y-saeson 25/5/68. Prescoed 8/6/69. Usk (Park Wood) 9/6/69. Hendre Wds 26/5/70. Slade Wds 1/6/71. Redding's Inclosure 15/6/71. Wyndcliff 22/6/71. Trelleck Bog 16/6/73. Piercefield Park 18/7/73. Usk, garden m.v.l. trap, 20/6/79. Hael Wds 26/6/85. Llansoy 2/7/85. Dinham 4/6/87. Magor Marsh 24/7/90. (MEA) : Pant-yr-eos 13/5/82, Mescoed Mawr 31/5/85, Risca 5/6/85, Nant Gwyddon 31/5/88, Abercarn 16/6/88.

Sq. **SO** 20,30,40,50,21,41,51. **ST** 28,48,29,39,49,59.

1694 *Scopula ternata* Schr.   Smoky Wave

Occurs locally on heaths and moorland in the west and north of the county during June and July. It sometimes flies by day and at night comes sparingly to light.

Abertillery dist. 1911 and 1912 (Rait-Smith 1912, 1913). Blorenge Mountain 24/6/71, plentiful by day (GANH). Cwm Tyleri 28/6/86 to m.v.l. (MEA) : Pontllanfraith 3/7/87, Trefil 3/7/87, Nant Gwyddon 25/6/88, Risca 20/6/93.

Sq. **SO** 20,11,21. **ST** 19,29.

1699 *Idaea vulpinaria* H.-S.   Least Carpet
     ssp. *atrosignaria* Lempke

1991: Wyndcliff 29/7/91, one to m.v.l., (MEA) (1st v-c. rec. and new to Wales) a probable vagrant, (in colln. GANH). BS wrote (in litt.) "1991 was a bumper year for *vulpinaria*, it turned up in many new sites as well as being abundant in many of its known haunts."

Sq. **ST** 59.

1702 *Idaea biselata* Hufn.   Small Fan-footed Wave

Common and widespread in gardens, woodland and hedgerows from late June until the end of August.

Wye Valley, 5/8/1906, "female deposited an ovum on apple leaf in nature" (Bird 1907).

Usk, garden m.v.l. trap, noted annually from 1967 to 1983 and often plentiful eg. 1974, 8/7 – 11/8. Ed. 23/6/75. Ld. 3/9/72. St Pierre's Great Wds 23/7/68. Wentwood 27/8/68. Wyndcliff 13/7/69. Rhyd-y-maen 18/7/70. Tintern 1/8/70. Magor Marsh 19/7/71. Redding's Inclosure 20/7/71. Usk (Park Wood) 23/8/72. Slade Wds 29/8/72. Hael Wds 17/7/73. Piercefield Park 18/7/73. Pont-y-saeson 10/8/74. Cwm Tyleri 29/6/76. Llangeview 28/7/79. Hendre Wds 11/7/86. Llansoy 13/7/86. Clydach 18/7/89.

(MEA) : Llantarnam 10/6/76, Ynysyfro 4/7/76, Mescoed Mawr 11/7/77, Upper Llanover 18/7/83, Risca 4/8/84 etc. inc. ab. *fimbriolata* Steph. 20/8/91, Cwm Merddog 8/8/86, Ochrwyth 9/8/86, Sor Brook 17/7/86, Cwmfelinfach 13/8/87.

Sq. **SO** 10,20,30,40,50,21,41,51. **ST** 28,48,19,29,39,49,59.

1707 *Idaea seriata* Schr. Small Dusty Wave
Usk, 1/8/73 (1) (1st. v-c. rec.) (GANH). Also single moths on 20/9/76, 20/7/77 and 23/7/79. All specimens were found resting on walls of house and out-buildings. No other records but moth probably overlooked.

Sq. **SO** 30.

1708 *Idaea dimidiata* Hufn. Single-dotted Wave
Flies in damp woods and marshy places from late June to mid-August. Local but fairly frequent where found.
Usk, garden m.v.l. trap, 6/7/68 and recorded annually until 1983 eg. 1975, (15) 21/6 – 4/8. Ed. 21/6/73. Ld. 22/8/68. Magor Marsh 11/7/69, 19/7/71. Wentwood 2/8/69. Slade Wds 23/7/79, 11/8/81 etc. (MEA) : Ynysyfro 28/6/76, Mescoed Mawr 11/7/77, Risca 13/7/82, Wyndcliff 20/7/83.

Sq. **SO** 30. **ST** 28,48,29,49,59.

1709 *Idaea subsericeata* Haw. Satin Wave
Occurs sparingly during June and July in damp open woodland and at night comes to light in small numbers. Local and mainly in the eastern half of the county.
Wye Valley, June 1906 (Bird 1907).
Usk, garden m.v.l. trap, recorded sporadically eg. 22/7/68; 1971, (2); 1979, (4) 22/6 – 30/6; 1983, 8/7. Pont-y-saeson 13/6/67, several to m.v.l. Magor Marsh 22/7/69. Prescoed 8/6/69. Redding's Inclosure 15/6/71. Hendre Wds 24/6/79. Llansoy 1/7/85. Slade Wds 19/6/85. (MEA) : Mescoed Mawr 4/9/77, Wyndcliff 29/6/78, Tintern 30/6/78, Risca 19/6/85, Nant Gwyddon 7/6/88, Abercarn 16/6/88.

Sq. **SO** 30,40,41,51. **ST** 48,29,49,59.

1711 *Idaea trigeminata* Haw. Treble Brown Spot
This moth is scarce and local, frequenting woodlands mainly in the east of the county and comes to light. It flies in June and early July with occasional, probably second generation moths, being seen at the end of July.
Usk, garden m.v.l. trap, 5/7/68 and noted annually until 1983 eg. 1978, (12) 20/5 – 14/7; 1979, (18) 10/6 – 7/7; 1983, (8) 22/6 – 11/7 and late moths

on 30/7 and 31/7. Wyndcliff 24/6/71. Slade Wds 26/6/80, 30/6/84, 27/6/85. Llansoy 4/7/85, 1986 (3). Hendre Wds 11/7/86. Kilpale 30/6/87. (MEA) : Risca 11/7/83, 5/7/87.

Sq. **SO** 30,40,41. **ST** 48,29,49,59.

1712 *Idaea emarginata* Linn. Small Scallop
"Monmouthshire" (Ince 1887). No other records.

1713 *Idaea aversata* Linn. Riband Wave
Common and widespread in open woodland, waste ground, hedgerows etc. from mid-June to mid-August, occasionally with late moths, probably representing a partial second generation, being seen in September. It is easily disturbed from bushes by day and at night comes readily to light.
    Abertillery dist. (Rait-Smith 1912, 1913).
    Usk, garden m.v.l. trap, 16/6/66 and noted annually until 1983 eg. 1971, abundant 30/6 – 23/8 with late individuals on 2/8 and 9/9. Ed. 11/6/75. Ld. 21/8/74. Pont-y-saeson 27/6/67. Prescoed 5/7/68. Wentwood 8/7/68. Wyndcliff 22/7/68. Usk (Park Wood) 23/7/68. Llangybi (Cae Cnap) 1/7/69. Hendre Wds 5/7/70. Hael Wds 5/7/70. Magor Marsh 10/7/70. Tintern 1/8/70. Cwm Tyleri 22/8/72. St Pierre's Great Wds 13/7/73. Piercefield Park 18/7/73. Trelleck Bog 16/7/79. Slade Wds 23/7/79. Llansoy 26/7/84. Kilpale 30/6/87. Sugar Loaf (St Mary's Vale) 8/7/87. (MEA) : Llantarnam 24/6/76, Ynysyfro 28/6/76, Mescoed Mawr 1/7/76, Risca 10/7/79, Newport Docks 11/7/84, Pant-yr-eos 6/7/85, Cwm Merddog 7/7/85, Ynysddu 12/7/85, Abercarn 16/7/85, Cwmfelinfach 15/7/86, Pontllanfraith 10/7/87, Cwmcarn 11/7/87, Ochrwyth 17/6/88, Dixton Bank 18/6/88.

Sq. **SO** 10,20,30,40,50,21,41,51. **ST** 28,38,48,19,29,39,49,59.

1715 *Idaea straminata* Borkh. Plain Wave
Local and scarce in and Monmouthshire. Found in woods in the east and moorland in the north of the county.
    Pont-y-saeson 13/6/67 to m v.l. (REMP, GANH) (1st v-c. rec.). Trelleck Common 14/6/67 and 21/6/68, disturbed from bilberry bushes by day (REMP, GANH). Llansoy one to m.v.l. 7/8/88. Trelleck Bog 25/7/90 (1) to m.v.l. (MEA) : Cwm Tyleri 28/7/91, plfl. to m.v.l.

Sq. **SO** 20,40,50. **ST** 59.

1716 *Rhodometra sacraria* Linn. The Vestal
A casual immigrant from S. Europe and Africa, usually seen in the autumn.
    Usk, 21/9/66, one on lighted door-panel (GANH) (1st v-c. rec.). Usk, garden m.v.l. trap, 1982, 18/9 (1); 1983, (9) 25/9 – 28/9. Llansoy, to u.v.l.,

30/8/87 (1), 6/9/89 (1). (MEA) : Risca 1/9/84, one to act. l.

Sq. **SO** 30,40. **ST** 29.

LARENTIINAE

1719 **Orthonama vittata** Borkh.   Oblique Carpet
Local and scarce in Monmouthshire, appearing only sporadically except at Magor Marsh where it has been encountered with some regularity from early June to late September.

Usk, garden m.v.l. trap, two records only from 1966 to 1983, viz. 3/6/66 (1st v-c. rec.) and 30/8/73. Rhyd-y-maen 6/6/70 (2). Magor Marsh 7/9/69 (12) to m.v.l., 2/6/70 (svl.), 11/8/71, 23/9/72, 3/7/73 (GANH); 2/9/87 (MEA).

Sq. **SO** 30,40. **ST** 48.

1720 **Orthonama obstipata** Fabr.   The Gem
An immigrant species seen mostly in the autumn and coming readily to light.

Usk (Plas Newydd), garden m.v.l. trap, 23/10/68 (1); 1969, singletons on 21/7, 26/7, 5/9 and 27/10 (all ♀). Usk Castle. m.v.l. trap (RBH), 1969, 21/10 ♂♂ (5), ♀ (1). Magor Marsh 3/7/73 (1) to m.v.l. Usk (Plas Newydd) m.v.l. trap, 18/10/77 ♂ (1), 8/11/83 ♀ (1).

Sq. **SO** 30. **ST** 48.

1722 **Xanthorhoe designata** Hufn.   Flame Carpet
Widespread and common in hedgerows and damp woodland, flying in May and June and as a second brood, in August and September.

Llandogo (Nesbitt 1893). Wye Valley (Barraud 1906). Wye Valley (Bird 1907). Aberbeeg (Rait-Smith 1912, 1913). Abertillery dist. (Rait-Smith 1915). Tintern (Bird 1913).

Usk, garden m.v.l. trap, annually 1966 – 1983 eg. 1973, 1st brood (190) 16/5 – 12/7, 2nd brood (14) 30/7 – 13/9. Ed. 21/4/72. Ld. 22/9/78. Gwehelog (Camp Wood) 30/4/67. Llantrisant (Nant-y-banw) 25/5/68. Wentwood (Darren) 24/8/68. Prescoed 8/6/69. Usk (Park Wood) 9/6/69. Wentwood (Foresters' Oaks) 19/8/69. Wyndcliff 2/9/69. Hendre Wds 26/5/70. Pont-y-saeson 2/6/70. Rhyd-y-maen 6/6/70. Craig-y-Master 7/5/71. Redding's Inclosure 15/6/71. Cwm Coed-y-cerrig 27/5/73. Cwm Tyleri 1/7/73. Newport (Coldra) 27/6/75. Newport Docks 17/8/76. Hael Wds 30/8/77. Llansoy 26/5/85. Kilpale 30/6/87. (MEA) : Mescoed Mawr 3/9/77, Risca 6/9/81, Slade Wds 8/7/82, Pant-yr-eos 6/7/85, Cwmfelinfach 12/8/86, Abercarn 27/6/87, Lower Machen (Park Wood) 13/5/88, Nant Gwyddon 10/6/88, Ochrwyth 17/6/88.

Sq. **SO** 20,30,40,50,41,51,22. **ST** 28,38,48,19,29,39,49,59.

1724   *Xanthorhoe spadicearia* D. & S.   Red Twin-spot Carpet
This moth, widespread and common in the eastern half of Gwent but less so in the west, frequents hedgerows and open woods from early May to early June and, as a second brood, from mid-July to early September. It is easily disturbed from bushes and herbage by day and at night comes to light.

Wye Valley (Bird 1907. 64).

Tintern and Redding's Inclosure 31/5/68 (JMC-H 1969). Usk, garden trap, recorded every year from 1966 to 1983 eg. 1969, 1st brood (7) 5/5 – 3/6, 2nd brood (11) 19/7 – 31/8. Ed. 26/4/82. Ld. 8/9/82. Gwehelog (Camp Wood) 30/4/67. Llantrisant (Nant-y-banw) 25/5/68. Prescoed 30/5/68. Tredean Wds 31/5/68. Wentwood 19/6/68 (JMD). Rhyd-y-maen 23/5/70. Slade Wds 30/5/71. Redding's Inclosure 7/6/71. Bica Common 5/6/72. Trelleck Bog 1/6/73. Runston 6/8/74. Hael Wds 10/8/74. Pont-y-saeson 30/5/75 (REMP). Cwm Coed-y-cerrig 24/5/76. Magor Marsh 8/8/76. Hendre Wds 29/8/78. Llansoy 11/8/84. Dinham 4/6/87. (MEA) : Wyndcliff 26/5/77, Risca 9/8/74, Cwmfelinfach 24/5/87, Nant Gwyddon 15/8/88.

Sq. **SO** 30,40,50,41,51,22. **ST** 48,19,29,39,49,59.

1725   *Xanthorhoe ferrugata* Cl.   Dark-barred Twin-spot Carpet
Widespread and fairly common in gardens, open woodland, waste-ground etc., this species is double-brooded, flying in May and June and the second generation in July, August and early September.

Llandogo dist. (Nesbitt 1892). Wye Valley (Bird 1907). Aberbeeg (Rait-Smith 1913). Abertillery dist. (Rait-Smith 1912, 1915).

Usk, garden m.v.l. trap, every year 1967 – 83 eg. 1971, 1st brood (8) 9/5 – 30/6, 2nd brood (16) 18/7 – 8/9. Ed. 9/5/71. Ld. 15/9/82. Glascoed 17/5/66. Wyndcliff 12/6/67 (REMP). Bica Common 23/8/68. Prescoed 8/6/69. Magor Marsh 22/7/69. Rhyd-y-maen 23/5/70. Redding's Inclosure 27/8/71. Cwm-mawr 30/8/71. Usk (Park Wood) 23/8/72. Llanfrechfa 26/5/73. Newport Docks 17/8/76 (5) to m.v.l. Llansoy 8/5/87. Hendre Wds 9/5/87. (MEA) : Allt-yr-yn 23/5/76, Wentwood 3/9/79, Risca 7/7/81, Mescoed Mawr 28/7/84, Ochrwyth 6/6/86, Trelleck Bog 29/8/86, Hael Wds 31/8/86, Cwmfelinfach 23/5/87, Pontllanfraith 16/8/87, Nant Gwyddon 21/5/88, Uskmouth 16/8/88.

Sq. **SO** 20,30,40,50,41,51. **ST** 28,38,48,19,29,39,49,59.

1727   *Xanthorhoe montanata* D. & S.   Silver-ground Carpet
    ssp. *montanata* D. & S.

Flying from late May to mid-July in gardens, hedgerows, rough ground, woodland rides and margins etc., this moth is readily disturbed from undergrowth and herbage by day and at night comes to light. Probably the commonest and most widespread Geometer in the county.

Tintern dist.: "one of our most abundant species", ab. *degenerata* Prout; very scarce (Bird 1905). Wye Valley (Barraud 1906). Abertillery dist. (Rait-Smith 1912,1913). Coed-y-Bynydd 5/6/68 (JMC-H 1969).

Usk, garden trap, recorded every year from 1966 to 1983 and often plentiful. Ed. 18/5/80. Ld. 15/7/81. Usk (Park Wood) 5/6/67. Wyndcliff 6/6/67. Pont-y-saeson 13/6/67. Trelleck Common 14/6/67. Craig-y-Master 21/5/68. Prescoed 31/5/68. Wentwood 19/6/68. Bica Common 20/6/68. Rhyd-y-maen 20/5/70. Magor Marsh 2/6/70. Slade Wds 30/5/71. Cwm-mawr 31/5/71. Hendre Wds 5/6/71. Redding's Inclosure 15/6/71. Cwm Tyleri 30/7/72. Cwm Coed-y-cerrig 27/5/73. Tredean Wds 5/6/73. Trelleck Bog 16/6/73. Mynyddislwyn 10/6/74. St Pierre's Great Wds 24/5/75. Newport (Coldra) 15/6/75. Hael Wds 27/6/78. Tal-y-coed 9/9/83. Llansoy 22/6/86. Dinham 4/6/87. Kilpale 30/6/87. (MEA) : Rhiwderin 21/5/76, Allt-yr-yn 23/5/76, Llantarnam 10/6/76, Ynysyfro 22/6/76, Mescoed Mawr 1/7/76, Black Vein 12/6/81, Risca 20/6/83, Pant-yr-eos 8/6/84, Ochrwyth 1/6/85, Cwm Merddog 7/7/85, Cwmfelinfach 10/6/87, Abercarn 27/6/87, Pontllanfraith 3/7/87, Fleur-de-lis 6/7/87, Dixton Bank 18/6/88, Nant Gwyddon 25/6/88.

Sq. **SO** 10,20,30,40,50,21,31,41,51,22. **ST** 28,38,48,19,29,39,49,59.

1728 *Xanthorhoe fluctuata* Linn.   Garden Carpet
     ssp. *fluctuata* Linn.

Common and widespread in gardens, waste ground, marshy places etc. flying from early May to October.

Tintern dist. ab. (Bird 1905). Aberbeeg (Rait-Smith 1912,1913).

Usk, garden m.v.l. trap, noted annually from 1966 to 1983 eg. 1971, (30) 18/5 – 22/9. Ed. 2/5/68. Ld. 11/10/78. Wyndcliff 14/6/67 (REMP). Llangwm 24/5/68. Llantarnam 25/6/68. Wentwood 19/8/69. Trelleck Bog 18/9/72. Pont-y-saeson 26/5/73. Cwm Tyleri 31/5/74. Magor Marsh 30/6/75. Newport Docks 17/8/76. Cwm Coed-y-cerrig 5/6/78. Hael Wds 27/6/78. Hendre Wds 29/5/79. Slade Wds 9/9/82. Llansoy 4/7/85. (MEA) : Allt-yr-yn 20/6/76, Mescoed Mawr 25/6/76, Risca 19/6/79, Ochrwyth 27/6/86, Cwmfelinfach 5/9/86, Pontllanfraith 20/6/87.

Sq. **SO** 20,30,40,50,41,22. **ST** 28,38,48,19,29,39,49,59.

1732 *Scotopteryx chenopodiata* Linn.   Shaded Broad-bar

Widespread and not uncommon in Monmouthshire frequenting grassy hillsides and embankments in July and August. Readily disturbed from the herbage by day and occasionally seen flying freely on sunny days. At night it comes sparingly to light.

Wye Valley (Bird 1907).

Usk, garden m.v.l. trap, recorded every year from 1966 to 1983 eg. 1969, (12) 18/7 – 18/8; 1973, (11) 5/7 – 22/8. Ed. 5/7/73. Ld. 22/8/73.

Glascoed 1965. Pont-y-saeson 1965, 10/8/68. Runston 23/7/68. St Pierre's Great Wds 11/8/68. Wentwood 2/8/69. Magor Marsh 2/8/70. Undy (sea-wall) 13/7/73. Newport Docks 17/8/76. Slade Wds 8/8/80. Wyndcliff 6/7/81. Usk 15/7/82, plentiful on old railway line. (MEA) : Risca 7/8/81, Dixton Bank 21/7/87, Pontllanfraith 15/8/87, Uskmouth 16/8/88.

Sq. **SO** 30,50,51. **ST** 38,48,19,29,49,59.

1733 *Scotopteryx mucronata* Scop.   Lead Belle
ssp. *umbrifera* Heyd.

Locally common in open heathy woods in the eastern half of the county from mid-May to mid-June and often flying by day.

Abertillery dist. (Rait-Smith 1912, 1913). Redding's Inclosure 31/5/68 (JMC-H 1969).

Glascoed 17/5/65. Wyndcliff 7/6/66 to m.v.l. Usk, garden m.v.l. trap, 11/5/67. Usk (Park Wood) 15/6/67. Trelleck Common 14/6/67 plentiful (by day). Trelleck Bog 2/6/70 (plentiful in daytime). Bica Common 13/6/72.

Sq. **SO** 20,30,50,51. **ST** 49,59.

1734 *Scotopteryx luridata* Hufn.   July Belle
ssp. *plumbaria* Fabr.

Not uncommon in Monmouthshire, but of local occurrence, mainly in the hills and moorlands of the north-west of the county and flying from mid-June to late July.

Llandogo dist. (Nesbitt 1892, 1893) as *Eubolia plumbaria*, "but I think probably *mucronata*" comments JMC-H (1966). However, I believe "*luridata*" may well be correct as it does occur in the vicinity (GANH pers. obs.).

Usk, garden m.v.l. trap, occurred sporadically over the years 1966 – 83 eg. 1969, 28/6, 8/7; 1973, 1/7; 1974, 17/7, 23/7; 1981, 14/7. Pont-y-saeson 13/6/67. Craig-yr-iar 28/6/70 Blorenge 25/6/71. Cwm Tyleri 26/6/76. Trelleck Bog 24/7/86 (one to m.v.l.). (MEA) : Sugar Loaf (St Anne's Vale) 7/7/71, Blorenge Mountain 3/7/76 & 7/7/84, Wentwood 22/6/78, Risca 16/7/86, Dinham 7/7/87.

Sq. **SO** 20,30,40,50,21. **ST** 29,49,59.

1735 *Catarhoe rubidata* D. & S. Ruddy Carpet

Local and scarce in Gwent and, as far as I am aware, recorded from one station only.

Wyndcliff 12/6/67 (1) to m.v.l. (REMP); 10/7/67 (1) (GANH); 7/7/84 (1), 6/7/86 (1), (MEA); 3/7/91 (2) (GANH).

Sq. **ST** 59.

1736   *Catarhoe cuculata* Hufn.   Royal Mantle
Two Monmouthshire records only viz. Usk, garden m.v.l. trap, 31/7/69 (1) (GANH) (1st v-c. rec.). Risca 8/7/83 (1) to act.l. (MEA).

Sq. **SO** 30. **ST** 29.

1737   *Epirrhoe tristata* Linn.   Small Argent & Sable
This species, flying in June and July, occurs locally in open woods and heaths in the east of the county though it is more common in the hills and moorland of the west and north of Monmouthshire. It may sometimes be seen flying freely in afternoon sunshine.

Aberbeeg (Rait-Smith 1912, 1913). Tintern "fairly common by day in 1947 and a few in 1946 and 1948" (Blathwayt 1967). Tintern, (Mere 1965)

Wyndcliff 23/6/67, to m.v.l. (GANH). Wentwood 18/6/69 (JMD), 2/6/70 (GANH). Trelleck Bog 12/7/70 (MWH in litt.). Mynyddislwyn 21/6/74. Cwmfelinfach 20/6/87. Trelleck (Fedw Fach) 27/5/88. (MEA) : Risca 14/7/83, Trefil 4/7/87.

Sq. **SO** 20,40,50,11. **ST** 19,29,49,59.

1738   *Epirrhoe alternata* Müll.   Common Carpet
   ssp. *alternata* Müll.
Widespread and common, frequenting gardens, waste ground, hedgerows, open woods etc. Flying mainly in June and July and again in August and September, it has been recorded from early May to early October.

Wye Valley (Bird 1907). Abertillery dist. 1911, "common everywhere" (Rait-Smith 1912).

Usk, garden m.v.l. trap, noted every year 1966 – 1983 eg. 1968, (21) 2/5 – 12/9; 1979, (22) 14/5 – 19/9. Ed. 2/5/68. Ld. 2/10/69. Tintern 31/5/68. Pont-y-saeson 8/6/88.m Wentwood 10/6/68. Usk (Park Wood) 10/6/68. Bica Common 20/6/68. St Pierre's Great Wds 11/8/68. Magor Marsh 24/8/68. Prescoed 8/6/69. Wyndcliff 13/7/69. Rhyd-y-maen 6/6/70. Hael Wds 5/7/70. Craig-yr-iar 28/6/70. Redding's Inclosure 7/6/71. Cwm Tyleri 9/7/73. Piercefield Park 18/7/73. Llangybi Park 26/8/73. Newport Docks 17/8/76. Hendre Wds 27/5/78. Slade Wds 28/5/78. Cwm Coed-y-cerrig 5/6/78. Trelleck Bog 8/7/78. Penallt 1/6/79. Llansoy 4/6/85. Kilpale 30/6/87. (MEA) : Mescoed Mawr 1/7/76, Risca 28/8/81, Pant-yr-eos 13/5/82, Ynysddu 12/7/85, Cwm Merddog 14/7/86, Cwmfelinfach 15/7/86, Ochrwyth 9/8/86, Gaer Hill (Newport) 13/6/87, Ebbw Vale 22/7/87, Nant Gwyddon 25/6/88.

Sq. **SO** 10,20,30,40,50,41,51,22. **ST** 28,38,48,19,29,39,49,59.

1739   *Epirrhoe rivata* Hb.   Wood Carpet
Only once, over the years, have I encountered this species in Monmouthshire so I find the 1911 record somewhat surprising.

Abertillery dist. 1911, "common everywhere" (Rait-Smith 1912).

Slade Wds, 15/7/80, one to m.v.l. (GANH).

Sq. **SO** 20. **ST** 48.

1740 *Epirrhoe galiata* D. & S.   Galium Carpet
Occurs in the hills and moorlands in the west and north of Gwent where it flies from June to August but not common.

Abertillery dist. (Rait-Smith 1912). Cwm Tyleri 1972, plentiful to m.v.l., 28/7 to 22/8. Also 1973, 74, 75, 76, 77 and 86. Ed. 31/5/74. Ld. 22/8/72. (MEA) : Blorenge 7/7/84, Risca 6/7/84 & 7/7/86, Ochrwyth 27/6/86.

Sq. **SO** 20,21. **ST** 28,29.

1742 *Camptogramma bilineata* Linn.   Yellow Shell
ssp. *bilineata* Linn.
Widespread and fairly common, flying from mid-June to mid-August in open woods, hedgerows and bushy places generally. It often flies by day when it is easily disturbed from herbage and bushes but at night it comes only sparingly to light.

Usk, garden m.v.l. trap, noted occasionally eg. 21/8/66, 23/9/68, 1977 (3) 15/8 – 28/8. Wyndcliff 12/6/67 (REMP). Pont-y-saeson 27/6/67, plentiful by day. Hael Wds 12/7/74. Magor Marsh 1974 (MJL), 13/7/83. Cwm Tyleri 20/6/76. Hendre Wds 9/7/78. Llansoy 28/6/87. Kilpale 30/6/87. Dixton Bank 18/6/88. Clydach 18/7/89. (MEA) : Llantarnam 24/6/76, Mescoed Mawr 25/6/76, Ynysyfro 4/7/76, Wentwood 10/7/77, Black Vein 3/7/81, Risca 8/7/83, Newport Docks 11/7/84, Gaer Hill (Newport) 17/7/86, Cwmfelinfach 6/7/87, Ebbw Vale 22/7/87, Pontllanfraith 16/8/87, Uskmouth 16/8/88.

Sq. **SO** 10,20,30,40,50,21,41,51. **ST** 28,38,48,19,29,39,49,59.

1744 *Entephria caesiata* D. & S.   Grey Mountain Carpet
Flying during June and July, this moth occurs locally in the hills of north-west Gwent.

Abertillery dist. two only (Rait-Smith 1912).

Cwm Tyleri (1,400 ft.) 1972, 28/7 (2), 30/7 (4); 1973, single moths on June 3, 17, 19 and July 9 and several on July 1; 1976, plentiful on June 26, 27, 29 and July 8. Also recorded 1974, 75, 77. (MEA) : Blorenge 7/7/84, Cwm Tyleri 1985, 86, 87.

Sq. **SO** 20,21.

1745 *Larentia clavaria* Haw.   The Mallow
Very scarce in Monmouthshire.

Usk, garden m.v.l. trap, 1967, 24/9 (1) (1st v-c. rec.); also single specimens on 1/10/68, 6/10/67, 16/9/81, 1/10/83. No other records.

Sq. **SO** 30.

1746   *Anticlea badiata* D. & S.   Shoulder Stripe

Common and widespread in Monmouthshire, frequenting woods, hedgerows, gardens etc. from mid-March to mid-May and coming readily to light.

Wye Valley (Bird 1907). Tintern (Bird 1913). Cwmyoy 1965 (SHR).

Usk, garden m.v.l. trap, recorded every year from 1967 to 1984 eg. 1971, (17) 23/3 – 19/4; 1980, (7) 30/3 – 12/5. Ed. 9/3/84. Ld.13/5/79. Coed-y-Fferm 21/4/70. Wentwood 6/5/70. Hadnock Quarries 30/3/71. Redding's Inclosure 7/3/72. Pont-y-saeson 20/3/72. Prescoed 16/4/72. St Pierre's Great Wds 23/3/73. Hendre Wds 27/4/79. Llansoy 4/5/85. Wyndcliff 28/4/87. (MEA) : Mescoed Mawr 27/2/77, Llantarnam 18/3/77, Slade Wds 21/4/78, Cwmfelinfach 14/4/87, Risca 7/3/88.

Sq. **SO** 30,40,41,51,32. **ST** 48,19,29,39,49,59.

1747   *Anticlea derivata* D. & S.   The Streamer

Widespread and fairly common in hedges, open woodland and bushy places flying from mid-March to mid-May.

Wye Valley (Bird 1907). Tintern (Bird 1913). Gilwern, one, 1911 (Rait-Smith 1912). Cwmyoy 1965 (SHR).

Usk, garden m.v.l.trap, a few most years from 1967 – 84 eg. 1976, (3) 19/4 – 2/5; 1980, (7) 15/4 – 29/4. Ed. 3/4/74. Ld. 22/5/78. Usk (Park Wood) 26/4/68. Craig-y-Master 7/5/71. Wyndcliff 26/4/76. Hendre Wds 21/4/80. Slade Wds 22/4/80. Llansoy 4/5/85, 18/3/90. Hael Wds 8/5/87. (MEA) : Mescoed Mawr 16/4/78, Risca 12/5/82, Pant-yr-eos 18/5/84, Abercarn 26/5/87.

Sq. **SO** 30,40,50,21,41,32. **ST** 29,49,59.

1748   *Mesoleuca albicillata* Hb.   Beautiful Carpet

This woodland moth, flying during June and July, is widely distributed but at low density in Monmouthshire. It is readily disturbed from bushes by day and at night comes sparingly to light. Ed. 26/5/70 (Pont-y-saeson). Ld. 1/8/70 (Tintern) and a very late specimen on 18/9/70 (Pont-y-saeson).

Near Tintern (Piffard 1859), Wye Valley 1892 (Nesbitt). Wye Valley (Bird 1907). Abertillery dist. (Rait-Smith 1912, 13, 15).

Usk (Park Wood) 27/6/67, 7/7/69. Pont-y-saeson 27/6/67. Wentwood 8/7/68. Wyndcliff 13/7/69. Usk, garden m.v.l. trap, 14/7/79. Hael Wds 5/7/70. Tintern 1/8/70. Hendre Wds 17/7/79. Slade Wds 23/7/79 (2), 24/6/80. Redding's Inclosure 4/7/81. (MEA) : Mescoed Mawr 3/7/76, Ynysyfro 4/7/76, Cwmfelinfach 15/7/86, Nant Gwyddon 25/6/88.

Sq. **SO** 20,30,50,41,51. **ST** 28,48,19,29,49,59.

1750   *Lampropteryx suffumata* D. & S. Water Carpet

Widespread and often plentiful in woodland rides and clearings from mid-April to early June.

Tintern dist. (Bird 1905). Wye Valley (Barraud 1906). Wye Valley (Bird 1907). Aberbeeg (Rait-Smith 1913). Pontllanfraith (Rait-Smith 1915).

Usk, garden m.v.l. trap, noted occasionally eg. 17/4/67; 1977, (3) 10/5 – 26/5; 22/4/84. Ed. 6/4/81 (Redding's Inclosure). Ld. 26/6/85 (Hael Wds). Usk (Park Wood) 28/4/67 (abdt.). Gwehelog (Camp Wood) 29/4/67. Llangybi (Cae Cnap) 23/4/68. Pont-y-saeson 26/5/70. Usk (Cockshoots Wood) 29/4/71. Cwm-mawr 30/5/71. Prescoed 16/4/72. Redding's Inclosure 17/4/72, 6/4/81.Trelleck Bog 22/5/73. Hendre Wds 27/5/79. Cwm Coed-y-cerrig 5/6/78. Hael Wds 21/5/84. Llansoy 19/5/85. Clydach 11/5/90. (MEA) : Wentwood 7/5/77, Wyndcliff 21/5/77, Black Vein 8/5/81, Ochrwyth 7/6/86, Risca 14/6/86, Cwmfelinfach 24/5/87, Nant Gwyddon 10/6/88.

Sq. **SO** 20,30,40,50,21,41,51,22. **ST** 28,19,29,39,49,59.

1751 *Lampropteryx otregiata* Metc. Devon Carpet
In Monmouthshire, occurs at one site only, an area of damp woodland and alder carr in the north of the county.

Northern Monmouthshire 1973, one to m.v.l. 27/5 (GANH) (1st v-c. rec.), 5/6/73 (4).

Sq. **SO** 22.

1752 *Cosmorhoe ocellata* Linn. Purple Bar
Widespread and fairly common, this moth inhabits open woodland, heathy areas and also the hills and moorland in the north-west of the county. It flies in June and again, as a partial second generation, in late August and the first half of September.

Llandogo dist. (Nesbitt 1892). Abertillery dist. (Rait-Smith 1912, 1913).

Usk, garden m.v.l. trap, noted sporadically eg. 1968, 1st brood (1) 12/6; 2nd brood (3), 25/8 – 15/9. Usk (Park Wood) 3/9/69. Cwm Tyleri 30/7/72, 22/8/72 and annually to 1977. Trelleck Bog 16/6/73. Wentwood 29/6/73. St Pierre's Great Wds 13/7/73. Piercefield Park 18/7/73. Hendre Wds 21/5/79. Slade Wds 21/8/79. Redding's Inclosure 23/6/81. Wyndcliff 3/8/81. Hael Wds 31/7/82. Llansoy 26/6/86. Dinham 4/6/87. Kilpale 30/6/87. (MEA) : Risca 19/7/79, Ochrwyth 27/6/86, Cwmfelinfach 15/6/86, Abercarn 16/5/87, Pontllan-fraith 22/6/87, Trefil 3/7/87, Nant Gwyddon 25/6/88.

Sq. **SO** 20,30,40,50,11,41,51. **ST** 28,48,19,29,49,59.

1753 *Nebula salicata* Hb. Striped Twin-spot Carpet
   ssp. *latentaria* Curt.
Local and scarce in Monmouthshire and to date recorded only from two sites in the moorland areas of the north.

Cwm Tyleri, 26/5/73, one to m.v.l. (GANH) (1st v-c. rec.).; 31/5/74, (10) to m.v.l. Trefil, 31/8/90 (MEA).

Sq. **SO** 20,11.

1754 *Eulithis prunata* Linn.   The Phoenix
Local and at low density in Gwent, flying in gardens and open woodland during July and August and coming to light.

Wye Valley (Bird 1907). Tintern, not uncommon (Bird 1913).

Usk 18/8/65. Usk, garden m.v.l. trap, a few recorded every year from 1966 to 1983 eg. 1969, (5) 15/7 – 3/8; 1978, (7) 13/7 – 9/9. Ed. 1/7/75. Ld. 9/9/78. Hael Woods 10/7/73, 6/8/74, 2/8/81. Hendre Wds 20/7/79. Wyndcliff 2/8/83. Llansoy 12/7/86, 31/7/86, 14/8/87. (MEA) : Wentwood 10/7/77, 12/8/81; Risca 12/8/81; Cwmfelinfach 13/8/87.

Sq. **SO** 30,40,50,41. **ST** 19,29,49,59.

1755 *Eulithis testata* Linn.   The Chevron
This species, flying from late July to late September, occurs at low density in open woodland and heathy localities in eastern Monmouthshire but is much more common in the hilly moorland areas in the north of the county.

Llandogo dist. (Nesbitt 1892). Wye Valley (Bird 1907). Abertillery dist. (Rait-Smith 1912, 1913).

Usk, garden m.v.l. trap, occurred sporadically and sparingly eg. 1968, 20/8 (1), 12/9(1); 1972, 13/9 (1); 1975, 11/8 (5). Cwm Tyleri, to m.v.l., 1982, 28/7, 22/8, 28/8; 1973; 1974; 15/9/76, abundant on herbage at night. Trelleck Bog 18/9/72, 4/9/73. Slade Wds. Blorenge 1986, abundant on heather at night 16/9 and 20/9. Llansoy 2/8/90 (1) to m.v.l. (MEA) : Hael Wds 31/8/86, Trefil 15/8/87, Pontllanfraith 1/9/87, Nant Gwyddon 7/8/88.

Sq. **SO** 20,30,40,50,11,21. **ST** 48,19.

1756 *Eulithis populata* Linn.   Northern Spinach
This moth, flying from early July to late September, is widespread and fairly common in the eastern half of the county in open woodland and heathy localities wherever whinberry (bilberry) occurs, but is abundant in the hilly moorland areas in the north of the county.

Wye Valley (Bird 1907). Abertillery dist. (Rait-Smith 1913).

Usk, garden m.v.l.trap, recorded occasionally eg. 19/7/68, 28/7/79. Pont-y-saeson 27/6/67. Llangybi (Coed-y-Fferm) 4/7/68. Wentwood 8/7/68. Trelleck Common 9/7/68. Usk (Park Wood) 7/7/69. Wyndcliff 13/7/69. Hael Wds 5/7/70. Cwm Tyleri, to m.v.l., 28/7/72 (abdt.), and annually 1973 – 1977; 5/8/74, many hundreds resting on flower-heads of rushes in moonlight (many in cop.). Blaenavon Mountain 25/8/73. Newport (Coldra) 27/6/75. Trelleck Bog 22/7/78. Blorenge 20/9/86, abundant on bilberry and heather at night. (MEA) : Risca 27/7/84, Pontllanfraith 20/7/87, Trefil 15/8/87.

Sq. **SO** 20,30,50,11,21,51. **ST** 38,19,29,39,49,59.

1757   *Eulithis mellinata* Fabr.   The Spinach
This species, flying during July, frequents gardens and open woods, but is very sparsely distributed in Monmouthshire.
    Abertillery dist. (Rait-Smith 1912, 1913).
    Usk, garden m.v.l. trap, 1967 30/6 (1), 12/7 (1); 1969 19/7 (1) – the only records in the eighteen years fron 1966 to 1983. Wentwood 11/7/69 (GANH), 9/7/84 (MEA). Caerwent (Brockwells) 3/8/71 (CT). Magor Marsh 1974 (MJL). (MEA) : Mescoed Mawr 1/7/76; Risca 1981, 7/7 and 9/7, also 1986, 87 and 88; Nant Gwyddon 7/8/88.

Sq. **SO** 20,30. **ST** 48,29,49.

1758   *Eulithis pyraliata* D. & S.   Barred Straw
This moth, flying from late June to early August, is widespread and locally common in the eastern half of Monmouthshire frequenting hedgerows, bushy places and woodland rides and margins.
    Monmouthshire (Ince 1887). Abertillery dist. (Rait-Smith 1912).
    Usk, garden m.v.l. trap, recorded every year from 1966 – 83 and usually plentiful eg 1973, (19) 5/7 – 1/8; 1978, (53) 27/6 – 8/8. Ed. 20/6/75. Ld. 8/8/78. Prescoed 5/7/68. Usk (Park Wood) 7/7/69. Magor Marsh, plentiful, 11/7/69, 13/7/83 etc. Hael Wds 3/7/71. Redding's Inclosure 20/7/71. Undy (sea-wall) 13/7/73. Cwm Tyleri 9/7/73. Piercefield Park 18/7/73. Trelleck Bog 16/7/79. Llansoy 4/7/85. (MEA) : Risca 13/7/83.

Sq. **SO** 20,30,40,50,51. **ST** 48,29,39,59.

1759   *Ecliptopera silaceata* D. & S.   Small Phoenix
Widespread and very common in gardens, waste ground, open woods, commons etc., flying in two broods, the first in May and June, the second in the latter half of July, August and September. The form *insulata* Haw. occurs frequently
    Tintern dist. (Piffard 1859). Wye Valley (Bird 1907.64). Tintern (Bird 1913). Tintern 31/5/68 (JMC-H 1969).
    Usk, garden m.v.l. trap, plentiful every year from 1966 – 83 eg. 1969, 1st brood (19) 21/5 – 20/6; 2nd brood (27) 19/7 – 14/9. Ed. 28/4/82. Ld. 24/9/70, 12/10/78. Usk (Park Wood) 26/4/68. Rhyd-y-maen 21/5/68. Prescoed 31/5/68. Wentwood 8/7/68. Wyndcliff 22/7/68. Bica Common 23/8/68. Hael Wds 25/5/70. Pont-y-saeson 26/5/70. Hendre Wds 26/5/70. Redding's Inclosure 7/6/71. Cwm-mawr 2/7/71. Tredean Wds 5/6/73. Cwm Coed-y-cerrig 5/6/73. Mynyddislwyn 10/6/74. Slade Wds 27/5/78. Magor Marsh 24/8/81. Llansoy 11/8/84. Trelleck Bog 18/7/86, (MEA) · Llantarnam 10/6/76, Ynysyfro 20/6/76, Mescoed Mawr 11/6/78, Risca

8/6/80, Pant-yr-eos 13/5/82, Ochrwyth 1/6/85, Cwmfelinfach 5/9/86, Abercarn 16/5/87, Pontllanfraith 16/8/87, Nant Gwyddon 15/8/88.

Sq. **SO** 20,30,40,50,41,51,22. **ST** 28,48,19,29,39,49,59.

1760 *Chloroclysta siterata* Hufn. Red-green Carpet
Scarce in Monmouthshire.
    Tintern, at ivy bloom, a few, Oct. 14 1876 (Jones 1876). Tintern, two 1911; at ivy, 1912 (Bird 1913).
    Cwmyoy, 1965 Mar/Apr, one to m.v.l. (SHR). Hilston Park (Skenfrith) 1986 4/10 (1) to m.v.l. (MEA). Llansoy 1986, 15/10 (1); 1990, (4) 10/10 – 16/11. Hael Wds 1987, 8/5 (1) to m.v.l. (GANH). Wyndcliff 1992, 6/11 (1) (MEA).

Sq. **SO** 40,50,41,32. **ST** 59.

1761 *Chloroclysta miata* Linn. Autumn Green Carpet
Rare in Monmouthshire, very few records.
    Llandogo (Nesbitt 1892). Wye Valley (Bird 1907). Tintern (Bird 1913). Cwmyoy, Oct. 1965 (SHR).
    Usk, garden m.v.l. trap, 4/10/68 (1) (GANH).

Sq. **SO** 30,50,32. **ST** 59.

1762 *Chloroclysta citrata* Linn. Dark Marbled Carpet
    ssp. *citrata* Linn.
This moth, of sporadic occurrence and sparse distribution in Monmouthshire, flies in open woodland during July and August.
    Wye Valley (Bird 1907).
    Usk, garden m.v.l. trap, 1968, 21/8; 1976, 19/7; 1983, 30/8 and 31/8, all single moths. Wentwood 10/8 and 27/8/68. Usk (Park Wood) 23/8/72. Hael Wds 14/8/78. Trelleck Bog, 25/7/90 (2). Llansoy 3/8/90. (MEA) : Ochrwyth 9/8/86, Cwmfelinfach 13/8/87, Nant Gwyddon 8/9/88.

Sq. **SO** 30,40,50. **ST** 28,19,29,49.

1764 *Chloroclysta truncata* Hufn. Common Marbled Carpet
This very common and variable moth inhabits woodland, gardens, hedgerows etc. flying in late May and June and as a second brood from late August to early November.
    Tintern dist.(Bird 1905). Wye Valley (Bird 1907). Abertillery dist. (Rait-Smith 1912).
    Usk, garden m.v.l. trap, noted annually from 1966 to 1983 and usually plentiful eg. 1968, 1st brood (9) 29/5 – 28/6, 2nd brood (70) 20/8 – 24/10; 1978, 1st brood (9) 25/5 – 29/6, 2nd brood (95) 23/8 – 4/11. Ed. 20/5/76. Ld. 4/11/78. Usk (Park Wood) 4/6/68. Llangybi (Coed-y-Fferm) 6/6/68. Magor

Marsh 23/8/68. Pont-y-saeson 25/8/68. Wentwood 27/8/68. St Pierre's Great Wds 21/9/68. Bica Common 24/9/68. Prescoed 8/6/69. Wyndcliff 2/9/69. Rhyd-y-maen 6/6/70. Hendre Wds 5/6/71. Redding's Inclosure 7/6/71. Slade Wds 29/8/72. Trelleck Bog 18/9/72. Cwm Coed-y-cerrig 5/6/73. Llansoy 13/6/85. Dinham 4/6/87. Kilpale 30/6/87. (MEA) : Llantarnam 10/6/76, Mescoed Mawr 25/6/76, Risca 8/10/79, Black-Vein 12/6/81, Pant-yr-eos 6/7/85, Cwm Merddog 20/6/86, Ochrwyth 27/6/86, Cwmfelinfach 15/7/86, Pontllanfraith 22/6/87, Nant Gwyddon 7/6/88, Dixton Bank 18/6/88.

Sq. **SO** 10,20,30,40,50,41,51,22,32. **ST** 28,48,19,29,39,49,59.

1765 *Cidaria fulvata* Forst.   Barred Yellow

Widely distributed and common in Gwent frequenting open woods and bushy places and flying from mid-June to mid-July.

Wye Valley (Bird 1907). Abertillery dist. (Rait-Smith 1912, 1913).

Usk, garden m.v.l. trap, recorded in small numbers most years from 1966 to 1982 eg. 1968, (8) 5/7 – 20/7; 1975, (5) 23/6 – 6/7. Ed. 15/6/76. Ld. 20/7/68 and a very late moth on 30/8/70. Prescoed 5/7/68. Magor Marsh 1/7/69 and plentiful in the years 1975 – 1978. Usk (Park Wood) 7/7/69. Wyndcliff 13/7/69. Hendre Wds 5/7/70. Newport (Coldra) 27/6/75. Slade Wds 21/7/79. Llansoy 1/7/85. Kilpale 30/6/87. (MEA) : Ynysyfro 22/6/76, Llantarnam 24/6/76, Mescoed Mawr 1/7/76, Sugar Loaf (St Anne's Vale) 7/7/77, Tintern 30/6/78, Risca 19/6/82, Dinham 7/7/87, Pontllanfraith 10/7/87, Nant Gwyddon 25/6/88.

Sq. **SO** 20,30,40,21,41. **ST** 28,38,48,19,29,39,49,59.

1766 *Plemyria rubiginata* D. & S.   Blue-bordered Carpet
      ssp. *rubiginata* D. & S.

Scarce in Monmouthshire, flying from early July to mid-August in marshy localities and damp woodland.

Wye Valley (Bird 1907). Llanock Wood, a few, 1911 (Rait-Smith 1912) [= Llanerch Wood, Crumlin. GANH].

Usk, garden m.v.l. trap, recorded sporadically eg. 1966, (1) 12/7; 1973, (5) 19/7 – 9/8, 1983, (2) 9/7 and 16/7. Ed. 2/7/80. Ld. 19/8/68. Magor Marsh 5/7/77 (MEA), 11/7/78 (GANH). Hael Wds 31/7/82 (GANH). (MEA) : Mescoed Mawr 1/7/76; Cwmfelinfach 12/8/86, 13/8/87; Cwm Coed-y-cerrig 17/8/87.

Sq. **SO** 30,50,22. **ST** 48,19,29.

1767 *Thera firmata* Hb.   Pine Carpet

Flies from late June to late October in woods and plantations containing pines. Locally common in Monmouthshire and comes to light.

Usk, garden m.v.l. trap, noted most years from 1967 to 1983 eg. 1968, (8) 6/7 – 28/10; 1979, (23) 22/6 – 19/10. Ed. 22/6/79. Ld. 28/10/68. Prescoed

5/7/68. Trelleck Bog 18/9/72. Magor Marsh 23/9/72. Slade Wds 19/6/85. Llansoy 9/7/85. (MEA) : Cwmfelinfach 26/9/86, Risca 30/9/86, Ochrwyth 17/9/88.

Sq. **SO** 30,40,50. **ST** 28,48,19,29,39.

1768   *Thera obeliscata* Hb.   Grey Pine Carpet
Double-brooded, flying in conifer plantations from mid-May to the end of June and from mid-September to late October. Common in Gwent and often abundant.

Cwmyoy, Oct. 1965 (SHR). Redding's Inclosure 31/5/68 (JMC-H 1969).

Usk, garden m.v.l. trap, noted annually from 1966 – 83 and usually common eg. 1968, 1st brood (18) 14/5 – 29/6, 2nd brood (140) 13/9 – 25/10. Ed. 14/5/68. Ld. 28/10/69. Usk (Park Wood) 4/6/68. Wentwood 27/8/68. Bica Common 24/9/68. Hendre Wds 26/5/70. Wyndcliff 28/9/70. Cwm-mawr 31/5/71. Slade Wds 24/10/71. Magor Marsh 23/9/72. Trelleck Bog 16/6/73. St Pierre's Great Wds 13/7/73. Llansoy 27/6/86. (MEA) : Mescoed Mawr 5/9/77, Pant-yr-eos 16/6/83, Risca 23/9/83, Cwmfelinfach 26/9/86, Ochrwyth 28/9/86, Abercarn 27/6/87, Dixton Bank 18/6/88, Nant Gwyddon 25/6/88.

Sq. **SO** 20,30,40,50,41,51,32. **ST** 28,48,19,29,49,59.

1769   *Thera britannica* Turner   Spruce Carpet
       = *variata* auct.
Common in spruce plantations, flying in two broods, the first from May to July and the second from late August to November.

Cwmyoy, Oct 1965 (SHR). Redding's Inclosure 31/5/68 (JMC-H 1969).

Usk, garden m.v.l. trap, occurred commonly every year from 1966 to 1983 eg. 1979, 1st brood (13) 30/5 – 23/7, 2nd brood (39) 30/8 – 5/11. Ed. 21/4/72. Ld. 12/11/71. Usk (Park Wood) 23/7/68. Rhyd-y-maen 6/6/69. Wyndcliff 28/9/69. Trostrey 25/10/69, at ivy bloom. Hendre Wds 5/6/71. Redding's Inclosure 19/10/71. Hael Wds 22/10/71. Bica Common 13/6/72. Slade Wds 29/8/72. Trelleck Bog 18/9/72. Tredean Wds 5/6/73. Llandogo 19/5/78 (CGMdW). Wentwood 26/5/80. Llansoy 12/8/85. (MEA) : Tintern 30/6/78, Risca 8/7/83, Ynysddu 12/7/85, Ochrwyth 27/6/76, Cwmfelinfach 15/7/86, Abercarn 27/6/87, Lower Machen (Park Wood) 13/5/88.

Sq. **SO** 30,40,50,41,51,32. **ST** 28,48,19,29,49,59.

1773   *Electrophaes corylata* Thunb.   Broken-barred Carpet
This widespread and common moth inhabits hedgerows, woodland and bushy places. It flies from mid-May to early July and comes to light.

Wye Valley (Barraud 1906). Wye Valley (Bird 1907). Abertillery dist. (Rait-Smith 1912, 1913).

Usk, garden m.v.l. trap, noted every year from 1966 to 1983 eg. 1979

(29) 15/5 – 6/7. Ed. 15/5/79. Ld. 10/7/67. Wyndcliff 7/6/66. Usk (Park Wood) 5/6/67. Pont-y-saeson 13/6/67. Prescoed 8/6/69. Wentwood 21/6/69. Bica Common 28/6/69. Hendre Wds 5/6/71. Redding's Inclosure 15/6/71. Cwm-mawr 2/7/71. Hael Wds 3/7/71. St Pierre's Great Wds 1/6/75. Slade Wds 28/5/78. Llansoy 15/6/85. Cwm Tyleri 28/6/86. Trelleck Bog 25/5/87. Dinham 4/6/87. (MEA) : Llantarnam 10/6/76, Allt-yr-yn 20/6/76, Mescoed Mawr 25/6/76, Risca 31/5/80, Cwm Merddog 14/7/86, Cwmfelinfach 15/7/86, Ochrwyth 27/6/86, Abercarn 26/5/87, Nant Gwyddon 25/6/88.

Sq. **SO** 10,20,30,40,50,41,51. **ST** 28,48,19,29,39,49,59.

1775 *Colostygia multistrigaria* Haw.   Mottled Grey
This species, flying in March and April, occurs at low density in open woodland in eastern Monmouthshire but is rather more plentiful in the moorland areas of the west and north of the county.

Abertillery dist. inc. ab. *virgata* Tutt and ab. *nubilata* Tutt (Rait-Smith 1912, 1913).

Usk, garden m.v.l. trap, noted sporadically, eg. 1967, (1) 17/4; 1969, (1) 9/5; 1971, (3) 6/4 – 13/4. Wentwood 26/4/69, 11/4/70 (JMD). Cwm-mawr 6/4/71, 27/3/73. Llansoy 8/4/88. (MEA) : Mescoed Mawr 16/4/78, Blorenge 27/3/82, Risca 25/4/82 and 5/4/87, Pontypool 12/4/87, Ochrwyth 13/4/87, Cwm Merddog 26/4/87.

Sq. **SO** 10,20,30,40,21. **ST** 28,29,49.

1776 *Colostygia pectinataria* Knoch   Green Carpet
Widespread and common in open woods, hedgerows and on rough ground, this moth flies from the end of May to mid-July while second generation moths have occasionally been noted in August and September.

Wye Valley (Bird 1907). Abertillery dist. (Rait-Smith 1912, 1913).

Usk, garden trap, a few in most years from 1966 to 83 eg. 1973, (4) 21/6 – 8/7; 1979, (4) 29/6 – 12/7, and a second brood moth on 12/8; 1982 second brood moths on 10/9 and 18/9. Pont-y-saeson 13/6/67, 30/5/75. Trelleck Common 9/7/68. Wentwood 12/7/68. Usk (Park Wood) 7/7/69. Wyndcliff 13/7/69. Hael Wds 5/7/70. Hendre Wds 7/7/70. Redding's Inclosure 7/6/71. Prescoed 21/6/71. Cwm-mawr 2/7/71. Cwm Tyleri 30/7/72. Trelleck Bog 16/6/73. Piercefield Park 18/7/73. Cwm Coed-y-cerrig 5/6/78. Slade Wds 28/7/72. Llansoy 16/6/86. Dinham 4/6/87 (plfl.). Kilpale 30/6/87. (MEA) : Sugar Loaf (St Anne's Vale) 7/7/77, Tintern 30/6/78, Mescoed Mawr 3/7/78, Risca 12/8/80, Black Vein 3/7/81, Blorenge 7/7/84 Newport Docks 11/7/84, Cwm Morddog 7/7/85, Ynysddu 12/7/85, Ochrwyth 27/6/86, Cwmfelin-fach 15/7/86, Trefil 3/7/87, Dixton Bank 18/6/88, Nant Gwyddon 25/6/88.

Sq. **SO** 10,20,30,40,50,11,21,41,51,22. **ST** 28,38,48,19,29,39,49,59.

1777   *Hydriomena furcata* Thunb.   July Highflyer
Common and widespread, flying from early July to early September, in fens, woodland, hedgerows and bushy places and coming readily to light. The smaller moorland forms are abundant in the hills of north-west Gwent and sometimes are to be seen, in the small hours, congregated in huge numbers, many in cop., settled on the flowers of rushes and on other herbage.

Wye Valley (Bird 1907). Abertillery dist. (Rait-Smith 1912, 1913).

Usk, garden m.v.l. trap, recorded every year from 1966 to 1983 and plentiful. eg. 1975, 30/6 – 26/8; 1979, 7/7 – 3/9. Pont-y-saeson 1/8/70. Redding's Inclosure 20/7/71. Usk (Park Wood) 23/8/72. Slade Wds 29/8/72. Cwm Tyleri 9/7/73; 5/8/74 (abundant in the moonlight on rushes together with numerous *Eulithis populata* and *Perizoma didymata*). Hael Woods 10/7/73. Trelleck Bog 4/9/73. Cwm Coed-y-cerrig 10/8/74. Magor Marsh 2/7/76. Hendre Wds 20/7/79. Llangeview (Allt-y-bela) 28/7/79. Wyndcliff 3/8/81. Llansoy 28/7/84. Clydach 18/7/89. (MEA) : Llantarnam 24/6/76, Ynysyfro 28/6/76, Mescoed Mawr 3/7/76, Wentwood 10/7/77, Risca 14/7/80, Abercarn 16/7/85, Sor Brook (Llandegfedd) 17/7/86, Cwmfelinfach 12/8/86, Pontllanfraith 20/7/87, Trefil 15/8/87, Nant Gwyddon 7/8/88.

Sq. **SO** 20,30,40,50,11,21,41,51,22. **ST** 28,48,19,29,39,49,59.

1778   *Hydriomena impluviata* D. & S.   May Highflyer
Fairly common in damp woods and river valleys flying from late May to early July and coming to light.

Abertillery dist. (Rait-Smith 1912, 1913). Pontllanfraith (Rait-Smith 1915). Tintern, "several both by day and at light in all three years 1946 – 48" (Blathwayt 1967).

Usk, garden m.v.l. trap, occurred sparingly in most years from 1966 – 83 eg. 1973, (3) 22/5 – 30/5; 1975, (6) 27/5 – 27/6. Usk (Park Wood) 6/6/66 (several). Rhyd-y-maen 6/6/70. Hael Wds 27/5/75, 8/5/87. Newport (Coldra) 15/6/75. Cwm Coed-y-cerrig 24/5/76. Hendre Wds 23/6/79. Llansoy 13/6/85. (MEA) : Mescoed Mawr 6/6/78 (melanic f.), Black Vein 8/5/81, Cwmfelinfach 23/5/87, Trelleck Bog 25/5/87, Abercarn 26/5/87, Lower Machen (Park Wood) 13/5/88.

Sq. **SO** 20,30,40,50,41,22. **ST** 28,38,19,29,59.

1781   *Horisme vitalbata* D. & S.   Small Waved Umber
Occurs locally in the eastern half of Monmouthshire mainly on the Carboniferous Limestone in the south and also, but more infrequently, on the Old Red Sandstone. It flies during the second half of May and in June and, as a second brood, in August. Easily disturbed by day from traveller's joy growing in hedges and bushy places and at night it comes to light.

Usk, garden m.v.l. trap, recorded sporadically and in small numbers eg. 1970, 10/8 (1), 12/8 (1); 1972, 2/9 (2); 1983, 18/6 (1). Wentwood 1969 (1)

(JMD). Slade Wds 1/6/71, 19/6/85 etc. Magor Marsh 11/8/71. Hael Wds 6/8/84. Wyndcliff 8/8/84. Dinham 4/6/87.

Sq. **SO** 30,50. **ST** 48,49,59.

1782 *Horisme tersata* D. & S.   The Fern

This moth occurs mainly in the south-east where traveller's joy (*Clematis vitalba*) flourishes on the Carboniferous Limestone. It flies during June and July in woods, hedgerows, and bushy places and comes readily to light.

Usk, garden m.v.l. trap, a few most years from 1966 to 83 eg. 1968, (1) 30/6; 1973, (5) 11/6 – 13/7; 1983, (2) 20/6, 26/7. Ed. 7/6/66. Ld. 2/8/70. Wyndcliff 7/6/66, 25/7/83, 19/6/87. Usk Castle 2/6/66 (1) (REMP). Llangwm 24/6/68, Trelleck Bog 16/7/79. Slade Wds 23/7/69, 21/7/83. Llansoy 25/6/86. Kilpale 30/6/87. (MEA) : Wentwood 10/7/77, Abercarn 27/6/87, Dixton Bank 18/6/88.

Sq. **SO** 30,40,50,51. **ST** 48,29,49,59.

1784 *Melanthia procellata* D. & S.   Pretty Chalk Carpet

This moth, flying from early June to late August, is found fairly commonly in the east and south of the county in hedgerows, open woods and bushy places. During the day it can be disturbed from traveller's joy and at night it comes to light. Earliest recorded date 6/6/67 (Wyndcliff), latest date 29/8/72 (Slade Wds).

Monmouthshire (Ince 1887). Wye Valley (Bird 1907).

Usk, garden m.v.l. trap, recorded most years from 1966 to 1983 but not more than one or two per season eg. 1968, (2) 25/7, 27/8; 1983, (1)5/7. Wyndcliff 7/6/66, 10/8/83. Wentwood 8/7/68 (4) to m.v.l., 29/6/73 (plfl.). Magor Marsh 22/7/69. Hendre Wds 5/7/70. St Pierre's Great Wds 22/8/72. Slade Wds 29/8/72 (plfl.). Piercefield Park 18/7/73. Redding's Inclosure 4/7/81 (3) to m.v.l. Hael Wds 5/7/82. Kilpale 30/6/87. (MEA) : Allt-yr-yn 20/6/76, Risca 12/8/80, Dixton Bank 18/6/88.

Sq. **SO** 30,50,41,51. **ST** 28,48,29,49,59.

1786 *Spargania luctuata* D. & S.   White-banded Carpet
A rare vagrant or immigrant.

Usk, one to garden m.v.l. trap 17/5/75 (GANH). The only record, but unfortunately not captured due to surprise to the author. Identity not in doubt.

Sq. **SO** 30.

1787 *Rheumaptera hastata* Linn.   Argent & Sable
      ssp. *hastata* Linn.
Wye Valley (Nesbitt 1892). Wye Valley (Barraud 1906). Wye Valley, larvae on birch (Bird 1907). Abertillery dist. (Rait-Smith 1912, 1913). Monmouth

dist. (Charles c. 1937).

Trelleck (Fedw Fach) 27/5/88 (1) (GANH). Croes Robert 6/6/91 (2) (RG).

Sq. **SO** 20,40. **ST** 59.

1788   *Rheumaptera cervinalis* Scop.   Scarce Tissue
Tintern, one at light, May 1912 (Bird 1913). The only Monmouthshire record.

Sq. **SO** 50.

1789   *Rheumaptera undulata* Linn.   Scallop Shell
Occurs in woods mainly in the eastern half of the county but is local and scarce. It flies from late June to early August and comes to light.

Usk, garden m.v,l, trap, appeared sporadically eg. 1967, (1) 13/7; 1969, (1) 20/7. A total of (5) recorded from 1966 to 1983. Pont-y-saeson 27/6/67 one beaten from bushes. Wentwood 11/7/69 (1), 21/7/69 (1) to m.v.l.

Wyndcliff 13/7/69. Slade Wds 2/8/80, 28/7/82, 1983 (5), 8/7 – 23/7. (MEA) : Risca 26/6/90 one to act.l., Cwm Tyleri 28/7/91 one to m.v.l.

Sq. **SO** 20,30. **ST** 48,29,49,59.

1790   *Triphosa dubitata* Linn.   The Tissue
This moth, flying in open woodland in August and September and in the late Spring after hibernation, is local and scarce in Gwent occurring mainly in the south-east of the county.

Wye Valley (Bird 1907). Abertillery dist. (Rait-Smith 1912, 1913).

Wyndcliff, to m.v.l., 15/9/66 (GANH); 12/9/77, 1/9/79, 12/8/83 (MEA). Went-wood, to m.v.l. trap, 1969 singletons on 3/5, 11/7, 21/7 (JMD). Usk, garden m.v.l. trap, 12/5/70. Magor Marsh 2/8/70. Llansoy 11/9/86, 28/8/87.

Sq. **SO** 20,30,40. **ST** 48,49,59.

1791   *Philereme vetulata* D. & S.   Brown Scallop
Local and rare in Monmouthshire, flying in July in woodland on the Carboniferous Limestone in the south of the county.

South Monmouthshire 23/7/79, ♀ (1) to m.v.l. (GANH) (1st v-c. rec.). Also 29/7/81 ♂ (1); 1983, 21/7 (2), 23/7 (1). No other records.

Sq. **ST** 48.

1792   *Philereme transversata* Hufn.   Dark Umber
         ssp. *britannica* Lempke
Llandogo (Nesbitt 1892, as *Scotosia rhamnata*).

No recent records.

Sq. **SO** 50.

1793 *Euphyia biangulata* Haw.   Cloaked Carpet
Very local in Monmouthshire, being found in some of the Wye Valley woods during July and August.
    Wye Valley (Bird 1907).
    Hael Wds, to m.v.l., 6/8/74 (1), 4/7/81 (1) (GANH). Llansoy 31/7/90, one to u.v.l.. (MEA) : Tintern 1/7/78 (1), Wyndcliff 6/7/86 (1).

Sq. **SO** 40,50. **ST** 59.

1794 *Eyphyia unangulata* Haw.   Sharp-angled Carpet
Widely distributed in the eastern half of Gwent but local and at low density. It flies in open woods in June, July and early August and is easily disturbed from bushes by day and at night comes to light.
    Tintern dist. (Piffard 1859). Llandogo dist. (Nesbitt 1892). Wye Valley (Bird 1907). Tintern, "a few at light in 1947" (Blathwayt 1967).
    Whitebrook 2/7/69 (SC). Usk, garden m.v.l. trap, 1973, 16/8 (1); 1980, 2/8 (1); 1981, (1); 1982, (1); 1983, (2), 17/7 and 26/7. Usk (Park Wood) 6/6/66 (1), 5/6/67 (1). Pont-y-saeson (by day), 13/6/67 (1), 27/6/67 (1), 14/7/68 (1). Wentwood 25/7/69 (JMD), 13/8/81 (MEA). Trelleck Bog 16/6/73, 16/7/79. Llansoy 2/7/85, 17/8/86, 1987 (2), 1990 (2). Kilpale 30/6/87. Dixton Bank 18/6/88. (MEA) : Upper Llanover 2/8/79, Trelleck (Fedw Fawr) 19/7/83.

Sq. **SO** 20,30,40,50,51. **ST** 49,59.

1795 *Epirrita dilutata* D. & S.   November Moth
Flying in October and November, this species is widespread and common in woods, hedgerows and bushy places and comes readily to light.
    Tintern, 20/10/1912 (Bird 1913). Abertillery dist. (Rait-Smith 1912, 1913). Cwmyoy 1965 (SHR). Brockwells (Caerwent) 1/12/66 (CT in litt.).
    Usk, garden m.v.l. trap, plentiful every year from 1966 to 1983 eg. 1978, (110) 10/10 – 11/11. Ed. 8/10/79. Ld. 11/11/78. Rhyd-y macn 2/11/70. Redding's Inclosure 19/10/71 (abdt.). Slade Wds 24/10/71. Hendre Wds 25/10/71. Magor Marsh 13/10/78. Pont-y saeson 17/10/78 (pltl.). Llansoy 13/10/85. (MEA) : Wentwood 19/10/78, Risca 14/10/84, Pontypool 10/10/86, Cwmfelinfach 9/10/86.

Sq. **SO** 20,30,40,50,41,51,32. **ST** 48,19,29,49,59.

1796 *Epirrita christyl* Allen   Pale November Moth
This woodland moth, flying in October, occurs locally in Gwent.
    Cwmyoy 1965 (SHR, det. AR).
    Wyndcliff 28/9/70 (1) to m.v.l. Redding's Inclosure 18/10/71 (svl.). Hendre Wds 25/10/71. (MEA) : Wentwood 19/10/78.

Sq. **SO** 41,51,32. **ST** 49,59.

1797   *Epirrita autumnata* Borkh.   Autumnal Moth
Occurs locally in woods, flying during October and early November.

Usk, garden m.v.l. trap, 22/10/66 and recorded sporadically eg. 1977, (16) 19/10 – 27/10; 1979, (25) 9/10 – 23/10; 1980, (4) 25/10. Llantrisant 8/11/65. Hendre Wds 25/10/71, 10/10/78. Slade Wds 24/10/71. Pont-y-saeson 7/10/77. Magor Marsh 6/10/79 (1). Llansoy 16/10/85.

Sq. **SO** 30,40,41. **ST** 48,39,59.

1798   *Epirrita filigrammaria* H.-S.   Small Autumnal Moth
Flies in September on the hills and moorlands of northern Gwent.

Cwm Tyleri (1,500 ft.), 15/9/86 one to m.v.l. and several settled on rushes (1st v-c. rec.) (GANH, MEA). Blorenge Mountain 16/9/86 and 20/9/86, abundant on heather after dark (GANH, MEA). Trefil 9/9/88, 31/8/90 (MEA).

Sq. **SO** 20,11,21.

1799   *Operophtera brumata* Linn.   Winter Moth
Widespread and abundant in gardens, hedgerows, woodland, and bushy places from November to January. The female has vestigial wings and is flightless but the male comes to light.

Tintern dist., "larvae on Cotoneaster" (Bird 1905). Wye Valley, larvae (Barraud 1906). Tintern (Bird 1913). Tintern, "larvae on apple", 31/5/68 (JMC-H 1969).

Usk, to garden m.v.l. trap, noted most years from 1966 to 1983 eg. 1971, 25/11 (4), 27/11 (12), 5/12 (4), 25/12 (1). 1966, Usk dist. (abdt.). 1971, Devauden, Shirenewton, Llanfrechfa, Coed-y-paen, Llanbadoc. 1975, 23/12 abundant at Magor Marsh and Wentwood. (MEA) : Risca 18/12/87.

Larvae (bred) (GANH): Pont-y-saeson 1969, on sallow; 5/6/78, on sallow and hazel. Trelleck Bog 1/6/75, on birch, crab apple and mountain ash. St Pierre's Great Wds 1/6/75 on birch and common lime (*Tilia europaea*).

Sq. **SO** 30,40,50,41. **ST** 48,29,39,49,59.

1800   *Operophtera fagata* Scharf.   Northern Winter Moth
This moth, flying in November and December, has occasionally been noted in Monmouthshire but is scarce. The female has small under-developed wings and is flightless but the male comes to light.

Wye Valley (Bird 1907). Tintern (Bird 1913).

Usk, garden m.v.l. trap, recorded sporadically eg. 1966, several, 11/11 – 12/12; 1971, (17) 25/11 – 5/12. Prescoed, Nov. 1971.

Sq. **SO** 30. **ST** 39,59.

1801   *Perizoma taeniata* Steph.   Barred Carpet
Tintern dist. 1859 (Piffard 1859). Barrett (*The Lepidoptera of the British Islands* 1902 **8**:239) states "excessively local in these Islands" and gives, among other sites, "Tintern, Monmouthshire", probably citing Piffard's 1859 record. 1972, "Between Tintern and Chepstow" . . . three to m.v.l. 18/7. (Withers 1972).

Sq. **ST** 59.

1802 *Perizoma affinitata* Steph.   The Rivulet
Widespread and fairly common, inhabiting waysides, woodland margins and rides etc. and flying from the end of May to early September.

Wye Valley (Barraud 1906). Tintern and Llandogo, "one of the commonest moths" (Bird 1909).

Usk, garden m.v.l. trap, recorded annually 1966 – 1983 eg. 1974, (4) 18/6 – 16/8; 1981, (7) 1/6 – 18/8. Ed.1/6/68. Ld. 14/9/72, 20/8/80. Wyndcliff 7/6/66. Usk (Park Wood) 15/6/67. Pont-y-saeson 13/6/67, 25/8/68. Prescoed 8/6/69. Tintern 1/8/70. Magor Marsh 2/8/70. Gwehelog (Camp Wood) 28/8/70. Slade Wds 1/6/71. Redding's Inclosure 15/6/71. Hael Wds 3/7/71. Cwm Coedycerrig 5/6/73. Wentwood 29/6/73. Cwm Tyleri 1/7/73. St Pierre's Great Wds 13/7/73. Piercefield Park 18/7/73. Hendre Wds 23/5/80, 1/8/80. Dinham 4/6/87. Llansoy 13/7/87. (MEA) : Llantarnam 10/6/76, Mescoed Mawr 23/6/78, Risca 9/7/79, Pant-yr-eos 16/6/83, Trelleck 19/7/83, Ochrwyth 12/7/86.

Sq. **SO** 20,30,40,50,41,51. **ST** 28,48,29,39,49,59.

1803   *Perizoma alchemillata* Linn.   Small Rivulet
Widespread and common in lanes, woodland rides and margins, waste ground etc., flying from early June to late August.

Wye Valley (Bird 1909).

Usk, garden m.v.l. trap, noted every year from 1966 to 1983 eg. 1971, (20) 4/6 – 2/8; 1981, (25) 25/7 – 28/8. Ed. 3/6/68. Ld. 12/8/70, 3/9/72, 28/8/80. Usk (Park Wood), 6/6/66; Sept. 1966, larvae abundant in seed capsules of hempnettle (*Galeopsis tetrahit*), bred Wyndcliff 10/7/67. Prescoed 5/7/68. Magor Marsh 22/7/69. Pont-y-saeson 26/5/73. Hael Wds 10/7/73. Piercefield Park 18/7/73. Cwm Tyleri 12/7/77. Hendre Wds 20/7/79. Slade Wds 21/8/79. Redding's Inclosure 23/6/81. Llansoy 28/6/86. Trelleck Bog 19/7/86. Dinham 4/6/87. (MEA) : Mescoed Mawr 28/6/76, Risca 7/8/81, Wentwood 13/8/81, Cwmcarn 11/7/87, Pontllanfraith 20/7/87, Cwmfelinfach 13/8/87, Nant Gwyddon 15/8/88.

Sq. **SO** 20,30,40,50,41,51. **ST** 48,19,29,39,49,59.

1804   *Perizoma bifaciata* Haw.   Barred Rivulet
Usk, garden m.v.l. trap, 1978 one on 23/8 (1st v-c. rec.) (GANH); 1981 one

in "mint" condition on 7/8. Believed to be the only Gwent records.

Sq. **SO** 30.

1806 *Perizoma blandiata* D. & S.   Pretty Pinion
ssp. *blandiata* D. & S.

Very local and scarce in Monmouthshire, flying during July and August.

Panta Arch (Angidy Valley), larvae swept from eyebright (*Euphrasia* spp.) Sept. 1966, 24/9/68 and 3/9/69, bred (REMP). Whitebrook 2/7/69 several to m.v.l. (SC). Usk, garden m.v.l. trap, single moths on 5/7/73, 13/8/76 and 17/7/77. Hael Wds, to m.v.l., 10/7/73 (2), 26/6/86 (1) (GANH).

Sq. **SO** 30,50. **ST** 59.

1807 *Perizoma albulata* D. & S.   Grass Rivulet
ssp. *albulata* D. & S.

Widely distributed and locally common in Monmouthshire, flying from mid-May to early August. Readily disturbed from herbage by day and at night comes to light.

Llandogo (Nesbitt 1893). Wye Valley (Barraud 1906). Abertillery dist. (Rait-Smith 1912, 1913).

Usk, garden m.v.l. trap, appeared sporadically in small numbers, eg. 1967, (1) 11/5; 1968, 29/5 (1), 1/6 (2); 1970, (12) 24/5 – 3/6; 1976, (1) 11/8. Pont-y-saeson 13/6/67, 25/5/76. Usk (Park Wood) 23/8/72 (several). Redding's Inclosure 17/6/74. Wyndcliff 2/7/74. Cwm Tyleri 26/6/76. Hael Wds 26/6/86, 8/5/87. Hendre Wds 9/5/87, 1/6/87. Llansoy 1/6/87. Pontllanfraith 20/6/87, 17/6/88 (abdt.), by day. (MEA) : Risca 5/6/85.

Sq. **SO** 20,30,40,50,41,51. **ST** 19,29,59.

1808 *Perizoma flavofasciata* Thunb.   Sandy Carpet

Occurs at low density in Monmouthshire, mainly in the east of the county, flying in hedgerows, waysides and open woodland from late May through June and July.

Wye Valley (Barraud 1906). Wye Valley (Bird 1907).

Usk, garden m.v.l. trap, recorded annually 1966 to 1983 eg. 1968, (5) 28/5 – 12/7; 1973, (9) 25/6 – 15/7; 1983, (11) 19/6 – 14/7. Ed. 22/5/77. Ld. 24/7/72. Prescoed 8/6/69. Usk (Park Wood) 9/6/69, 7/7/69. Redding's Inclosure 15/6/71. Pont-y-saeson 18/7/72. Hael Wds 10/7/73. Llansoy 26/6/86. Dinham 4/6/87. Hendre Wds 14/5/88. (MEA) : Llantarnam 10/6/76, Mescoed Mawr 25/6/76, Wentwood 10/7/77, Tintern 30/6/78, Risca 13/7/82, Wyndcliff 13/6/86.

Sq. **SO** 30,40,50,41,51. **ST** 29,39,49,59.

1809   *Perizoma didymata* Linn.   Twin-spot Carpet
        ssp. *didymata* Linn.

Flying from late June until the end of August, this species is local but widespread and at low density in the central and eastern districts of Monmouthshire, while on the hills and moorlands in the north of the county it is abundant.

Wye Valley (Bird 1907). Abertillery dist. (Rait-Smith 1913).

Trelleck Common 27/4/68, larvae on bilberry, bred (REMP). Wentwood, to m.v.l. 2/8/69 (1), 1/9/69 (1) (JMD). Cwm Tyleri (1,400 ft.) 7/8/72, 6/8/73 (plfl.); 1974, abundant on 5/8, many hundreds sitting at night on the flowers of rushes with many in cop.; 29/7/75 (abdt.). Usk, garden m.v.l. trap, 2/8/78 (1), 13/8/80 (1), the only records in the eighteen years from 1966 to 1983. (MEA) : Mescoed Mawr 3/7/76, Ynysyfro 28/6/76, Blorenge Mountain 30/8/86, Trefil 7/8/88, Nant Gwyddon 15/8/88.

Sq. **SO** 20,30,50,11,21. **ST** 28,29,49.

1811   **Eupithecia tenuiata** Hb.   Slender Pug

Widespread and fairly common among sallows in damp woods and marshy places this moth flies in June and July and comes to light.

Tintern dist. (Piffard 1859).

Bred from sallow catkins collected at Usk 1976, Hendre Wds 6/4/80 and 7/4/81, and Llangybi (Coed-y-Ffern) 12/4/81 (GANH). Imagines to m.v.l.: Slade Wds 30/6/84, Hael Wds 14/6/86. (MEA) : Wyndcliff 6/7/84, Risca 27/6/86, Dixton Bank 18/6/88.

Sq. **SO** 30,50,41,51. **ST** 48,29,39,59.

1813   *Eupithecia haworthiata* Doubl.   Haworth's Pug

Common where traveller's joy (*Clematis vitalba*) flourishes, especially in central and south-east Gwent. It flies in July and August, and occasionally in June, and comes to light.

Wye Valley (Bird 1907).

Usk, garden m.v.l. trap, every year from 1966 to 1983 and often plentiful eg. 1982, 8/7 – 6/8. Ed. 8/7/70. Ld. 12/8/81. Hendre Wds 20/7/79. Slade Wds 28/7/80, 19/6/85, etc. Wyndcliff 6/7/81. Llansoy 14/7/86.

Sq. **SO** 30,40,41. **ST** 48,59.

1814   *Eupithecia plumbeolata* Haw.   Lead-coloured Pug

Flies in June and the first half of July and comes to light, being found locally in open woods and heaths where cow-wheat grows.

Tintern dist. (Piffard 1859). Tintern dist. 1965 (Mere 1965).

Trelleck Common 31/7/74, larvae on cow-wheat (*Melampyrum pratense*) bred, (REMP). Kilpale 30/6/87. (MEA) : Wyndcliff, to m.v.l.

6/7/86; Pontllanfraith 1987, several 20/6 – 10/7; Dinham 7/7/87.

Sq. **SO** 50. **ST** 19,49,59.

1816 *Eupithecia linariata* D. & S.   Toadflax Pug
Uncommon in Monmouthshire.

Wentwood (Darren Wood) 9/7/77 to m.v.l. (MEA) (1st v-c. rec.). Risca, to act.l., 1980, 13/7, 8/8, 28/8, 29/8; 1981, (4) 12/7 – 28/8 (MEA).

Sq. **ST** 29,49.

1817 *Eupithecia pulchellata* Steph.   Foxglove Pug
   ssp. *pulchellata* Steph.

This species is widespread and common in hedgerows, open woods, heaths and moorland, wherever foxglove flourishes. It flies from late May to mid-July and comes to light.

Tintern dist. (Piffard 1859). Wye Valley, larvae in foxglove flowers, reared (Bird 1907).

Usk, garden m.v.l. trap, recorded annually from 1966 to 1983 and often plentiful eg, 1981 31/5 – 28/7. Ed. 26/5/82. Ld. 30/8/80, 3/9/82. Pont-y-saeson 13/6/67. Wyndcliff 14/6/67. Usk (Park Wood) 15/6/67. Prescoed 8/6/69. Wentwood 2/6/70. Hael Wds 5/7/70. Redding's Inclosure 15/6/71. Trelleck Bog 22/5/73. Cwm Tyleri 1/7/73. Magor Marsh 11/7/78. Newport Docks 17/8/76. Cwm Coedycerrig 5/6/77. Llansoy 2/7/85. Hendre Wds 11/7/86. Kilpale 30/6/87. (MEA) : Tintern 30/6/78, Risca 14/7/80, Mescoed Mawr 8/7/84, Cwm Merddog 7/7/85, Ynysddu 12/7/85, Ochrwyth 27/6/86, Cwmfelinfach 15/7/86, Abercarn 16/5/87, Pontllanfraith 10/7/87, Nant Gwyddon 25/6/88.

Sq. **SO** 10,20,30,40,50,41,51. **ST** 28,38,48,19,29,39,49,59.

1818 *Eupithecia irriguata* Hb.   Marbled Pug
This species, described by Skinner as "very local in southern England", occurs in one wooded locality in the east of the county.

East Monmouthshire 1979, one to m.v.l. 29/5 (GANH) (1st v-c. rec.). Also 29/4/80 (2); 1981, (1); 1982, (9) 21/4 – 28/4; 1983 (1); 1987, (2).

Elsewhere in Wales it has apparently been recorded from only two other sites viz. Aberystwyth 1965-1970 and Newtown, Powys 1988 and 89 (per AMR in litt.).

Sq. **SO** 41.

1819 *Eupithecia exiguata* Hb.   Mottled Pug
   ssp. *exiguata* Hb.
Widespread and common in woods, hedgerows and bushy places, this species

flies in May, June and early July and comes to light.

Usk, garden m.v.l. trap, noted every year from 1966 to 1983 and often numerous eg. 1979, (54) 4/6 – 9/7. Ed. 6/5/70. Ld. 7/7/83. Prescoed 8/6/69. Usk (Park Wood) 9/6/69. Rhyd-y-maen 23/5/70. Hendre Wds 26/5/70. Wentwood 2/6/70. Slade Wds 1/6/71. Redding's Inclosure 7/6/71. Magor Marsh 19/6/71. Wyndcliff 22/6/71. Hael Wds 3/7/71. Trelleck Bog 22/5/73. Pont-y-saeson 30/5/75. Llansoy 4/6/85. (MEA) : Llantarnam 10/6/76, Mescoed Mawr 11/6/78, Risca 16/6/83, Ynysddu 12/7/85.

Sq. **SO** 30,40,50,41,51. **ST** 48,19,29,39,49,59.

1822   *Eupithecia pygmaeata* Hb.   Marsh Pug
Crosskeys, 1989, one flying by day 11/6 (MEA) : (1st v-c. rec.).

Sq. **ST** 29.

1823   *Eupithecia venosata* Fabr.   Netted Pug
       ssp. *venosata* Fabr.
Three records only viz. Usk, garden m.v.l. trap, one, 23/6/75 (GANH) (1st v-c. rec.). Dinham 4/6/87, one netted at dusk (MEA). Wyndcliff 28/6/91, one to m.v.l. (BS).

Sq. **SO** 30. **ST** 49,59.

1824   *Eupithecia egeneria* H.-S.   Fletcher's Pug (Pauper Pug)
A very local moth, first recorded and confirmed as a resident British species in 1962 by Messrs Bradley, Mere, and Pelham-Clinton when they found it near Tintern in woodland containing the small-leaved lime (*Tilia cordata*) and the large-leaved lime (*T. platyphyllos*). The previous year Messrs Fletcher and Bradley had searched for it in the same area but without success. Since its discovery it has been noted regularly during June and July in the lower Wye Valley both in Monmouthshire and in Gloucestershire. Elsewhere in the country it is known only from one locality in south-west Norfolk.

Tintern dist., 1962, (8) to m.v.l 15/6 and 16/6 (JDB, RMM, ECP-C). 18/6/65, several (RMM). Frequently since, eg. 1967, June 12, 16, 23. 1969, 13/7 (svl.). 1984, 1987, 1991.

Sq. **ST** 59.

1825   *Eupithecia centaureata* D. & S.   Lime-speck Pug
This species, flying from June to August, is sparsely distributed and at low density in Gwent. Ed. 4/6/87. Ld. date 26/8/70.

Tintern, one, 1911 (Bird 1912).

Usk, garden m.v.l. trap, 1970, 27/7 (1), 26/8 (1); 1971, 11/6 (1) and single moths in 1972, 75 and 83. Newport Docks 17/8/76 (4) to m.v.l. Pont-y-saeson 18/9/79, larvae on goldenrod (*Solidago virgaurea*), bred (REMP).

(MEA) : Dinham 1987, 4/6 (1), 8/7 (1); Risca 8/7/81 (1), 1984 (1), 1986 (1), 1987 (2); Uskmouth 16/8/88.

Sq. **SO** 30. **ST** 38,29,49,59.

1826 *Eupithecia trisignaria* H.-S.   Triple-spotted Pug
Occurs locally in damp woodland.

Larvae feeding on seed-heads of angelica (*Angelica sylvestris*), Usk (Park Wood), Sept 1966, reared (REMP) (1st v-c. rec.) and Pont-y-saeson 18/9/79, reared (REMP). Larvae on angelica, St Pierre's Great Wds 21/9/68 and Devauden 1/10/68, reared (GANH). Usk, garden m.v.l. trap, 26/6/79, imago (1), Slade Wds 1989, imagines to m.v.l., 20/7 (1) and 8/8 (numerous).

Sq. **SO** 30. **ST** 48,49,59.

1827 *Eupithecia intricata* Zett.   Freyer's Pug
         ssp. *arceuthata* Freyer
This moth, which comes readily to light, flies during June and July in gardens containing cypresses (*Cupressus* spp.).

Risca, 1984, one to act. l. 18/6 (1st v-c. rec.). 1985, (3) 15/6 – 10/7; 1986, (12) 16/6 – 8/7; 1987, (17) 15/6 – 7/7 (MEA). Llansoy 1987, (5) 23/6 – 6/7 (GANH).

Sq. **SO** 40. **ST** 29.

1830 *Eupithecia absinthiata* Cl.   Wormwood Pug
Widespread and not uncommon in Gwent, flying from early May to September in gardens, woods, waste ground etc.

Tintern dist. (Piffard 1859). Wye Valley 1906 (Bird 1907).

Usk, garden m.v.l. trap, occurred sporadically eg. 1970, 12/5, 30/5; 1974, (3) 9/7 – 3/8. Ed. 3/5/69. Ld. 11/8/81. Pont-y-saeson 18/9/79, larvae on goldenrod (Solidago virgaurea), reared (REMP).

(MEA) : Rhiwderin 21/5/76; Mescoed Mawr 14/8/81; Wentwood 3/9/79, 17/8/83; Risca 13/7/83, 1985 (4) 13/7 – 27/8, 1986, 1987; Wyndcliff 12/8/83; Cwmfelinfach 5/9/86; Pontllanfraith 16/8/87.

Sq. **SO** 30. **ST** 28,19,29,49,59.

1831 *Eupithecia goossensiata* Mab.   Ling Pug
A moorland moth recorded from the north of the county.

Cwm Tyleri (1,500 ft.) 12/7/77 one to m.v.l. (DSF) (1st v-c. rec.).

Sq. **SO** 20.

1832   *Eupithecia assimilata* Doubl.   Currant Pug
Monmouthshire records are few but moth probably overlooked.
    Wye Valley (Bird 1907). Usk, garden m.v.l. trap, 1981, 15/6 (1), 14/7 (1). Llansoy 9/7/87 (1). (MEA) : Risca 6/8/89.

Sq. **SO** 30,40. **ST** 29.

1833   *Eupithecia expallidata* Doubl.   Bleached Pug
This species flying during July and August, occurs locally in open woods and woodland rides and clearings. Wyndcliff 12/6/67 (REMP), 3/8/81 and 10/8/83 (GANH). Usk, m.v.l. trap, 17/7/72, 14/7/81. Hael Wds 14/8/78. Slade Wds 2/8/80. (MEA) : Wentwood 16/8/84, Risca 29/6/85.
    Usk (Park Wood) 1966, larvae on goldenrod (*Solidago virgaurea*), reared (REMP); also Pont-y-saeson 18/9/79 and Slade Wds 14/9/80, reared, (GANH).

Sq. **SO** 30,50. **ST** 48,29,49,59.

1834   *Eupithecia vulgata* Haw.   Common Pug
    ssp. *vulgata* Haw.
This pug, widespread and common in gardens, woods etc. comes readily to light, flying in May and June and, as a second brood, in July and August.
    Wye Valley (Bird 1907). Abertillery dist. (Rait-Smith 1912, 1913). Deri-fach 2/6/68 (JMC-H 1969).
    Usk, garden m.v.l.trap, recorded annually 1966 – 83 eg. 1973, 1st brood, (21) 24/5 – 29/6, 2nd brood, (12) 9/7 – 20/8. Ed. 5/5/71. Ld. 31/8/74. Wyndcliff 12/6/67. Prescoed 8/6/69. Hendre Wds 26/5/70. Rhyd-y-maen 6/6/70. Slade Wds 1/6/71. Magor Marsh 6/7/71. Hael Wds 3/7/71. Redding's Inclosure 18/7/71. Trelleck Bog 16/6/73. Cwm Tyleri 1/7/73. St Pierre's Great Wds 13/7/73. Usk (Park Wood) 31/5/75. Newport (Coldra) 27/6/75. Cwm Coed-y-cerrig 24/5/76. Pont-y-saeson 24/5/76. Llansoy 18/5/85. Dinham 4/6/87. (MEA) : Allt-yr-yn 23/5/76, Wentwood 9/7/77, Mescoed Mawr 11/6/78, Risca 19/6/79, Pant-yr-eos 13/5/82, Ochrwyth 1/6/85, Cwm Merddog 20/6/86, Abercarn 16/5/87, Cwmfelinfach 24/5/87, Pontllanfraith 27/5/88, Nant Gwyddon 10/6/88.

Sq. **SO** 10,20,30,40,50,21,41,51,22. **ST** 28,38,48,19,29,39,49,59.

1835   *Eupithecia tripunctaria* H.-S.   White-spotted Pug
Widespread and fairly common in damp woods and marshes where wild angelica grows, flying from late April until the end of August.
    Usk, garden m.v.l. trap, recorded most years from 1971 to 1983 eg. 1971 (2) 20/8 and 24/8; 1979 (5) 29/5 – 20/6. Ed. 27/4/80. Ld. 30/8/77. Wyndcliff 14/6/67 (REMP). Pont-y-saeson 30/5/75 (REMP). Usk (Park Wood) 31/5/75. Hael Wds 30/8/77, 31/8/86. Magor Marsh 1/6/79. Llansoy 1986 to m.v.l., (4) 31/5 – 31/8; 3/5/90 (1), 7/7/90 (1). (MEA) : Mescoed

Mawr 11/9/78, Risca 10/8/87, Pontllanfraith 16/8/87.

Usk (Park Wood), Sept. 1966, larvae on seed-heads of angelica, bred (REMP). Larvae on angelica, St Pierre's Great Wds 24/9/68 and Pont-y-saeson 24/9/68 and 18/9/79, reared (GANH).

Sq. **SO** 30,40,50. **ST** 48,19,29,59.

1837   *Eupithecia subfuscata* Haw.   Grey Pug

Widely distributed and fairly common, flying from May to early July, and occasionally in August, in gardens, woodland etc.

Wye Valley (Barraud 1906). Wye Valley, "larva in seed-vessel of sweet William, reared." (Bird 1907).

Wyndcliff 12/6/67, 22/6/71. Wentwood 8/7/68. Usk, garden m.v.l. trap, 4/7/75, 2/6/78. Cwm Tyleri 12/7/77. Hael Wds 27/6/78. Slade Wds 2/8/80. Hendre Wds 11/7/86. Llansoy 14/7/90, ab. *obscurissima* (1) to u.v.l. (MEA) : Mescoed Mawr 11/6/78, Risca 29/6/85, Pant-yr-eos 6/7/85, Cwm Merddog 7/7/85, Ynysddu 12/7/85, Ochrwyth 12/7/86, Trelleck Bog 25/5/87, Cwmfelinfach 10/6/87, Pontllanfraith 20/6/87, Nant Gwyddon 10/6/88.

Sq. **SO** 10,20,30,40,50,41. **ST** 28,48,19,29,49,59.

1838   *Eupithecia icterata* Vill.   Tawny Speckled Pug
         ssp. *subfulvata* Haw.

Widespread and not uncommon in Gwent, flying during August and September on wayside embankments, waste ground and in rough open woodland.

Wye Valley (Bird 1907).

Usk, m.v.l. trap, a few most years 1966 – 83 eg. 1966, 21/8 (2); 1979, (5) 12/8 – 7/9. Ed. 1/8/69. Ld. 16/9/72. Hael Wds 6/8/74. Cwm Tyleri 8/7/76. Newport Docks 17/8/76 (3) to m.v.l. Magor Marsh 12/8/87. (MEA) : Risca 8/9/79, Wyndcliff 12/8/84, Wentwood 16/8/84, Pontllanfraith 16/8/87.

Sq. **SO** 20,30,50. **ST** 38,48,19,29,49,59.

1839   *Eupithecia succenturiata* Linn.   Bordered Pug

Very few records for this moth which flies in July and August.

Usk, garden m.v.l. trap, 19/7/69 (1), 2/7/75 (1); 1979, (2) 29/7 and 1/8; 1982, 20/7 (1), 5/8 (1). Usk Castle 2/8/80 one to m.v.l. (REMP). Slade Wds 2/8/80 (GANH). (MEA) : Risca to act.l., single moths on 14/7/83, 15/7/88, 27/7/88.

Sq. **SO** 30. **ST** 48,29.

1842   *Eupithecia simpliciata* Haw.   Plain Pug
Risca, 2/8/91, one to act.l. (1st v-c. rec.) (MEA).

Sq. **ST** 29.

1843 *Eupithecia distinctaria* H.-S.   Thyme Pug
    ssp. *constrictata* Geun.
Clydach, a few to m.v.l. 17/6/89 (1st v-c. rec.) (MEA), several also on 1/7, and 9/7; 31/5/90 (1). Risca, 23/6/92, one to act.l. (MEA).

Sq. **SO** 21. **ST** 29.

1844 *Eupithecia indigata* Hb.   Ochreous Pug
Local and scarce in Monmouthshire. It flies in May, frequenting pine plantations, and comes to light.
    Pont-y-saeson, one to m.v.l. 2/5/72 (GANH) (1st v-c. rec.). Llansoy 31/5/86 (2). Trelleck Bog, (6) to m.v.l. 2/5/87 (MEA, GANH).

Sq. **SO** 40,50. **ST** 59.

1846 *Eupithecia nanata* Hb.   Narrow-winged Pug
    ssp. *angusta* Prout
Flying from April to August, this species is widespread in open heathy woods in the eastern half of Gwent but is much more plentiful on the hills and moorlands in the west and north of the county.
    Abertillery dist. (Rait-Smith 1912, 1913).
    Usk, garden m.v.l. trap, recorded most years from 1966 to 1983 eg. 1972, 26/4 (1), 25/7 (1); 1978, (6) 14/7 – 7/8. Wentwood 2/8/69 (JMD). Magor Marsh 11/8/71. Usk (Park Wood) 23/8/72. Cwm Tyleri 17/6/73. Trelleck Bog 16/6/73. Hael Wds 17/7/73. Pont-y-saeson 30/5/75 (REMP). Hendre Wds 29/8/78. (MEA) : Sugar Loaf (St Anne's Vale) 7/7/77, Blorenge Mountain 7/7/84, Risca 10/6/84, Ochrwyth 11/7/86, Trefil 15/8/87 7/7/84, Risca 10/6/84, Ochrwyth 11/7/86, Trefil 15/8/87, Nant Gwyddon 25/6/88.

Sq. **SO** 20,30,50,11,21,41. **ST** 28,48,29,49,59.

1851 *Eupithecia virgaureata* Doubl.   Golden-rod Pug
A few records from the east of the county.
    1968 St Pierre's Great Wds 21/9, larvae on flowers of goldenrod (*Solidago virgaurea*), reared (GANH) (– ab. *nigra* Lempke). 1979, Pont-y-saeson 18/9, larvae on goldenrod, bred (REMP). Hendre Wds 1/8/80, one to m.v.l.

Sq. **SO** 41. **ST** 59.

1852 *Eupithecia abbreviata* Steph.   Brindled Pug
This moth, widespread and very common in oakwoods, flies during April and May and comes readily to light. The melanic form ab. *hirschkei* Bastelburger is very common in Gwent.
    Wye Valley "very variable" (Bird 1907).

Usk, garden m.v.l. trap, recorded annually 1966 – 84 eg. 1971, (29) 28/3 – 18/5 inc. (18) ab. *hirschkei*. Llangybi (Coed-y-Fferm) 21/4/70.
Cwm-mawr 6/4/71. Craig-y-Master 17/5/71. Pont-y-saeson 3/4/72 (plfl.). Redding's Inclosure 24/4/72 (abdt.). Hael Wds 26/3/73. Trelleck Bog 22/5/73. Cwm Coed-y-cerrig 27/5/73. Usk (Park Wood) 31/5/75. Wyndcliff 26/4/76. Wentwood 4/5/76. Hendre Wds 23/4/79. Llansoy 7/5/85. (MEA) : Mescoed Mawr 19/4/85, Risca 8/5/85, Cwmfelinfach 14/4/87, Cwm Merddog 26/4 87, Abercarn 3/5/87.

Sq. **SO** 10,30,40,50,21,41,51,22. **ST** 48,19,29,39,49,59.

1853   *Eupithecia dodoneata* Guen.   Oak-tree Pug
Redding's Inclosure 1981, one to m.v.l. 21/4 (GANH) (detn. confirmed by DSF).

Sq. **SO** 51.

1854   *Eupithecia pusillata* D. & S.   Juniper Pug
    ssp. *pusillata* D. & S.
Flies during July and August in gardens containing junipers and other cultivated conifers.
Usk, garden m.v.l. trap, 1/8/80 (2) (1st v-c. rec.) (REMP, GANH), 2/8/80, 7/8/80; 1981, (6) 19/7 – 4/8; 1982, (12) 8/7 – 3/8; 1983, (8) 10/7 – 7/8.

Sq. **SO** 30.

1856   *Eupithecia lariciata* Freyer   Larch Pug
Common in larch plantations throughout Monmouthshire, flying during May, June and early July and readily coming to light. The melanic form ab. *nigra* Prout is frequent.
Wye Valley (Bird 1907). Abertillery dist. (Rait-Smith 1912, 1913).
Wyndcliff 7/6/66. Usk, garden trap, 28/5/66. Prescoed 8/6/69. Usk (Park Wood) 9/6/69. Hendre Wds 26/5/70. Slade Wds 1/6/71. Redding's Inclosure 7/6/71. Trelleck Bog 22/5/73. Wentwood 29/6/73. Hael Wds 27/7/75. Cwm Tyleri 8/7/76. Sugar Loaf (St Mary's Vale) 8/7/87. (MEA) : Mescoed Mawr 25/6/76, Ynysyfro 28/6/76, Ochrwyth 11/7/86, Cwmfelinfach 14/7/86, Abercarn 16/5/87, Lower Machen (Park Wood) 13/5/88, Nant Gwyddon 10/6/88.

Sq. **SO** 20,30,50,21,41,51. **ST** 28,48,19,29,39,49,59.

1857   *Eupithecia tantillaria* Boisd.   Dwarf Pug
Locally common in conifer plantations during May and June.
Usk, garden m.v.l. trap, 8/6/66, 22/5/68, 13/5/79 etc. Usk (Park Wood) larvae on spruce 15/6/67, reared (REMP). Prescoed 31/5/68. Redding's

Inclosure 31/5/68 (JMC-H 1969). Trelleck Bog 22/5/73. Pont-y-saeson 27/5/78. Hendre Wds 27/5/78. Slade Wds 28/5/78. Hael Wds 26/6/85. Clydach 11/5/90. (MEA) : Ochrwyth 17/6/88.

Sq. **SO** 30,50,21,41,51. **ST** 28,48,39,59.

1858 *Chloroclystis v-ata* Haw.　　　The V-Pug

This species inhabiting open woodland, hedgerows and bushy places is widespread and common in Monmouthshire, flying in two broods, the first from late April to June and the second from mid-July until mid-September.

Usk, garden m.v.l. trap, 18/9/66 and recorded every year until 1983 eg. 1969, 1st brood, 22/5 – 6/6. 2nd brood, 11/7 – 4/8. Ed. 2/5/70. Ld. 24/9/70. Wyndcliff 12/6/67. Wentwood 12/7/68. Hael Wds 3/7/71, 26/4/82. Redding's Inclosure 20/7/71. Prescoed 16/4/72. Cwm Tyleri 8/7/76. Magor Marsh 21/8/76. Pont-y-saeson 18/9/79, larvae on goldenrod, reared (REMP). Hendre Wds 19/5/80. Slade Wds 2/8/82. Llansoy 26/7/84. (MEA) : Mescoed Mawr 3/7/76, Black Vein 8/5/81, Abercarn 26/5/87, Pontllanfraith 10/7/87, Lower Machen (Park Wood) 13/5/88, Clydach 18/7/89.

Sq. **SO** 20,30,40,50,21,41,51. **ST** 28,48,19,29,39,49,59.

1860 *Chloroclystis rectangulata* Linn.　　Green Pug

Widespread and fairly common in gardens, woodland etc. during June and July. Comes to light.

Wye Valley (Bird 1907).

Usk, garden m.v.l. trap, 29/6/68 and most years until 1983 eg. 1982, (11) 31/5 – 30/7. Ed. 31/5/82. Ld. 2/8/72. Hael Wds 3/7/71. Piercefield Park 18/7/73. Wyndcliff 6/7/81. Slade Wds 16/9/82. Llansoy 4/7/85. Hendre Wds 11/7/86. (MEA) : Tintern 30/6/78, Risca 9/7/79, Mescoed Mawr 8/7/84, Wentwood 9/7/84, Cwmfelinfach 14/7/86, Abercarn 27/6/87, Dinham 7/7/87, Dixton Bank 18/6/88.

Sq. **SO** 30,40,50,41,51. **ST** 48,19,29,49,59.

1861 *Chloroclystis debiliata* Hb.　　Bilberry Pug

This moth, found in the east of the county, is very scarce and local, occurring in some of the woods where bilberry grows. It flies in June and July and comes to light.

"Monmouthshire", (South, *Moths of the British Isles* 1961 ed.).

Llandogo 1953, "larvae plentiful on bilberry" (GMH per REMP).

Pont-y-saeson 13/6/67, one to m.v.l. (REMP). Hael Wds, 4/7/87 several to m.v.l. and 7/7/87 (1) (GANH, MEA).

Sq. **SO** 50. **ST** 59.

1862   *Gymnoscelis rufifasciata* Haw.   Double-striped Pug

Widespread and common in gardens, hedgerows, woods etc., flying from April to mid-June, with a second brood in July and August. Comes to light.

Tintern "1947, fairly common, both by day and by light at night" (Blathwayt 1967).

Usk, garden m.v.l. trap, noted every year from 1966 to 1983 eg. 1978, 1st brood, 24/5 to 8/6; 2nd brood, 18/7 to 5/8. Ed. 30/3/76. Ld. 5/8/78. Wyndcliff 12/6/67 (REMP). Craig-y-Master 7/5/71. Redding's Inclosure 4/4/72. Trelleck Bog 22/7/73. Hael Wds 10/7/73. Usk (Park Wood) 31/5/75. Magor Marsh 11/7/78. Hendre Wds 20/7/79. Slade Wds 23/7/79. Pont-y-saeson 27/4/80. Llansoy 24/8/90. (MEA) : Wentwood 25/5/77, Risca 7/8/81, Pant-yr-eos 13/5/82, Abercarn 16/5/85, Lasgarn Wds 20/5/87.

Sq. **SO** 20,30,40,50,41,51. **ST** 48,29,49,59.

1864   *Chesias legatella* D. & S.   The Streak

Fairly common in Monmouthshire, flying from late September to early November in open woods and heathy places where broom grows.

Llandogo (Nesbitt 1892). Wye Valley (Bird 1907). Tintern, "larvae very common on broom in 1946, bred" (Blathwayt 1967). Hael Wds 1976, larvae on broom 30/6 (MWH in litt.).

Usk, garden m.v.l. trap, 14/10/62 (2) and recorded annually until 1983 eg. 1969, (28), 3/10 – 27/10; 1970, (13), 26/9 – 1/11. Ed. 26/9/70. Ld. 7/11/83. Redding's Inclosure 19/10/71. Hendre Wds 10/10/78. Magor Marsh 16/10/79. Llansoy 9/8/85. (MEA) : Wentwood 19/8/78, Risca 28/10/84, Cwmfelinfach 9/10/86.

Sq. **SO** 30,40,50.41,51. **ST** 48,19,29,49,59.

1865   *Chesias rufata* Fabr.   Broom-tip
         ssp. *rufata* Fabr.

Fairly common, flying from mid-April to early August in open woods etc. where broom grows.

Llandogo dist. (Nesbitt 1892). Wye Valley (Bird 1907). Tintern, "bred" (Bird 1913). Monmouth dist. (Charles c. 1937). Tintern, "a few by day in 1946" (Blathwayt 1967).

Usk, garden m.v.l. trap, recorded frequently eg. 1970, (5) 2/5 – 12/7. Ed. 15/4/67. Ld. 21/7/78. Usk (Park Wood) 26/4/67 Wentwood 2/6/70. Rhyd-y-maen 7/5/71. Redding's Inclosure 20/7/71. Hael Wds 6/8/74. Trelleck Bog 18/7/86. (MEA) : Risca 22/4/72, Pant-yr-eos 13/5/82, Cwmfelinfach 15/7/86.

Sq. **SO** 30,40,50,51. **ST** 19,29,49,59.

1867   *Aplocera plagiata* Linn.   Treble-bar
        ssp. *plagiata* Linn.

This moth fairly common and widespread in Monmouthshire, inhabits open woods and rough places where St John's wort grows. Double-brooded, it flies in late May and June and again from mid-August to early October. It is readily disturbed from herbage by day and at night it comes to light.

Llandogo dist. (Nesbitt 1892). Wye Valley (Barraud 1906). Wye Valley (Bird 1907). Tintern and Llandogo (Bird 1909). Abertillery dist. (Rait-Smith 1912).

Usk, garden m.v.l. trap, recorded every year from 1966 to 1983 eg. 1968, 1st brood (6), 22/5 – 10/6, 2nd brood (18), 15/8 – 4/10. Ed. 20/5/70. Ld. 5/10/69. Pont-y-saeson 17/6/58, 16/8/76. Wentwood 19/6/68. Blca Common 24/9/68. Slade Wds 30/5/71. Redding's Inclosure 7/6/71. St Pierre's Great Wds 6/7/71. Hael Wds 20/9/74. Newport (Coldra) 27/6/75. Newport Docks 17/8/76. Hendre Wds 24/6/79. Wyndcliff 15/9/81. Llansoy 1/6/85. Dinham 4/6/87. Kilpale 30/6/87. Dixton Bank 18/6/88. (MEA) : Ynysyfro 4/7/76, Mescoed Mawr 5/9/77, Risca 3/6/80, Ochrwyth 27/6/86, Cwmfelinfach 23/5/87, Abercarn 26/5/87, Pontllanfraith 20/6/87.

Sq. **SO** 20,30,40,50,41,51. **ST** 28,38,48,19,29,39,49,59.

1868   *Aplocera efformata* Guen.   Lesser Treble-bar

This double-brooded species frequents open woods and rough ground amongst St John's-wort but is far less common and widespread in Monmouthshire than is *A. plagiata*.

Usk, garden m.v.l.trap, 7/9/66 and recorded sporadically eg. 1969, (5) 25/8 – 5/9; 1979 (5) 1/9 – 29/9. Rhyd-y-maen 6/6/70. Redding's Inclosure 15/6/71. Hendre Wds 6/6/78, several flying around clumps of St John's-wort in late afternoon. (MEA) : Risca, to act.l., 7/7/84, 9/7/84.

Sq. **SO** 30,40,41,51. **ST** 29.

1870   *Odezia atrata* Linn.   Chimney Sweeper

A day-flying moth very scarce and local in Gwent. All but one of the following records refer to the western half of the county.

"1911, swarmed in one spot at Crumlin and another at Abertillery" (Rait-Smith 1912). Abertillery dist. (Rait-Smith 1913).

Pontllanfraith, 1987, plentiful 3/7 and 4/7 (MEA, GANH). Llansoy 6/7/87, a single vagrant drifted by, carried on a strong N.E. breeze (GANH). Cwm Merddog 13/6/88 (2) (CT). "Occurs in the Grwyne Fawr valley" (CT 1989).

Sq. **SO** 10,20,40,22. **ST** 19.

1872   *Discoloxia blomeri* Curt.   Blomer's Rivulet
This species flies from late May to early September in woods containing wych elms and is locally plentiful in eastern Monmouthshire but scarce elsewhere in the county. Although patently less numerous than it was prior to the ravages of Dutch Elm Disease in the 1970s it still (1990) appears to remain fairly well-established.

Tintern dist. (Piffard 1859). Wye Valley (Bird 1907). Bettws Newydd (Crowther 1935). Tintern dist. 1965, "to m.v.l. 19/6" (Mere 1965). Wyndcliff 7/6/66 and plentiful most years eg. 2/9/69, 3/8/81, 31/7/91. Wentwood 23/8/68. Hael Wds 3/7/71, 29/5/82, 19/6/88. Redding's Inclosure 20/7/71, 4/7/81. St Pierre's Great Wds 13/7/73. Piercefield Park 18/7/73. Usk, garden m.v.l. trap, 21/6/74. Cwm Coed-y-cerrig 5/6/78. Kilpale 30/6/87. Croes Robert 6/6/91 (1) (RG). (MEA) : Trelleck Common 11/6/82, Clydach 17/6/89.

Sq. **SO** 30,40,50,21,51,22. **ST** 49,59.

1873   *Venusia cambrica* Curt.   Welsh Wave
Very scarce and local in Monmouthshire, occurring in the east of the county in open heathy woodland where mountain ash grows.

Trelleck Common, 1983, to m.v.l. 20/7 (MEA) (1st v-c. rec.). Trelleck Bog 1986, 19/7 (2), 24/7 (3), 17/8 (3) (GANH).

Sq. **SO** 50.

1874   *Euchoeca nebulata* Scop.   Dingy Shell
Frequents damp woods and river valleys from late May to the end of August but local and at low density in Monmouthshire.

Tintern dist. (Piffard 1859). Wye Valley (Bird 1907). Abertillery dist. (Rait-Smith 1912, 1913, 1915).

Usk, garden m.v.l. trap, 1968 11/6 (1), 27/8 (1) and noted sporadically in subsequent years eg. 12/6/74, 12/7/83. Usk (Park Wood) 6/6/67. Cwm-mawr 2/7/71. Hendre Wds 17/7/79. Hael Wds 29/5/82. Clydach 18/7/89. Cwm Coed-y-cerrig 26/6/93. (MEA) . Mescoed Mawr 16/6/78, Pant-yr-eos 16/6/83, Cwmcarn 11/7/87, Dixton Bank 21/7/87, Abercarn 16/6/88.

Sq. **SO** 20,30,50,21,41,51,22. **ST** 29.

1875   *Asthena albulata* Hufn.   Small White Wave
A widespread and common woodland moth, flyng at dusk from late May to mid-July and readily disturbed from undergrowth by day but rarely coming to light.

Tintern dist. (Bird 1905). Wye Valley (Barraud 1906).

Pont-y-saeson 13/6/67 to m.v.l. (REMP), 8/6/68 plentiful by day (GANH). Wyndcliff 13/6/67. Slade Wds 3/6/72. Cwm Coed-y-cerrig 27/5/73. St Pierre's Great Woods 1/6/75. Hendre Wds 28/5/80. Usk, garden m.v.l.

trap, 1/6/82. Hael Wds 14/6/86. Clydach 15/6/90. (MEA) : Mescoed Mawr 24/5/77, Tintern 30/6/78, Trelleck Common 11/6/82, Risca 8/7/85, Cwm Tyleri 10/7/85, Ochrwyth 7/6/86, Llandegfedd 17/7/86, Dixton Bank 18/6/88.

Sq. **SO** 20,30,50,21,41,51,22. **ST** 28,48,29,39,49,59.

1876   *Hydrelia flammeolaria* Hufn.   Small Yellow Wave
Widespread and common, frequenting hedgerows and woodland during June and July and coming readily to light.
   Tintern dist. (Piffard 1859). Wye Valley (Bird 1907).
   Usk, garden m.v.l. trap, 9/7/67 and every year until 1983 eg. 1973, (20) 25/6 – 28/7; 1983, (12) 25/6 – 20/7. Wyndcliff 12/6/67. Usk (Park Wood) 15/6/67. Hael Wds 3/7/71. Pont-y-saeson 21/5/73. St Pierre's Great Wds 13/7/73. Piercefield Park 18/7/73. Newport (Coldra) 27/6/75. Cwm Coed-y-cerrig 5/6/78. Slade Wds 2/8/80. Hendre Wds 11/7/86. Llansoy 29/6/93. (MEA) : Wentwood 18/6/78, Black Vein 15/7/82, Risca 11/7/83, Mescoed Mawr 28/7/84, Ynysddu 12/6/85, Cwmfelinfach 5/7/86, Abercarn 27/6/87.

Sq. **SO** 30,40,50,41,22. **ST** 38,48,19,29,49,59.

1877   *Hydrelia sylvata* D. & S.   Waved Carpet
A scarce woodland moth which occurs locally in the eastern half of Monmouthshire. It flies in June and July and comes to light.
   Tintern dist. (Piffard 1859). Wye Valley (Bird 1907).
   Pont-y-saeson 1967, one to m.v.l. 13/6, two beaten from bushes by day 27/6 (GANH). Trelleck Common 9/7/68 (1). Whitebrook 2/7/69 (SC). Wyndcliff 13/7/69 (10) to m.v.l. Usk, garden m.v.l. trap, single moths on 4/6/71 and 23/7/72. Hael Wds 3/7/71, also 1973, 74, 76 and 87. Redding's Inclosure 17/6/74. (MEA) : Tintern 30/6/78.

Sq. **SO** 30,50,51. **ST** 59.

1878   *Minoa murinata* Scop.   Drab Looper
Occurs locally in the eastern half of Monmouthshire, flying by day during May and June, and occasionally in August, in woodland where the wood spurge (*Euphorbia amygdaloides*) grows. Often abundant, especially when wood spurge grows in profusion following clear-felling of woods.
   Tintern (Piffard 1859). Wye Valley (Barraud 1907). Wye Valley (Bird 1907). Tintern, Aug. 1911 (Bird 1912). Redding's Inclosure and Tintern 31/5/68 (JMC-H 1969).
   Pont-y-saeson 7/6/66, 10/8/68 and often abundant eg. 26/5/70 and 1/6/81. Usk (Lady Hill Wood) 6/6/67, a single moth seen flying by day in several acres of Wood Spurge in a part of the wood recently felled but the following year on 10/6/68 the species "swarmed" there. Tredean Wds 31/5/68. Prescoed 8/6/69, one to m.v.l. St Pierre's Great Wds 24/6/69. Slade

Wds 30/5/71, 2/6/81. Hendre Wds 6/6/79. Dixton Bank 18/6/88.

Sq. **SO** 30,41,51. **ST** 48,39,49,59.

1879   *Lobophora halterata* Hufn.   The Seraphim
A woodland moth, scarce in Monmouthshire, found locally during May and June in damp deciduous woods in the eastern half of the county.

Wye Valley (Bird 1907). Usk, garden m.v.l. trap, (8) specimens recorded in the years 1966 – 1983, inc. 25/5/68 (1), May 1978 (2), 1979 (2), 1982 (1). Tredean Wds 31/5/68, one by day. Hael Wds 30/5/78, 14/6/86. Hendre Wds 27/5/78 (4) to m.v.l., 14/6/79.

Sq. **SO** 30,50,41. **ST** 49.

1881   *Trichopteryx carpinata* Borkh.   Early Tooth-striped
A common and widespread woodland moth, flying from late March to mid-May and coming to light.

Wye Valley 1906, imagines, also larvae on sallow, reared (Bird 1907). Llandogo 1912 (Bird 1913).

Usk, garden m.v.l. trap, 17/4/67 and recorded every year until 1984 eg. 1968, (4) 8/4 – 2/5; 1982, (4) 2/4 – 28/4. Ed. 30/3/76. Ld. 17/5/78. Gwehelog (Camp Wood) 9/4/67. Llangybi (Cae Cnap) 23/4/68. Usk (Park Wood) 3/5/69 (plfl.). Hendre Wds 25/3/72. Pont-y-saeson 3/4/72. Redding's Inclosure 4/4/72. Hael Wds 25/3/72. Wentwood 26/5/80. Llansoy 15/4/87. (MEA) : Slade Wds 21/4/78, Mescoed Mawr 19/4/85, Risca 8/5/85, Cwm-felinfach 14/4/87, Cwm Merddog 26/4/87, Wyndcliff 27/4/87, Abercarn 3/5/87.

Sq. **SO** 10,30,40,50,41,51. **ST** 48,19,29,39,49,59.

1882   *Pterapherapteryx sexalata* Retz.   Small Seraphim
Scarce and local in Monmouthshire, flying in damp woods and marshy places during June and July.

Pont-y-saeson 13/6/67, to m.v.l. (REMP) (1st v-c. rec.). Whitebrook 2/7/69 (SC). Usk, garden m.v.l trap, 1972, 16/6 (1), 20/7 (3), and a further (18) specimens recorded sporadically up to 1983 inc. 1979, (7) 4/6 – 10/7. Ed. 4/6/79. Ld. 20/7/72. Magor Marsh 5/7/77 (MEA); 19/6/79, 2/7/89 (GANH). Wyndcliff 2/7/85 (MEA). Llansoy 26/6/86 (GANH).

Sq. **SO** 30,40,50. **ST** 48,59.

1883   *Acasis viretata* Hb.   Yellow-barred Brindle
Widespread but at low density in Monmouthshire, flying in woods and bushy places during May and June and as a second generation in August and September. It comes to light.

Usk, garden m.v.l. trap, 4/6/67 and recorded most years until 1983 eg. 1979, 1st brood, (8) 13/5 – 30/5; 2nd brood, (1) 20/8. Ed. 2/5/70. Ld.

14/9/69. Usk (Park Wood) 6/6/67. Hendre Wds 4/9/78. Hael Wds 6/6/79. Slade Wds 28/8/79. Trelleck Bog 25/5/87. (MEA) : Risca 8/9/79, Black Vein 12/6/81, Wyndcliff 6/7/84, Ochrwyth 1/6/85, Cwmfelinfach 12/8/86, Abercarn 25/6/87.

Sq. **SO** 30,50,41. **ST** 28,48,19,29,59.

ENNOMINAE

1884 *Abraxas grossulariata* Linn.   The Magpie

Widespread and common in gardens, hedgerows, margins of woods, bushy places etc., flying during July and August and often in September.

Wye Valley, larvae (Barraud 1906). Wye Valley (Bird 1907). Abertillery (Rait-Smith 1912).

Croesyceiliog 30/7/27, plentiful (GANH). Usk dist. 1965 (plentiful). Usk, garden m.v.l. trap, annually 1966 to 1983 and often plentiful eg. 1969, 19/7 – 1/9. Ed. 17/7/73. Ld. 6/9/68, 23/9/72. Wentwood 10/8/68. Prescoed 16/8/68. Magor Marsh 23/8/68. Bica Common 23/8/68. Pont-y-saeson. Usk (Park Wood) 3/9/69. Tintern 1/8/70. Cwm Tyleri 1/7/73. Hendre Wds 29/8/78. Slade Wds 23/7/79. Wyndcliff 3/8/81. Tal-y-coed 9/9/83. Llansoy 28/7/84. Clydach 18/7/89. (MEA) : Risca 12/8/80, Mescoed Mawr 14/8/81, Uskmouth 16/8/88.

Sq. **SO** 20,30,40,50,21,31,41. **ST** 38,48,29,39,49,59.

1885 *Abraxas sylvata* Scop.   Clouded Magpie

Flying from early June to mid-August this woodland moth occurs locally where wych elm flourishes, mainly in the eastern half of the county, and is sometimes abundant. Although less numerous than it was prior to the appearance of Dutch elm disease, it is still plentiful at some sites in the Wye Valley. Earliest date recorded 5/6/73 (Cwm Coed-y-cerrig), latest date 19/8/69 (Wentwood).

Near Tintern (Piffard 1859). Llandogo dist. 18/5/92 (Nesbitt 1892). Wye Valley (Barraud 1906). Wye Valley (Bird 1907). "This species does not occur in the Abertillery district at all" (Rait-Smith 1912). Tintern dist. 19/6/65 to m.v.l. (Mere 1965).

Wyndcliff 7/6/66, 13/7/69 (plfl.). Pont-y-saeson 13/6/67, 18/7/72 (abdt.). Magor Marsh 22/7/69. Wentwood 11/7/69. Hael Wds 5/7/70, 5/7/82 (abdt.) to m.v.l. with (25) at a time on the sheet. Usk, garden m.v.l. trap, 27/7/68 (1), 1969 (2), 1973 (2), 1979 (1), 1981 (2), 1983 (2), (the only records from 1966 to 83). Tintern 1/8/70. Rodding's Inclosure 7/6/71. Cwm Coed-y-cerrig 5/6/73 (plfl.), 5/6/78. St Pierre's Great Wds 13/7/73. Clydach 11/5/90 (1). (MEA) : Mescoed Mawr 1/7/76, Risca (Dan-y-Graig) 1/7/87, Cwmfelinfach 6/7/87, Dinham 7/7/87, Ochrwyth 17/6/88.

Sq. **SO** 30,50,21,51,22. **ST** 28,48,19,29,49,59.

1887 *Lomaspilis marginata* Linn.   Clouded Border
Flying from mid-May to late August this species is widespread and common in hedgerows, woodland and damp bushy places.

Llanock Wood [= Llanerch Wood, Crumlin] (Rait-Smith 1912, 1913, 1915).

Usk, garden m.v.l. trap, every year from 1966 to 1983 and plentiful eg. 1968, (13) 24/5 – 15/8; 1979, (21) 29/5 – 20/8. Ed. 15/5/70. Ld. 15/8/68, 20/8/79. Wyndcliff 7/6/66, 23/6/86. Pont-y-saeson 7/6/66. Prescoed 31/5/68. Usk (Park Wood) 4/6/68. Wentwood 8/7/68. Bica Common 28/6/69. Hendre Wds 26/5/70, 11/7/86. Trelleck Bog 2/6/70, 18/7/86. Tintern 1/8/70. Magor Marsh 2/8/70. Redding's Inclosure 7/6/71. Cwm Coed-y-cerrig 5/6/73. Piercefield Park 18/7/73. Hael Wds 28/6/74. Newport (Coldra) 27/6/75. Slade Wds 28/5/78. Llansoy 28/7/84. Clydach 18/7/89. (MEA) : Risca 9/7/79, Mescoed Mawr 21/6/83, Cwm Merddog 14/7/86, Pontllanfraith 2/7/87, Sugar Loaf (St Mary's Vale) 8/7/87, Cwmfelinfach 15/7/86, Nant Gwyddon 25/6/88.

Sq. **SO** 10,30,40,50,21,41,51,22. **ST** 38,48,19,29,39,49,59.

1888 *Ligdia adustata* D. & S.   Scorched Carpet
This double-brooded species frequenting woods, hedgerows, gardens etc. is widespread but at low density in the eastern half of Monmouthshire flying from early May to late August.

Near Bigsweir (Barraud 1906).

Usk, garden m.v.l. trap, a few recorded most years from 1966 to 1983 eg. 1968, (2) 29/5, 11/8; 1970, (5) 24/4 – 26/6. Ed. 26/4/82. Ld. 26/8/69. Usk (Park Wood) 5/6/67, 3/5/69. Wyndcliff 6/6/67. Llantrisant (Nant-y-banw) 25/5/68. Redding's Inclosure 20/7/71. Prescoed 18/5/73. Hael Wds 28/6/74. Newport Docks 17/8/76. Magor Marsh 21/8/76. Slade Wds 28/5/78. St Pierre's Great Wds 25/7/78. Llangeview 28/7/79. Llansoy 26/5/85. Hendre Wds 1/7/86. Kilpale 30/6/87. (MEA) : Wentwood 10/7/77.

Sq. **SO** 30,40,50,41,51. **ST** 38,48,39,49,59.

1889 *Semiothisa notata* Linn.   Peacock Moth
Flies in woodland in June and July but local and scarce in Gwent.

Tintern dist., not uncommon (Bird 1905, 1907).

Hael Wds 10/7/73 (1) to m.v.l. and found here fairly regularly in subsequent years eg. 1974, 28/6 (10), 12/7 (2); 1978, 27/6 (10); 1982, 31/7 (1); 1987, 24/6 (1). Trelleck Bog 22/7/78 (1), 1979 (2), 1986 (2). (MEA) : Mescoed Mawr 25/6/76, 1984, 1985; Cwmfelinfach 6/7/87.

Sq. **SO** 50. **ST** 19,29.

1890 *Semiothisa alternaria* Hb.   Sharp-angled Peacock
Rare in Monmouthshire. Usk, garden m.v.l. trap, 14/8/76, one (1st v-c. rec.)

(GANH). Newport Docks (1) to m.v.l. 17/8/76, (GANH).

Sq. **SO** 30. **ST** 38.

1893 *Semiothisa liturata* Cl.   Tawny-barred Angle
Common and widespread in conifer plantations, flying from early June to mid-August. The melanic form *nigrofulvata* Collins occurs frequently.

Wye Valley (Nesbitt 1892a). Abertillery dist. (Rait-Smith 1912, 1913).

Usk, garden m.v.l. trap, recorded every year from 1966 to 1983 eg. 1968, (10) 3/6 – 12/8; 1979, (21) 8/6 – 13/8. Ed 3/6/68. Ld. 21/8/74. Wyndcliff 7/6/66. Usk (Park Wood) 15/6/67. Prescoed 5/7/68. Wentwood 8/7/68 (abdt.). Trelleck Common 9/7/68 (abdt.). Bica Common 28/6/69. Hendre Wds 5/7/70. Hael Wds 5/7/70. Tintern 1/8/70. Slade Wds 1/6/71. Pont-y-saeson 18/7/72. Trelleck Bog 16/6/73. St Pierre's Great Wds 13/7/73. Magor Marsh 1/7/75. Cwm Coed-y-cerrig 5/6/78. Redding's Inclosure 23/6/81. Llansoy 26/7/84. Ysguborwen (Lantrisant) 17/7/86. (MEA) : Mescoed Mawr 25/6/76, Risca 12/7/82, Cwm Meiddog 7/7/85, Ochrwyth 27/6/86, Cwmfelinfach 5/9/86, Abercarn 27/6/87.

Sq. **SO** 10,20,30,40,50,41,51,22. **ST** 28,48,19,29,39,49,59.

1894 *Semiothisa clathrata* Linn.   Latticed Heath
   ssp. *clathrata* Linn.
Scarce in Monmouthshire and with a patchy distribution, this moth frequents open woodland, embankments and rough ground from June to August. By day it flies in sunshine and at night comes to light.

Usk, garden m.v.l. trap, noted only twice in eighteen years, viz. 22/8/68 and 8/8/82. Wentwood (Darren Wood) 24/8/68. Magor Marsh, often noted at m.v.l. eg. 2/6/70; 1971; 1976, (6) 21/6 – 21/8; 1981. Runston 6/8/74. Newport Docks 17/8/76, particularly large specimens plentiful at m.v.l. on damp, bushy, waste ground. Llansoy 26/8/88. (MEA) : Uskmouth 15/7/88 (by day), 16/8/88 (plfl. to m.v.l.); Magor Pill (sea-wall) 6/8/89 (2).

Sq. **SO** 30,40. **ST** 38,48,49.

1897 *Semiothisa wauaria* Linn.   The V-Moth
Flies from late June to early August and found mainly in gardens. Local and scarce in Monmouthshire.

Abertillery dist. (Rait-Smith 1912).

Usk, garden m.v.l. trap, recorded most years from 1966 to 83 but in small numbers eg. 1966, 12/7 (2); 1968, (8) 27/6 – 22/7; 1979, (4) 11/7 – 29/7. Ed. 27/6/78. Ld. 3/8/77. Hael Wds 5/7/70. Llansoy 8/7/87. Clydach 18/7/89 (1) to m.v.l. (MEA) : Risca 30/7/84, 8/8/86, 23/7/87.

Sq. **SO** 20,30,40,50,21. **ST** 29.

1901   *Cepphis advenaria* Hb.   Little Thorn

This species, locally common in the eastern half of the county, flies in late May and June inhabiting open woodland where bilberry flourishes. Earliest recorded date 25/5/68 (Pont-y-saeson). Latest date 21/6/83 (Risca).

Llandogo dist. (Nesbitt 1892a). Llandogo 15/5 1892 (idem 1892). Wye Valley, May 1907 (Barraud 1907b). Wye Valley (Bird 1907). Tintern, 1946 – 1948, "quite common both by day and at light in all three years" (Blathwayt 1967). Tintern dist. (Mere 1965). Tintern 31/5/68 (JMC-H 1969).

Pont-y-saeson 7/6/66 to m.v.l. (REMP). Wyndcliff 6/6/67. Trelleck Common 14/6/67, many disturbed from bilberry bushes by day (REMP, GANH). Usk, garden m.v.l. trap, 11/6/68 (1), 25/5/76 (1). Prescoed 8/6/69 (2) to m.v.l. Slade Wds 30/5/71, 26/5/80. Hendre Wds 5/6/71. Bica Common 5/6/72. Trelleck Bog 16/6/73. Hael Wds 29/5/82. Croes Robert 6/6/91 (3) by day (RG). (MEA) : Risca, to act.l. 21/6/83; Ochrwyth 1/6/85.

Sq. **SO** 30,40,50,41. **ST** 28,48,29,39,49,59.

1902   *Petrophora chlorosata* Scop.   Brown Silver-line

This moth, flying from the end of April to mid-June, is widespread and common and found wherever bracken grows.

Llandogo dist. (Nesbitt 1892, 1893). Wye Valley (Barraud 1906). Wye Valley (Bird 1907). Aberbeeg (Rait-Smith 1913). Tintern 31/5/68 and Derifach 2/6/68 (JMC-H 1969).

Pont-y-saeson 13/6/67. Usk (Park Wood) 12/5/67. Graig-y-Master 21/5/68 (abdt.). Llantrisant (Nant-y-banw) 25/5/68. Tredean Wds 31/5/68. Prescoed 31/5/68. Wentwood (Cadira Beeches) 10/6/68. Whiteleye 9/6/69. Bica Common 28/6/69. Rhyd-y-maen 23/5/70. Hael Wds 25/5/70. Hendre Wds 26/5/70. Magor Marsh 2/6/70. Trelleck Bog 2/6/70 (abdt.). Slade Wds 30/5/71. Cwm-mawr 31/5/71. Redding's Inclosure 7/6/71. Cwm Tyleri 26/5/73. Mynyddislwyn 10/6/74. Trostrey Common 11/5/80. Pontypool (Mountain Air) 25/5/80. Llansoy 4/6/85. Dinham 4/6/87. (MEA) : Sugar Loaf (St Anne's Vale) 7/7/79, Risca 2/6/80, Mescoed Mawr 21/6/83, Cwm Merddog 31/5/85, Ochrwyth 7/6/86, Cwmfelinfach 24/5/87, Pontllanfraith 20/6/87, Lower Machen (Park Wood) 13/5/88, Nant Gwyddon 21/5/88.

Sq. **SO** 10,20,30,40,50,21,41,51. **ST** 28,48,19,29,39,49,59.

1903   *Plagodis pulveraria* Linn.   Barred Umber

This moth, flying in deciduous woodland from mid-May to mid-July, is widely but thinly distributed in the eastern half of Gwent.

Llandogo, "to light" 26/5/92 (Nesbitt 1892). Wye Valley (Barraud 1906). Wye Valley (Bird 1907). Bettws Newydd 31/5/35 (RBH). Tintern, "fairly common in 1948" (Blathwayt 1967).

Usk, garden m.v.l. trap, recorded most years from 1966 to 1983 eg. 1967, 10/5 (1), 28/5 (1); 1968, (6) 1/6 – 11/6; 1982, (5) 20/5 – 11/7. Ed.

10/5/67. Ld. 11/7/82. Pont-y-saeson 7/6/66. Usk (Park Wood) 12/5/67. Llangybi (Coed-y-Fferm) 7/6/68. Hendre Wds 26/5/70, 9/5/87. Trelleck Bog 2/6/70. Cwm Coed-y-cerrig 5/6/73. Magor Marsh 20/7/76 (3) to m.v.l. Redding's Inclosure 30/5/77. Hael Wds 29/5/82. Llansoy 15/6/86. Wyndcliff 23/6/86. Wentwood. (MEA) : Mescoed Mawr 14/5/78 (plfl.), Risca 2/6/80.

Sq. **SO** 30,40,50,41,51,22. **ST** 48,29,39,49,59.

1904 *Plagodis dolabraria* Linn.   Scorched Wing
Widespread and fairly common in the eastern half of the county, flying in deciduous woodland during June and the first half of July.

Bettws Newydd (Crowther 1933). Bettws Newydd 31/5/35 (RBH).

Usk, garden m.v.l. trap, recorded annually from 1966 to 1983 inc. 1969, (7) 24/6 – 12/7; 1983, (13) 21/6 – 7/7. Ed. 31/5/82. Ld. 13/7/68. Usk (Park Wood) 5/6/67. Wyndcliff 6/6/67. Pont-y-saeson 13/6/67. Wentwood 23/6/69. Rhyd-y-maen 6/6/70. Redding's Inclosure 7/6/71. Prescoed 21/6/71. Cwm-mawr 2/7/71. Hael Wds 3/7/71. Hendre Wds 27/5/78. Llansoy 4/6/85. Slade Wds 27/6/85. Trelleck Bog 25/5/87. Dinham 4/6/87. Kilpale 30/6/87. (MEA) : Ochrwyth 27/6/86.

Sq. **SO** 20,30,40,50,41,51. **ST** 28,48,39,49,59.

1906 *Opisthograptis luteolata* Linn.   Brimstone Moth
One of the commonest and most widespread of moths, flying from April to October in gardens, hedgerows, woodland and bushy places in general.

Wye Valley (Barraud 1906). Abertillery dist. (Rait-Smith 1912, 1913). Cwmyoy 1965 (SHR).

Usk, garden m.v.l. trap, noted annually from 1966 to 1983 and often abundant eg. 1968, 2/5- 4/10; 1978, 4/5 – 13/10. Ed. 15/4/80. Ld. 16/10/77. Gwehelog (Camp Wood) 29/4/67. Usk (Park Wood) 12/5/67. Wyndcliff 6/6/67. Pont-y-saeson 13/6/67. Nant-y-banw 25/5/68. Coed-y-Fferm 7/6/68. Wentwood 10/8/68. Magor Marsh 23/8/68. Risca Common 27/8/68. Prescoed 8/6/69. Hendre Wds 26/5/70. Rhyd-y-maen 6/6/70. Hael Wds 5/7/70. Undy (Collister Pill) 30/8/70. Cwm-mawr 31/5/71. Redding's Inclosure 20/7/71. St Pierre's Great Wds 15/5/73. Cwm Coed-y-cerrig 27/5/73. Trelleck Bog 16/6/73. Piercefield Park 18/7/73. Cwm Tyleri 13/7/75. Newport Docks 17/8/76. Slade Wds 28/5/78. Tal-y-coed 9/9/83. Llansoy 26/7/84. Dinham 4/6/87. Kilpale 30/6/87. St Mary's Vale 8/7/87. (MEA) : Rhiwderin 21/5/76, Llantarnam 10/6/76, Allt-yr-yn 20/6/76, Ynysyfro 4/7/76, Risca 5/7/79, Mescoed Mawr 14/8/81, Ochrwyth 27/6/86, Cwm Merddog 14/7/86, Cwmfelinfach 4/9/86, Hilston Park 4/10/86, Abercarn 26/5/87, Pontllanfraith 2/7/87, Lower Machen (Park Wood) 13/5/88, Dixton Bank 18/6/88, Uskmouth 16/8/88, Clydach 18/7/89.

Sq. **SO** 10,20,30,40,50,21,31,41,51,22,32. **ST** 28,38,48,19,29,39,49,59.

1907   *Epione repandaria* Hufn.   Bordered Beauty
Occurs locally and at low density in Monmouthshire, frequenting marshy places, river valleys and damp woods from July to September.

Wye Valley 1911 (Bird 1912).

Usk, garden m.v.l.trap, recorded in ten of the eighteen years from 1966 to 1983 but in small numbers eg. 1967, (2) 7/8 and 13/8; 1982, (5) 20/7 – 30/7. Ed. 20/7/82. Ld. 27/8/66. Tal-y-coed 9/9/83. Llansoy 27/8/86, 4/9/86. Trelleck Bog 30/8/86. (MEA) : Ynysyfro 4/7/76; Mescoed Mawr 14/8/81, 13/7/82; Wentwood 17/8/83, 16/8/84; Pontllanfraith 1/9/87.

Sq. **SO** 30,40,50,31. **ST** 28,19,29,49.

1909   *Pseudopanthera macularia* Linn.   Speckled Yellow
This day-flying moth, frequenting open woodland from mid-May to mid-June, occurs locally, mainly in the eastern half of the county, but is often abundant where found.

Llandogo (Nesbitt 1893). Wye Valley 1906 (Barraud 1906). Wye Valley 1906 (Bird 1907). Wye Valley, May 1907 (Barraud 1907b). Tintern 31/5/68 (JMC-H 1969).

Pont-y-saeson 17/6/58 (plfl.) and noted most years from 1966 to 1981 and often abundant eg. 25/5/68, 27/5/78. Prescoed 31/5/68. Wentwood (Cadira Beeches) 10/6/68. Usk (Park Wood) 10/6/68. Slade Wds 30/5/71, (plfl.). Hael Wds 30/5/78. Hendre Wds 6/6/78, 16/5/88. Trostrey Common 11/5/80. St Pierre's Great Wds 13/5/80. Trelleck (Fedw Fach) 14/5/88 (abdt.). (MEA) : Pant-yr-eos 17/5/82, Cwmfelinfach 9/5/87, Abercarn 21/5/88.

Sq. **SO** 30,40,50,41. **ST** 48,19,29,39,49,59.

1910   *Apeira syringaria* Linn.   Lilac Beauty
Occurs locally and at low density in Monmouthshire, flying in open woodland in late June and July.

Wye Valley (Nesbitt 1892a). Wye Valley 1906 (Bird 1907). Usk (Cockshoots Wood), July 1933 (RBH).

Devauden 5/7/61. Usk 30/6/64. Usk, garden m.v.l. trap, a few sporadic records eg. 18/7/72, 1973 (1), 75(1), 76(2), 78(2), 79(2), 81(1), 1983 (5) 5/7 – 14/7. Ed. 21/6/76. Ld. 29/7/78. Magor Marsh 10/7/70. Wentwood 13/7/71 (JMD). Llansoy 4/7/86. (MEA) : Llantarnam 24/6/76, Mescoed Mawr 25/6/76, Ynysyfro 28/6/86, Risca 7/7/81, Wyndcliff 6/7/84, Ynysddu 12/7/85.

Sq. **SO** 30,40. **ST** 28,48,19,29,39,49,59.

1912   *Ennomos quercinaria* Hufn.   August Thorn
Flying during August and early September in deciduous woodland, gardens etc, this moth is scarce and of sporadic occurrence, although in central Monmouthshire (Usk dist.) it appears with greater regularity.

Wye Valley 1906 (Bird 1907).

Usk, garden trap, recorded every year from 1966 to 1983 and in some years fairly frequent eg. 1968, (11) 9/8 – 3/9; 1971, (15) 29/7 – 24/8. Ed. 29/7/78. Ld. 7/9/73. Pont-y-saeson 10/8/74. Hael Wds 14/8/78. Wentwood. Newport Docks 17/8/76. Tal-y-coed 9/9/83. (MEA) : Risca 11/8/83, Wyndcliff 12/8/83, Cwmfelinfach 26/9/86.

Sq. **SO** 30,50,31. **ST** 38,19,29,49,59.

1913 *Ennomos alniaria* Linn.   Canary-shouldered Thorn

Common and widespread in Monmouthshire, in gardens, open woodland, marshes etc., flying from late July to mid-October.

Bettws Newydd 1934, "at light" 16/8 (RBH). Bettws Newydd (Crowther 1935). Brockwells (Caerwent) 6/10/65 (CT). Cwmyoy, Oct. 1965 (SHR).

Usk, garden m.v.l. trap, recorded every year from 1966 to 1983 and often plentiful eg. 1968, (60) 10/8 – 2/10. Ed. 21/7/76. Ld. 25/10/72. Prescoed 21/8/68. Bica Common 23/8/68. Wentwood 24/8/68. Magor Marsh 5/8/69. Usk (Park Wood) 3/9/69. Gwehelog (Camp Wood) 28/8/70. Trelleck Bog 4/9/73. Pont-y-saeson 7/9/73. Newport Docks 17/8/76. Hendre Wds 29/8/78. Slade Wds 28/8/79. Wyndcliff 15/9/81. Hael Wds 31/7/82. Llansoy 24/8/86. (MEA) : Mescoed Mawr 4/9/77, Risca 8/9/79, Cwm Tyleri 29/9/86, Pontypool 10/10/86, Cwmfelinfach 26/9/86.

Sq. **SO** 20,30,40,50,41,32. **ST** 38,48,19,29,39,49,59.

1914 *Ennomos fuscantaria* Haw.   Dusky Thorn

A woodland moth locally common in Monmouthshire, flying from late July to mid-October.

Tintern, 1911 (Bird 1912). Bettws Newydd 1934, (4) "at light" 8/8 (RBH). Bettws Newydd (Crowther 1935).

Usk, garden m.v.l. trap, recorded annually from 1966 to 1983 and often plentiful eg. 1968, (90) 11/8 – 5/10 inc. (14) on 21/9; 1973, (66) 30/7 – 24/9. Ed. 19/7/76, 26/7/82. Ld. 13/10/69, 23/10/72. Magor Marsh 23/8/68. Bica Common 23/8/68. Wyndcliff 2/9/69. Usk (Park Wood) 3/9/69. Gwehelog (Camp Wood) 28/8/70. Rhyd-y-maen 1/9/73. Hael Wds 18/8/75. Newport Docks 17/8/76. Hendre Wds 23/8/87. Pont-y-saeson 30/9/79. Slade Wds 28/7/82. Llansoy 30/8/87. (MEA) : Risca 1/9/84, Cwmfelinfach 9/10/86, Ochrwyth 17/9/88.

Sq. **SO** 30,40,50,41. **ST** 28,38,48,19,29,49,59.

1915 *Ennomos erosaria* D. & S.   September Thorn

This species flies from late July to mid-October in woods, gardens and bushy places. It is locally plentiful in central Monmouthshire (Usk dist.) but is sparsely distributed elsewhere in the county.

Usk, garden m.v.l. trap, noted every year from 1966 to 1983 and sometimes frequent eg. 1972, (32) 22/7 – 12/10. Ed. 20/7/73, 8/7/82. Ld. 13/10/78. Prescoed 3/8/69. Magor Marsh 2/8/70. Usk (Park Wood) 23/8/72. Hendre Wds 1/8/78. Wyndcliff 3/8/81. Llansoy 17/10/85. Wentwood. (MEA) : Risca 1/9/84, Cwmfelinfach 5/9/86, Pontllanfraith 1/9/87.

Sq. **SO** 30,40,41. **ST** 48,19,29,39,49,59.

1917 *Selenia dentaria* Fabr.   Early Thorn

A double-brooded moth very common and widespread, inhabiting open woodland, gardens and bushy places generally. The first brood flies from mid-March to the middle of May and the second from mid-July until early September.

Wye Valley (Bird 1907).

Usk, garden m.v.l. trap, recorded every year from 1966 to 1983 eg. 1969, (46) 1/4 – 31/8; 1972, (70) 2/4 – 2/9. Ed. 11/3/68, 18/3/73. Ld. 5/9/69, 31/8/79. Usk (Park Wood) 28/4/67. St Pierre's Great Wds 23/7/68. Bica Common 23/8/68. Wentwood 25/4/69. Magor Marsh 5/8/69. Coed-y-Fferm 21/4/70. Tintern 1/8/70. Cwm-mawr 6/4/71. Usk (Cockshoots Wood) 29/4/71. Rhyd-y-maen 7/5/71. Redding's Inclosure 20/7/71. Prescoed 16/4/72. Hendre Wds 22/4/72. Slade Wds 29/8/72. Wyndcliff 22/3/74. Hael Wds 6/4/74. Llantrisant 11/4/80. Llansoy 18/7/86. Trelleck Bog 30/8/86. (MEA) : Llanarth 2/5/76, Mescoed Mawr 3/7/76, Ynysyfro 4/7/76, Risca 12/8/81, Cwmfelinfach 14/4/87, Ochrwyth 24/4/87, Abercarn 29/4/87, Cwm Coed-y-cerrig 17/8/87, Nant Gwyddon 15/8/88.

Sq. **SO** 20,30,40,50,31,41,51,22. **ST** 28,48,19,29,39,49,59.

1918 *Selenia lunularia* Hb.   Lunar Thorn

Scarce and local in Monmouthshire, flying in deciduous woodland from late May to early July.

Llandogo dist. "to light" 26/5/92 (Nesbitt 1892). Whitebrook (Mere 1965). Usk Castle 6/6/66 (RBH). Pont-y-saeson 13/6/67 (2) to m.v.l. (GANH). Wyndcliff 16/6/67 (2). Cwm-mawr 2/7/71. Trelleck Bog 16/6/73. Usk, garden m.v.l. trap, 24/6/83 (the only record in the eighteen years 1966 – 1983). Llansoy 23/6/86, 1987 (2). 1988 (2). (MEA) : Black Vein 12/6/81, Risca 6/7/87.

Sq. **SO** 20,30,40,50. **ST** 29,59.

1919 *Selenia tetralunaria* Hufn.   Purple Thorn

A double-brooded woodland moth fairly common in Gwent, the first brood appearing in April and May and the second flying in July and August.

Usk, garden m.v.l. trap, recorded every year from 1966 to 1984 eg. 1974, 1st brood, (7) 10/4 – 22/5, 2nd brood, (7) 12/7 – 7/8; 1982, 1st brood, (10) 5/4 – 12/5, 2nd brood, (10) 13/7 – 3/8. Ed. 5/4/82. Ld. 20/8/71. Usk

(Park Wood) 28/4/67. Gwehelog (Camp Wood) 29/4/67. Wentwood 10/6/68. Prescoed 4/5/70. Caerwent 1/4/74 (CT). Pont-y-saeson 10/8/74. Wyndcliff 26/5/76. Hendre Wds 27/4/79. Slade Wds 22/4/80. Hael Wds 26/4/82. Llansoy 3/5/85. (MEA) : Risca 30/7/84, Mescoed Mawr 17/5/85, Cwm Merddog 26/4/87, Cwmfelinfach 8/5/88.

Sq. **SO** 10,30,40,50,41. **ST** 48,19,29,39,49,59.

1920 *Odontopera bidentata* Cl.   Scalloped Hazel
Widespread and common in gardens, woods etc. during May and June.

Llandogo dist. "to light" 26/5/92 (Nesbitt 1892). Tintern dist. (Bird 1905). Wye Valley 1906 (Barraud 1906). Wye Valley 1906 (Bird 1907). Abertillery dist. (Rait-Smith 1912). Bettws Newydd 31/5/35 (RBH).

Usk, garden m.v.l. trap, recorded every year from 1966 to 83 eg. 1968, (23) 9/5 – 13/6; 1982, (17) 12/5 – 9/6. Ed. 6/5/81. Ld. 18/6/79. Usk (Park Wood) 12/5/67. Wyndcliff 6/6/67. Pont-y-saeson 13/6/67. Prescoed 31/5/68. Wentwood 23/6/69 (JMD). Bica Common 28/6/69. Hael Wds 25/5/70. Hendre Wds 26/5/70. Cwm-mawr 31/5/71. Slade Wds 1/6/71 (plfl.) Redding's Inclosure 7/6/71. St Pierre's Great Wds 15/5/73. Trelleck Bog 22/5/73. Cwm Coed-y-cerrig 5/6/73, (plfl.) Mynyddislwyn 10/6/74. Llansoy 1/6/87. (MEA) : Rhiwderin 21/5/76, Mescoed Mawr 19/6/77, Pant-yr-eos 13/5/82, Risca 13/6/85, Ochrwyth 27/6/86, Cwmfelinfach 24/5/87, Abercarn 26/5/87, Lower Machen (Park Wood) 13/5/88, Nant Gwyddon 7/6/88.

Sq. **SO** 20,30,40,50,41,51,22. **ST** 28,48,19,29,39,49,59.

1921 *Crocallis elinguaria* Linn.   Scalloped Oak
Widespread and locally common in Monmouthshire, flying from early July to mid-August in fenny places, and deciduous woodland as well as the hills and moorlands in the north of the county.

Abertillery dist., series bred from larvae collected from heather, 1911 (Rait-Smith 1912).

Usk, garden trap, recorded every year from 1966 to 83 eg. 1978, (36) 14/7 – 13/8. Ed. 6/7/68. Ld. 16/8/74. Wentwood 22/7/68. Usk (Park Wood) 23/7/68. Bica Common 23/8/68. Tintern 1/8/70. Magor Marsh 2/8/70. Hael Woods 6/8/74 (plfl.). Pont-y-saeson 10/8/74. Cwm Tyleri 23/8/74. Newport Docks 17/8/76. Slade Wds 28/7/80. Wyndcliff 28/7/82. Llansoy 28/7/84. (MEA) : Risca 7/7/84, Dinham 7/7/87.

Sq. **SO** 20,30,40,50. **ST** 38,48,29,49,59.

1922 *Ourapteryx sambucaria* Linn.   Swallow-tailed Moth
Common and widespread, inhabiting gardens, woodland etc. and flying from late June to early August.

Llandogo, larvae on privet (Nesbitt 1893). Wye Valley 1906 (Bird

1907). Usk Castle July 1933 (2) (RBH).

Usk, garden m.v.l. trap, noted annually 1966 to 1983 and usually frequent eg. 1968, (22) 29/6 – 9/8; 1980, (17) 28/6 – 2/8. Ed. 25/6/73. Ld. 13/8/79. Wentwood 17/7/68. Usk (Park Wood) 7/7/69. Magor Marsh 11/7/69. Hendre Wds 5/7/70. Hael Wds 12/7/71. Redding's Inclosure 20/7/71. Pont-y-saeson 18/7/72. Cwm Tyleri 9/7/73. Piercefield Park 18/7/73. Slade Wds 23/7/79. Ysguborwen (Llantrisant) 17/7/86. Trelleck Bog 18/7/86. Llansoy 11/7/87. (MEA) : Ynysyfro 28/6/76, Mescoed Mawr 1/7/76, Risca 18/7/79, Ynysddu 12/7/85, Abercarn 16/7/85, Cwm Merddog 7/8/86, Cwmfelinfach 5/9/86, Dinham 7/7/87, Pontllanfraith 20/7/87.

Sq. **SO** 10,20,30,40,50,41,51. **ST** 28,48,19,29,49,59.

1923 *Colotois pennaria* Hb.   Feathered Thorn

Widespread and common in woodland, gardens, hedgerows, and bushy places generally, flying from mid-September to late November.

Tintern 1912 (Bird 1913).

Caerwent (Brockwells) 1/12/62 (CT). Llanbadoc Oct. 1963 (abdt.) (GANH). Cwmyoy 1965 (SHR). Usk, garden m.v.l. trap, recorded every year from 1966 to 82. eg. 1978, (67) 10/10 – 18/11. Ed. 3/10/69. Ld. 27/11/71. Wentwood 29/10/68 (JMD). Llantarnam 25/10, 26/10/69 (WTH). Trostrey 25/10/69. Wyndcliff 28/9/70. Nant-y-derry 2/11/70. Rhyd-y-maen 2/11/70. Redding's Inclosure 19/10/71. Hael Wds 22/10/71. Hendre Wds 25/10/71. Prescoed 1/11/71. Trelleck 2/11/71. Pont-y-saeson 18/9/79. Llansoy 12/10/85. (MEA) : Risca 15/10/84, Cwmfelinfach 9/11/86.

Sq. **SO** 30,40,50,41,51,32. **ST** 48,19,29,39,49,59.

1924 *Angerona prunaria* Linn.   Orange Moth

This moth flying from mid-June to mid-August, occurs locally in the eastern half of Monmouthshire frequenting open woodland and heathy places. It comes readily to light and is often numerous in its favoured haunts. The form f. *corylaria* Thunberg often occurs. Earliest recorded date 13/6/67 (Pont-y-saeson). Latest date 21/8/79 (Slade Wds).

Monmouthshire 1889 (Patten 1890). Wye Valley, 1906 (Bird 1907). Tintern, "fairly common at dusk and light in 1947" (Blathwayt 1967).

Pont-y-saeson 1965; 13/6/67, abundant both at dusk and to m.v.l. inc. f. corylaria; 18/7/72. Whitebrook 2/7/69 (SC). Hael Wds 3/7/71, plentiful to m.v.l. inc. f. *corylaria*; 4/7/87. Redding's Inclosure 20/7/71, 10/7/73 (f. *corylaria*), 24/7/83. Magor Marsh 8/7/77 (1). Slade Wds 1979 23/7, 21/8; 1983, 8/7/83, abdt. inc f. *corylaria*; 1986 22/7. Hendre Wds 1/7/86 (plfl.). Kilpale 30/6/87. Dixton Bank 18/6/88. (MEA) : Wyndcliff 12/7/87.

Sq. **SO** 50,41,51. **ST** 48,49,59.

1925 *Apocheima hispidaria* D. & S.   Small Brindled Beauty
This species, found locally in oakwoods in the eastern half of Gwent, is scarce. The female is wingless but the male, flying from late February to April, is readily attracted to light. Earliest recorded date 26/2/67 (Usk), latest date 21/4/80 (Hendre Wds).

Usk, garden m.v.l. trap, noted sporadically from 1967 – 83 eg. 1967, 26/2 (1); 1978, 10/3 (1); 1983, single moths on Mar. 3, 13, 14. Wentwood 7/3/69 (1) (JMD). Redding's Inclosure 7/3/72 (5) to m.v.l. Hendre Wds 15/4/79, 14/4/80 (2), 21/4/80(1).

Sq. **SO** 30,41,51. **ST** 49.

1926 *Apocheima pilosaria* D. & S.   Pale Brindled Beauty
Common and widespread in gardens and woodland. The female is wingless but the male comes to light and flies from late January to April. The melanic form f. *monacharia* Stdgr. is frequent.

Wye Valley 1906 (Bird 1907). Llandogo 1912 (Bird 1913). Abertillery dist. (Rait-Smith 1913), ab. *monacharia* (Rait-Smith 1912). Brockwells (Caerwent) 15/3/63 (CT). Cwmyoy 1965 (SHR).

Usk, garden m.v.l. trap, recorded every year from 1967 to 1984 eg. 1971, (8) 7/2 – 12/8 inc. ab. *monacharia* (1); 1973, (11) inc. f. *monacharia* (7); 1980, (7) 28/2 – 26/4. Ed. 27/1/73. Ld. 29/4/79. Wentwood 20/3/70 (JMD). Redding's Inclosure 1972, 27/2, 7/3 (inc f. *monacharia*). Hendre Wds 25/3/72. Pont-y-saeson 3/4/72. St Pierre's Great Wds 23/3/73. Trelleck Bog 27/3/73 (f. *monacharia*). Hael Wds 6/4/75. Slade Wds 26/3/83. (MEA) : Goetre 28/2/87, Cwmcarn 23/3/87, Cwmfelinfach 11/4/87, Risca 7/3/88.

Sq. **SO** 20,30,50,41,51,32. **ST** 48,19,29,39,49,59.

1927 *Lycia hirtaria* Cl.   Brindled Beauty
This species flying from late March to early May in gardens and woodland is not uncommon in Gwent though its appearance is rather sporadic. Earliest recorded date 19/3/65 (Caerwent). Latest date 16/5/82 (Usk).

Brockwells, Caerwent 19/3/65 (CT in litt). Usk, garden m.v.l. trap, 25/4/68 (1), 1970 (2) and one or two most years from 1976 to 1984. Prescoed 4/5/70 (2). Redding's Inclosure, 17/4/72 (5), 1/5/72 (1). Usk Castle 24/4/75 (1), 29/4/85 (2) (RBH). Pont-y-saeson, 11/6/77, larvae on alder, reared (GANH). Hendre Wds 14/4/80. Llansoy, to u.v.l., 1985 (2); 1986, 7/5 melanic male; 1987, (30) 13/4 – 29/4. Hael Wds 1/5/86. Wyndcliff 27/4/87. (MEA) : Risca 8/4/83, Cwmfelinfach 21/4/87.

Sq. **SO** 30,40,50,41,51. **ST** 48,19,29,39,59.

1930 *Biston strataria* Hufn.   Oak Beauty
This moth, frequenting deciduous woods and parkland from late February to mid-April, is widespread and locally abundant in Monmouthshire especially

in the woodlands of the Wye and Usk Valleys. The male comes readily to light.

Brockwells, Caerwent 9/2/61 (CT). Cwmyoy 1965 (SHR). Usk, garden m.v.l. trap, recorded every year from 1966 to 1984 eg. 1968 (35) 9/3 – 20/4 inc. (10) on 29/3, 1971 (41) 20/2 – 13/4. Ed. 18/2/80. Ld. 34/4/78. Llangybi (Coed-y-Fferm) 21/4/70. Wentwood 1970 (JMD). Redding's Inclosure to m.v.l., 27/2/72 (1), 20/3/72 (abdt.) (15 at a time on the sheet). Pont-y-saeson 3/4/72 (abdt.). St Pierre's Great Wds 23/3/73 (plfl.) Hael Wds 26/3/73, 19/3/83. Wyndcliff 22/3/74. Hendre Wds 14/4/80. Llansoy 30/3/87. (MEA) : Risca 19/4/85, Cwmfelinfach 11/4/87, Pontypool 12/4/87, Ochrwyth 13/4/87.

Sq. **SO** 30,40,50,41,51,32. **ST** 28,48,19,29,39,49,59.

1931   *Biston betularia* Linn.   Peppered Moth
This common and widespread moth flies from mid-May to late August in woods, gardens, bushy places etc. coming readily to light. The melanic form *carbonaria* Jordan and also f. *insularia* Thierry-Mieg are both common in Monmouthshire wherever the species occurs.

Tintern dist. a larva on *Spiraea ulmaria*, a larva on perrenial sunflower and a larva on thistle (Bird 1905). Abertillery dist. (Rait-Smith 1913). South, in *Moths of the British Isles* 1908 states "In Wales *carbonaria* is in the ascendant at Newport, Monmouth."

Usk, garden m.v.l. trap, recorded annually 1966 to 1983. Ed. 12/5/71. Ld. 30/8/72. Over the nine years 1973 to 1981 counts were made of the Type insect and of the two above-mentioned forms which came to the trap. Of a total of (534) moths almost exactly one half viz. (268) were of the Type, while there were (266) others of which (165) were f. *insularia* and (101) were the melanic f. *carbonaria*.

Llantrisant 17/7/56. Usk (Park Wood) 5/6/67. Wyndcliff 12/6/67. Pont-y-saeson 13/6/67. Prescoed 5/7/68. Wentwood 21/6/69. Bica Common 28/6/69. Magor Marsh 1/7/69. Hendre Wds 26/5/70. Rhyd-y-maen 6/6/70. Tintern 1/8/70. Cwm-mawr 31/5/71. Slade Wds 1/6/71. Redding's Inclosure 15/6/71. Hael Wds 3/7/71. Cwm Tyleri 1/7/73. St Pierre's Great Wds 13/7/73. Trelleck Bog 16/7/79. Llansoy 18/5/85. Kilpale 30/6/87. (MEA) : Mescoed Mawr 23/6/78, Risca 25/5/82, Cwm Merddog 7/7/85, Pant-yr-eos 16/7/85, Ochrwyth 27/6/86, Cwmfelinfach 15/7/86, Abercarn 27/6/87, Pontllanfraith 2/7/87, Dixton Bank 18/6/88.

Sq. **SO** 10,20,30,40,50,41,51. **ST** 28,38,48,19,29,39,49,59.

1932   *Agriopis leucophaearia* D. & S.   Spring Usher
Frequents oakwoods but is scarce and local in Monmouthshire. The female is wingless but the male comes to light and is on the wing from early in February to mid-April. The form ab. *marmorinaria* Esp. occurs.

Wye Valley (Bird 1907). Abertillery dist. inc. ab. *marmorinaria*, 1911,

(Rait-Smith 1912).

Usk, garden m.v.l. trap, 1966 – 1984 (6) only: 1971, 11/2; 1973, 2/2, 18/4; 1977, 13/2; 1979, 3/3; 1980, 1/3 (ab. *marmorinaria*). Redding's Inclosure 27/2/72 (1). Hendre Wds 4/3/79, one flying freely by day; 14/4//80 (1) to m.v.l. (MEA) : Wyndcliff 9/3/78 (1) to m.v.l.

Sq. SO 20,30,41,51. ST 59.

1933 *Agriopis aurantiaria* Hb. Scarce Umber

This largely woodland species is uncommon in Monmouthshire. The female's wings are vestigial but the male flies from October until mid-December and comes to light.

Tintern 15/11 1912 (Bird 1913). Abertillery dist. (Rait-Smith 1912). Cwmyoy 1965 (SHR).

Usk, garden m.v.l. trap, 1966, (3) 22/10 – 26/11; 12/2/69 (1); 7/11/83 (1). Wentwood 1968 (JMD). Hendre Wds. Pont-y-saeson 5/6/78, larvae on hazel and sallow (REMP, GANH). (MEA) : Risca 18/12/87 (1) to act.l.

Sq. SO 20,30,41,32. ST 29,49,59.

1934 *Agriopis marginaria* Fabr. Dotted Border

Common and widespread in deciduous woodland and bushy places generally, flying from late February to late April and occasionally well into May. The males come to light but the females have vestigial wings and are flightless.

Wye Valley (Bird 1907). Wye Valley, near Bigsweir, "on sallows" 31/3/07 (Barraud 1907b). Tintern 1912 (Bird 1913). Abertillery dist. (Rait-Smith 1912, 1913). Cwmyoy 1965 (SHR).

Usk, garden m.v.l. trap, noted every year from 1967 to 1984 eg. 1967 (2), 4/3 and 18/3; 1979, (13) 1/4 – 28/4. Ed. 25/2/80. Ld. 20/5/82. Wentwood 5/4/71 (JMD). Redding's Inclosure 27/2/72. Pont-y-saeson 29/2/72. Prescoed 10/3/73. St. Pierre's Great Wds 23/3/73. Cwm-mawr 27/3/73. Trelleck Bog 27/3/73. Wyndcliff 22/3/74. Llansoy 30/3/87. (MEA) : Mescoed Mawr 27/2/77, Risca 15/4/85.

Sq. SO 20,30,40,50,51,32. ST 29,39,49,59.

1935 *Erannis defoliaria* Cl. Mottled Umber

A very variable moth, widespread and common in deciduous woods and bushy places from October to December. The females are wingless but the males come to light.

Tintern dist., larvae on Cotoneaster (Bird 1905). Wye Valley, larvae (Barraud 1906). Wye Valley (Bird 1907). Tintern, melanic ab. 1911 (Bird 1912). Cwmyoy 1965 (SHR). Tintern 31/5/68, larva on apple; Deri-fach 2/6/68, larvae abundant on oak, also hawthorn (JMC-H 1969).

Usk, garden m.v.l. trap, a few most years from 1966 to 1983 eg. 1971,

(5) 12/11 – 5/12, 1980, (3) 19/10 – 22/11. Melanic ab. 25/11/66. Ed. 4/10/68. Ld. 12/12/69. Prescoed 22/10/70. Hendre Wds 25/10/71. Goetre 1971. Llansoy 1985, to u.v.l. trap, 20/10 ♂ (1), 27/10 ♂ (1); 23/11 a female specimen (trap sited beneath an oak). St Pierre's Great Wds, larvae on birch 1/6/75; Pont-y-saeson, larvae on hazel and oak 25/5/76, on alder 11/6/77, on sallow and blackthorn 5/6/78 (GANH). (MEA) : Mescoed Mawr 15/4/85, Risca 15/4/85, Cwmfelinfach 1986.

Sq. **SO** 30,40,50,21,41,32. **ST** 19,29,39,49,59.

1936 *Menophra abruptaria* Thunb.   Waved Umber
Widely distributed and fairly common in woods and gardens from late April to middle of June.

Llandogo 1892 (Nesbitt 1892b). Llandogo 1906, one (Bird 1907).

Usk, garden m.v.l. trap, noted in small numbers every year from 1967 to 1982 eg. 1978, (9) 7/5 – 3/6; 1980, (8) 27/4 – 24/5. Ed. 22/4/68. Ld. 18/6/79. Gwehelog (Camp Wood) 29/4/67. Rhyd-y-maen 9/5/70. Hael Wds 25/5/70. Magor Marsh 2/6/70. Wentwood 1970 (JMD). Redding's Inclosure 1/5/72. Slade Wds 28/5/78. Hendre Wds 31/5/79. Llansoy 18/5/85. Trelleck Bog 25/5/87. (MEA) : Wyndcliff 26/5/77, Risca 19/5/82, Pant-yr-eos 18/5/84, Abercarn 26/5/87, Lower Machen (Park Wood) 13/5/88.

Sq. **SO** 30,40,50,41,51. **ST** 28,48,29,49,59.

1937 *Peribatodes rhomboidaria* D. & S. Willow Beauty
Widespread and common in woods and gardens, flying from late June until the end of August with, in some years, a partial second brood in September and early October. It comes readily to light.

Wye Valley (Bird 1907). Abertillery dist. (Rait-Smith 1912, 1913, 1915).

Usk, garden m.v.l. trap, plentiful every year from 1966 to 1983 eg. 1971, 2/7 – 22/8; 1974, 28/6 – 31/8 and single moths on 5/9 and 19/9. Ed. 14/6/76. Ld. 19/9/74, 11/10/78. Wyndcliff 12/6/67. Pont-y-saeson 13/6/67. Usk (Park Wood) 27/6/67. Trelleck Common 9/7/68. Wentwood 12/7/68. Bica Common 23/8/68. Hendre Wds 5/7/70. Hael Wds 5/7/70. Tintern 1/8/70. Gwehelog (Camp Wood) 28/8/70. Redding's Inclosure 20/7/71. Slade Wds 29/8/72. St Pierre's Great Wds 13/7/73. Cwm Tyleri 8/7/76. Magor Marsh 3/8/76. Newport Docks 17/8/76. Trelleck Bog 8/7/78, 9/10/78. Llansoy 3/7/85. Kilpale 30/6/87. (MEA) : Mescoed Mawr 3/7/76, Risca 18/7/79, Ynysddu 12/7/85, Ochrwyth 12/7/86, Cwmfelinfach 5/9/86, Pontllanfraith 14/8/87, Uskmouth 16/8/88, Clydach 18/7/89.

Sq. **SO** 20,30,40,50,21,41,51. **ST** 28,38,48,19,29,49,59.

1940 *Deileptenia ribeata* Cl.   Satin Beauty
Locally frequent, sometimes abundant, in both deciduous and coniferous

woodland mainly in the eastern half of the county. Flying from mid-June, or earlier, until mid-August this moth readily comes to light. In my experience all Monmouthshire specimens are melanic forms.

South 1908 (*Moths of the British Isles*) gives "Monmouthshire".

Wentwood 10/8/68, in spruce plantation, abundant to m.v.l. and all melanic. Also 1969, 71, 74 etc. Wyndcliff 13/7/69, 3/8/81, plentiful in vicinity of yews and all specimens melanic. Usk, garden m.v.l. trap, 29/6/70 (1) and sporadically in small numbers eg. 1978, 27/6 (1); 1981, (4) 27/7 – 4/8. Hael Wds 29/8/78, 29/5/82 etc. Hendre Wds. Slade Wds 28/7/80. Llansoy 13/6/85. Ysguborwen (Llantrisant) 17/7/86. Kilpale 30/6/87. Redding's Inc. 2/8/90. (MEA) : Mescoed Mawr 8/7/84, Cwmfelinfach 13/8/87.

Sq. **SO** 30,40,50,41,51. **ST** 48,19,29,49,59.

1941 *Alcis repandata* Linn. Mottled Beauty
  ssp.*repandata* Linn.

A widespread and very variable species, often abundant in wooded and bushy places. Found also in the hills and moorland of the west and north of the county. It flies throughout June and July and, in some years, well into August, coming readily to light. Melanic forms occur, and the striking, banded variety ab. *conversaria* Hb. is especially common in some localities in the east of the county.

Tintern dist., including ab. *conversaria* (Bird 1905). Wye Valley 1906 (Bird 1907). Abertillery dist. (Rait-Smith 1912, 1913); ab. (idem 1912).

Usk, garden m.v.l. trap, noted every year from 1966 to 1983 eg. 1972, (35) 2/7 – 7/8 and a late moth on 1/9; 1981, (34) 12/6 – 17/7. Ed. 2/6/70. Ld. 27/8/68, 4/8/71. Devauden 5/7/61, several flying by day. Usk (Park Wood) 5/6/67. Wyndcliff 12/6/67, 16/6/67 (inc. ab. *conversaria*). Pont-y-saeson 13/6/67, 18/7/72 (plfl.). Prescoed 5/7/68. Wentwood 8/7/68 (abdt.) Trelleck Common 9/7/68. Bica Common 23/8/68. Magor Marsh 1/7/69, 24/6/75 ab. *conversaria*. Hendre Wds 5/7/70. Hael Wds 5/7/70. Tintern 1/8/70. Redding's Inclosure 15/6/71. Cwm-mawr 2/7/71. Slade Wds 29/8/72. St Pierre's Great Wds 13/7/73. Piercefield Park 18/7/73. Newport (Coldra) 25/6/75 (inc. melanic vars.). Cwm Tyleri 8/7/76, (inc. melanic vars.). Trelleck Bog 18/7/76 (inc. ab. *conversaria*), 16/7/79 (inc. melanic vars.). Llansoy 2/7/85. Kilpale 30/6/87 (plfl.). Clydach 18/7/89. (MEA) : Allt-yr-yn 20/6/76, Ynysyfro 22/6/76, Llantarnam 24/6/76, Mescoed Mawr 25/6/76, Sugar Loaf (St Anne's Vale) 7/7/77, Risca 9/7/79, Black Vein 3/7/81, Ynysddu 12/7/85, Ochrwyth 12/7/86, Cwmfelinfach 15/7/86, Abercarn 27/6/87, Cwmcarn 12/7/87.

Sq. **SO** 20,30,40,50,21,41,51. **ST** 28,38,48,19,29,39,49,59.

1943   *Boarmia roboraria* D. & S.   Great Oak Beauty

This species, scarce and local in Monmouthshire, occurs in some oakwoods in the Wye Valley and Forest of Dean.

   Tintern, "at light in 1947" (Blathwayt 1967).

   Hael Woods to m.v.l., 5/7/70 (4), 3/7/71 (1), 12/7/71 (1) (GANH). Redding's Inclosure 20/7/71 (2) (GANH).

Sq. **SO** 50,51.

1947   *Ectropis bistortata* Goeze   The Engrailed

A widespread and fairly common woodland species which comes readily to light. Double-brooded, it flies in March and April and again in July and early August while occasionally moths of a third brood have been noted late in the year. Melanic forms occur frequently.

   Llandogo dist. (Nesbitt 1892). Tintern, 2/3/1912 (Bird 1913). Abertillery dist. 1914, "a few, including an ab. on Apr. 24" (Rait-Smith 1915).

   Usk, garden m.v.l. trap, recorded sporadically from 1966 to 1983 eg. 1968, 2nd brood, (2) 19/7 and 20/7; 3rd brood, (1) 30/10; 1974, 1st brood (1), 11/4, 2nd brood, (5) 8/7 – 7/8; 1979, 3rd brood moth 14/10. Melanic specimens on 3/7/67, 18/7/68 etc. Usk (Park Wood) 28/4/67. Wyndcliff 23/6/67. Llangybi (Coed-y-Fferm) 21/4/70. Tintern 1/8/70. Hadnock Quarries 30/3/71. Wentwood 5/4/71 (JMD). Cwm-mawr 6/4/71. Hael Wds 3/7/71, 26/4/82 (melanic f.). Redding's Inclosure 17/4/71. Pont-y-saeson 20/3/72. Prescoed 16/4/72. St Pierre's Great Wds 23/3/73. Hendre Wds 23/4/77, 14/7/77 (melanic f.). Trelleck Bog 22/7/78. Slade Wds 23/7/79. Llansoy 22/4/87. (MEA) : Ynysyfro 4/7/76, Mescoed Mawr 16/4/78, Risca 4/4/82, Abercarn 3/5/87, Cwmfelinfach 14/4/87, Cwm Merddog 26/4/87.

Sq. **SO** 10,20,30,40,50,21,41,51. **ST** 28,48,19,29,39,49,59.

1948   *Ectropis crepuscularia* D. & S.   Small Engrailed

Common in open woodland from mid-May to mid-June it comes readily to light. Melanic forms are common.

   Llandogo dist. (Nesbitt 1892). Wye Valley, May 1906, melanic ab. bred (Bird 1907). Abertillery dist., inc. ab. nigra (Rait-Smith 1913).

   Usk, garden m.v.l. trap, recorded sporadically in small numbers eg. 1972, (1) 10/6; 1979, (5) 15/5 – 10/6 inc. one melanic form. Wyndcliff 7/6/66. Usk (Park Wood) 12/5/67. Pont-y-saeson 13/6/67. Graig-y-Master 21/5/68. Wentwood 21/6/69, 2/6/81. Bica Common 28/6/69. Hael Wds 25/5/70. Hendre Wds 8/6/79. Cwm-mawr 31/5/71. Redding's Inclosure 1/5/72. Trelleck Bog 22/5/73. St Pierre's Great Wds 14/5/73. Cwm Coed-y-cerrig 24/5/76. Slade Wds 28/5/78. Llansoy 7/5/85. (MEA) : Mescoed Mawr 19/5/76, Risca 7/7/84, Cwmfelinfach 10/6/87, Lower Machen (Park Wood)

13/5/88, Nant Gwyddon 10/6/88.

Sq. **SO** 30,40,50,41,51,22. **ST** 28,48,19,29,39,49,59.

1949 *Paradarisa consonaria* Hb.   Square Spot
This locally common woodland moth flies from late April to mid-June and comes to light. Earliest date recorded 22/4//68 (Park Wood, Usk). Latest date 20/6/67 (Wyndcliff). The form ab. *waiensis* Richardson occurs in Gwent.

Tintern, "fairly common, mainly at rest, in 1946 and 1948 including a few partially melanic specimens" (Blathwayt 1967). Tintern dist. 19/6/65, ab. *waiensis* to m.v.l. (Mere 1965).

Wyndcliff 7/6/66, 25/6/86. Usk (Park Wood) 28/4/67, 26/4/68 (plfl.). Pont-y-saeson 13/6/67. Wentwood 1970, 26/5/80. Graig-y-Master 7/5/71. Slade Wds 1/6/71. Hael Wds 21/5/73, 8/5/87 (plfl.). Hendre Wds 2/6/80. (MEA) : Mescoed Mawr 10/5/78, Cwmfelinfach 5/5/87.

Sq. **SO** 30,50,41. **ST** 48,19,29,49,59.

The attractive melanic form ab. *waiensis* Richardson was thought to be restricted to the Forest of Dean, Gloucestershire and the Wye Valley, Monmouthshire, however it also occurs in central Monmouthshire (Wentwood and Bica Common) and, more recently, in 1987, it was noted in the Ebbw Valley in the western half of the county by Dr. M. E. Anthoney.

ab. *waiensis*: Wentwood 24/5/69 (JMD). Pont-y-saeson 2/5/72. Bica Common 13/6/72; Hendre Wds 2/6/80; Hael Wds 8/5/87 (4) to m.v.l.; Trelleck Bog 25/5/87 (GANH). (MEA) : Abercarn 16/5/87.

Sq. **SO** 50,41. **ST** 29,49,59.

1950 *Paradarisa extersaria* Hb.   Brindled White Spot
In Monmouthshire this species is found in a few deciduous woods in the east of the county, flying mainly in June, but noted from late in May until early August. It comes to light. Although very local it is sometimes plentiful in its chosen haunts.

Wye Valley 1904, 1905 (Bird 1907).

Wyndcliff 7/6/66, 3/8/81, 23/6/84 (plfl.). Pont-y-saeson 13/6/67. Redding's Inclosure 7/6/71, Hael Wds 3/7/71, 29/5/82 (plfl.). Hendre Wds 4/6/79, 2/6/80.

Sq. **SO** 50,41.51. **ST** 59.

1951 *Aethalura punctulata* D. & S.   Grey Birch
This moth, flying from late April to mid-June and inhabiting heaths and open deciduous woodland, is widespread but at low density in Gwent.

Llandogo dist., "last week of March 1892" (Nesbitt 1892). Wye Valley, Bigsweir dist. 1907 (Barraud 1907b). Wye Valley 1906 (Bird 1907). Llandogo 1912 (Bird 1913). Llanock Wood [= Llanerch Wood, Crumlin

(GANH)] one 1912, Abertillery dist. (Rait-Smith 1913). Pontllanfraith (Rait-Smith 1915).

Usk, garden m.v.l. trap, a few most years from 1967 – 1983 eg. 1968, (3) 29/5 – 3/6; 1982, (4) 29/4 – 31/5. Gwehelog (Camp Wood) 29/4/67. Wyndcliff 12/6/67 (REMP). Usk (Park Wood) 4/6/68. Prescoed 8/6/69. Pont-y-saeson 26/5/70. Hendre Wds 26/5/70. Wentwood 2/6/70. Cwm-mawr 31/5/71. Bica Common 13/6/72. Hael Wds 21/5/73. Trelleck Bog 22/5/73. Cwm Coed-y-cerrig 5/6/73. Slade Wds 28/5/78. Llansoy 29/4/87. (MEA) : Mescoed Mawr 10/5/78, Abercarn 16/5/87, Cwmfelinfach 23/5/87.

Sq. **SO** 20,30,40,50,41,22. **ST** 48,19,29,39,49,59.

1952 *Ematurga atomaria* Linn.   Common Heath
ssp. *atomaria* Linn.

Widespread and fairly common, flying by day in May, June and early July and frequenting open heathy woods, and the hills and moorland of the west and north of the county.

Llandogo (Nesbitt 1893). Wye Valley (Barraud 1906). Abertillery dist. (Rait-Smith 1913).

Pont-y-saeson 7/6/66, 8/5/67, 28/6/70. Trelleck Common 14/6/67. Wentwood 19/6/68 (JMD). Whiteleye Bog 9/6/69. Trelleck Bog 2/6/70 (plfl.), 3/7/73. Llangeview (Craig-yr-iar) 28/6/70. Blorenge Mountain 24/6/71 (abdt.). Mynyddislwyn 21/6/74. Trostrey Common 11/5/80. Pontypool (Mountain Air) 25/5/80. (MEA) : Cwm Tyleri 10/7/85, Ochrwyth 13/6/86, Pontllanfraith 25/5/87, Nant Gwyddon 12/6/88.

Sq. **SO** 20,30,40,50,21. **ST** 28,19,29,49,59.

1954 *Bupalus piniaria* Linn.   Bordered White

Found in coniferous woods and plantations from mid-May to late July and sometimes abundant. The male is often seen flying freely by day and at night comes to light.

Usk, garden m.v.l. trap, frequently recorded in the years 1967 – 1983 eg. 1971, (6) 30/6 – 11/7; 1983, (16) 22/6 – 22/7. Ed. 18/5/80, Ld. 31/7/70, 1/8/77. Trelleck Common 14/6/67 (abdt.). Wentwood 3/7/68 (JMD). Magor Marsh 11/7/69. Wyndcliff 13/7/69. Slade Wds 1/6/71. Cwm-mawr 2/7/71. Hael Woods 3/7/71. Redding's Inclosure 20/7/71. Trelleck Bog 16/6/72. Llansoy 4/7/85. (MEA) : Mescoed Mawr 3/5/77, Risca 6/7/83, Cwm Tyleri 10/7/85, Ochrwyth 27/6/86, Cwmfelinfach 15/7/86, Pontllanfraith 10/7/87, Cwmcarn 11/7/87, Nant Gwyddon 10/6/88.

Sq. **SO** 20,30,40,50,51. **ST** 28,48,19,29,49,59.

1955 *Cabera pusaria* Linn.   Common White Wave

Common and widespread, flying from mid-May until the end of August in open woodland and bushy places generally.

Wye Valley (Barraud 1906). Wye Valley 1906 (Bird 1907). Abertillery dist. (Rait-Smith 1913, 1915).

Usk, garden m.v.l. trap, recorded annually 1966 – 1983 eg. 1973, (34) 18/6 – 31/8; 1975, (35) 29/5 – 1/8. Ed. 20/5/82. Ld. 31/8/73. Pont-y-saeson imagines 1966 and 25/5/68; larvae on alder 1/10/68, reared, (GANH). Prescoed 31/5/68. Usk (Park Wood) 4/6/68. Wentwood 8/7/68. Trelleck Common 9/7/68. St Pierre's Great Wds 23/7/68. Bica Common 23/8/68. Wyndcliff 13/7/69. Magor Marsh 22/7/69. Rhyd-y-maen 23/5/70. Hendre Wds 26/5/70. Hael Wds 5/7/70. Cwm-mawr 31/5/71. Redding's Inclosure 20/7/71. Slade Wds 29/8/72. Cwm Coed-y-cerrig 5/6/73. Trelleck Bog 16/6/73. Piercefield Park 18/7/73. Mynyddislwyn 10/6/74. Newport (Coldra) 27/6/75. Llansoy 31/5/86. Kilpale 30/6/87. (MEA) . Allt-yr-yn 20/6/76, Mescoed Mawr 25/6/76, Ynysyfro 28/6/76, Risca 8/6/80, Black Vein 12/6/81, Ynysddu 12/7/85, Abercarn 16/7/85, Cwm Merddog 20/6/86, Ochrwyth 27/6/86, Cwmfelinfach 15/7/86, Pontllanfraith 2/7/87, Sugar Loaf (St Mary's Vale) 8/7/87, Nant Gwyddon 25/6/88.

Sq. **SO** 10,20,30,40,50,21,31,41,51,22. **ST** 28,38,48,19,29,39,49,59.

1956 *Cabera exanthemata* Scop.   Common Wave
This moth, inhabiting damp woods, river valleys and marshy places, is widespread and fairly common in Gwent. It flies from the middle of May to early September and comes to light.

Wye Valley (Barraud 1906). Wye Valley 1906 (Bird 1907). Abertillery dist. (Rait-Smith 1913).

Usk, garden m.v.l. trap, recorded every year from 1966 to 1983 eg. 1973, 1st brood, (7) 18/5 – 12/7; 2nd brood, (16) 28/7 – 31/8. Ed. 13/5/82. Ld. 7/9/69, 1/9/79. Usk (Park Wood) 12/5/67. Wyndcliff 12/6/67 (REMP). Wentwood 3/7/68 (JMD). Pont-y-saeson 14/7/68; larva on sallow 1/10/68, reared. Magor Marsh 2/6/70. Prescoed 21/6/71. Hael Woods 21/5/73. Cwm Coed-y-cerrig 5/6/73. Slade Wds 15/6/73. Rhyd-y-maen 1/9/73. Llansoy 25/6/85. (MEA) : Allt-yr-yn 23/5/76, Mescoed Mawr 1/7/76, Ynysyfro 4/7/76, Risca 13/7/82, Pant-yr-eos 16/6/83, Cwm Merddog 8/8/86, Cwmfelinfach 23/5/87, Pontllanfraith 20/7/87.

Sq. **SO** 10,20,30,40,50,22. **ST** 28,48,19,29,39,49,59.

1957 *Lomographa bimaculata* Fabr.   White-pinion Spotted
This moth, frequenting hedgerows and open woodland throughout May and June, is widespread and common in Monmouthshire.

Usk, garden m.v.l. trap, seen most years from 1967 to 1983 eg. 1970, (5) 16/5 – 3/6; 1975, (8) 20/5 – 30/6. Ed. 4/5/72. Ld. 9/7/77. Gwehelog (Camp Wood) 29/4/67. Pont-y-saeson 25/5/68. Usk (Park Wood) 9/6/69. Cwm-mawr 31/5/71. Wyndcliff 24/6/71. Redding's Inclosure 15/5/72. Trelleck Bog 16/6/73. Hendre Wds 27/5/78. Slade Wds 28/5/78. Hael Wds

30/5/78. Cwm Coed-y-cerrig 5/6/78. Magor Marsh 19/6/79. Dinham 5/6/87. (MEA) : Allt-yr-yn 23/5/76, Llantarnam 10/6/76, Mescoed Mawr 21/5/78, Black Vein 8/5/81, Risca 3/6/84, Cwm Merddog 20/6/86, Cwmfelinfach 24/5/87, Abercarn 26/5/87.

Sq. **SO** 10,20,30,50,41,51,22. **ST** 28,48,19,29,39,49,59.

1958 *Lomographa temerata* D. & S.   Clouded Silver
This moth flying from mid-May to mid-July frequents woods, hedgerows and bushy places generally. It is common and widely distributed in Monmouthshire and comes to light.

Llandogo dist., "to light" 26/5/1892 (Nesbitt 1892). Wye Valley 1906, (Bird 1907).

Usk, garden m.v.l. trap, recorded most years from 1967 to 1983 eg. 1973, (10) 18/5 – 9/7. Ed. 18/5/73. Ld. 18/7/79. Usk (Park Wood) 6/6/66. Wyndcliff 7/6/66. Pont-y-saeson 7/6/66. Prescoed 31/5/68. Trelleck Common 9/7/68. Wentwood 12/7/68. Hendre Wds 26/5/70. Trelleck Bog 2/6/70. Rhyd-y-maen 6/6/70. Hael Wds 5/7/70, 26/6/85. Slade Wds 30/5/71, 28/5/77 (plfl.). Cwm-mawr 31/5/71. Cwm Coed-y-cerrig 5/6/78. Magor Marsh 9/7/78. Penallt 1/6/79. Llansoy 5/6/85. Dinham 5/6/87. Kilpale 30/6/87. Clydach 11/5/90 to m.v.l., a striking aberration (netted by MEA). (MEA) : Llantarnam 24/6/76, Mescoed Mawr 25/6/76, Risca 3/6/80, Cwm Merddog 14/7/86, Ochrwyth 27/6/86, Cwmfelinfach 15/7/86, Pontllanfraith 22/6/87, Dixton Bank 18/6/88, Nant Gwyddon 25/6/88.

Sq. **SO** 10,20,30,40,50,21,41,51,22. **ST** 28,48,19,29,39,49,59.

1960 *Theria primaria* Haw.   Early Moth
= rupicapraria auct.
This moth is locally common in woods and bushy places. The males are to be seen flying along hedgerows at dusk from January to mid-March and come to light, but the females are flightless.

Gilwern (Rait-Smith 1912). Wye Valley (Bird 1907). Wentwood 21/2/35 (RBH). Cwmyoy 1965 (SHR).

Usk, garden m.v.l. trap, occasional records eg. 1968, 12/3; 1971, Feb 8, 9, 20, Mar 9, 10; 1984 5/3. Llantrisant 31/1/67 (abdt.). Llanbadoc 31/1/67 (abdt.). Llanfrechfa 1971. Shirenewton 1971. Trelleck Bog 27/3/73. Caerwent dist. 18/1/74 (CT). Llansoy 13/3/91. (MEA) : Ochrwyth 28/2/77, Risca 28/2/87.

Sq. **SO** 30,40,50,21,32. **ST** 28,48,29,39,49.

1961 *Campaea margaritata* Linn.   Light Emerald
A common and widespread woodland moth, flying from mid-June until the end of July and coming readily to light.

Wye Valley 1906 (Bird 1907). Abertillery dist. (Rait-Smith 1912). Usk

2/7/33 (RBH).

Usk, garden m.v.l. trap, plentiful every year from 1966 to 83 eg. 1973, (31) 17/6 – 30/7; 1979, (20) 26/6 – 27/7. Ed. 14/6/76. Ld. 2/8/72. Wentwood 8/7/68 (abdt.). Pont-y-saeson 14/7/68. Usk (Park Wood) 7/7/69. Wyndcliff 13/7/69. Hendre Wds 5/7/70. Hael Wds 5/7/70. Magor Marsh 6/7/71. Redding's Inclosure 20/7/71, 24/7/83. St Pierre's Great Wds 13/7/73. Piercefield Park 18/7/73. Cwm Tyleri 8/8/73. Trelleck Bog 16/7/79. Slade Wds 15/7/80. Llansoy 27/6/86. Kilpale 30/6/87. Clydach 18/7/89. (MEA) : Mescoed Mawr 25/6/76, Ynysyfro 28/6/76, Risca 5/7/79, Pant-yr-eos 6/7/85, Cwm Merddog 7/7/85, Ynysddu 12/7/85, Abercarn 16/7/85, Ochrwyth 12/7/86, Cwmfelinfach 15/7/86, Sugar Loaf (St Mary's Vale) 8/7/87, Pontllanfraith 10/7/87, Dixton Bank 18/6/88, Nant Gwyddon 25/6/88.

Sq. **SO** 10,20,30,40,50,21,41,51. **ST** 28,48,19,29,49,59.

1962 *Hylaea fasciaria* Linn   Barred Red
Widespread but at low density in Monmouthshire, flying from late May to early August in coniferous woods and plantations.

Wye Valley (Nesbitt 1892).

Usk, garden m.v.l. trap, a few every year from 1967 to 1983 eg. 1969, (6) 29/6 – 21/7. Ed. 25/5/76 (Pont-y-saeson). Ld. 10/8/68 (Wentwood).

Wentwood 8/7/68. Tintern 1/8/70. Hael Wds 6/8/74. Pont-y-saeson 10/8/74. Cwm Tyleri 13/7/75. Slade Wds 28/5/78. Trelleck Bog 16/7/79. Hendre Wds 11/7/86. Llansoy 12/7/86. (MEA) : Mescoed Mawr 26/6/76, Wyndcliff 31/5/77, Cwmfelinfach 15/7/86, Risca 5/7/87, Pontllanfraith 20/7/87.

Sq. **SO** 20,30,40,50,41. **ST** 48,19,29,49,59.

1964 *Gnophos obscuratus* D. & S.   The Annulet
This moth, flying during July and August, is scarce in Monmouthshire, and is found mainly in the Carboniferous Limestone areas in the south-east and in the north of the county. Those found in old quarries in northern Gwent are comparatively dark forms.

Monmouth dist. (S.G. Charles c. 1937).

Cwm Tyleri, one to m.v.l. 28/7/72 and another 30/7/72 (GANH) (1st v-c. rec.). Piercefield Park 18/7/73 (1). Usk, garden m.v.l. trap, 20/8/74 (1). Slade Wds 23/7/79 (1). (MEA) : Risca, to act.l., 28/8/81 (1); Wyndcliff, to m.v.l., 12/8/83 (1); Clydach 9/7/89, 18/7/89 (3).

Sq. **SO** 20,30,21,51. **ST** 48,29,59.

SPHINGIDAE
SPHINGINAE

1972 *Agrius convolvuli* Linn.   Convolvulus Hawk-moth
An immigrant, occasionally recorded in Momouthshire.

Caerwood near Chepstow, Aug. 5th 1870, imago at petunia blossoms (Sellon, *Entomologist* **5**: 164). Abertillery 1911, ♂, (Rait-Smith 1912). Triley, near Abergavenny, 1928/29, one, (CMS-B pers. comm.). Monmouth, ♂ (1), 23/10/76 (handed to GANH by VCT). Bulwark (Chepstow) imago 28/9/80; Trewen (Caerwent), larva 13/9/84, reared (CT in litt.). Grosmont 18/9/92 (IR pers. comm.).

Sq. **SO** 20,31,51,42. **ST** 49,59.

1973 *Acherontia atropos* Linn. Death's-head Hawk-moth
An immigrant moth occurring only very sporadically in Gwent. The larval food-plant is the potato (*Solanum tuberosum*) and the insect is usually seen in potato-fields, either as a larva on the foliage, or as a pupa uncovered when the crop is lifted.

1865, "several larvae in gardens about Pontypool" (*Pontypool Free Press*, Aug. 8 1975, quoting an old issue in its column "1865 – 110 Years Ago "). Near Monmouth; Newport; (Parry teste Tutt, *Br. Lep.* **4**: 468).

Brockwells (Caerwent), larvae in potato-field Sept. 8; Caerwent, pupae in potato-field Sept. 15 (CT in litt.). 1976, Coed-y-paen, larvae in potato-field 27/7, 31/8, 17/9 and also a pupa 17/9 (GANH). Parc-Seymour, larvae 19/9 (CT). 1984, Trelleck, imago 31/7, (*Gwent Trust for Nature Conservation Newsletter No 51).* Kemeys Commander, full-fed larva 11/9/84 (GANH).

Sq. **SO** 30,50. **ST** 39,49,48.

1976 *Sphinx ligustri* Linn. Privet Hawk-moth
Occurs in Monmouthshire but very infrequently.

Llandogo 1905, larvae on ash, reared (Bird 1907).

Caerwent 29/9/62, larvae on lilac; Sudbrook 1/7/64, imago (CT in litt.). Wyndcliff, one to m.v.l. 16/6/67 (GANH). Slade Wds 1968, larvae on wild privet (RD). Wentwood, one to m.v.l. trap 27/5/70 (JMD). Usk, garden m.v.l. trap, 9/7/71 (1), 8/7/72 (1) (GANH). Usk Castle 16/7/71, one to m.v.l. trap, (RBH).

Sq. **SO** 30,50. **ST** 48,58,49,59.

1979 *Mimas tiliae* Linn. Lime Hawk-moth
Uncommon in Monmouthshire but occurs fairly regularly in one or two localities, notably Usk and the Wyndcliff.

Usk, to m.v.l. trap: 1966 7/6; 1967 1/6, 3/6; 1968 27/5, 30/6; 1971 7/6; 1973 22/6; 1976 29/5 (RBH); 1977 2/6, 3/6, 15/6; 1982 27/5; 1983 25/6, 29/6(2), 7/7. Wyndcliff: 1967 19/6 (CT), 23/6; 1971 22/6; 1974 2/7; 1986 13/6, 25/6; 1991 22/5 (8) to m.v.l. Wentwood: 1970 31/5 (JMD). Llansoy: 1990 25/5 (1). Magor Pill (sea-wall) 14/6/90 (1).

Sq. **SO** 30,40. **ST** 48,49,59.

1980   *Smerinthus ocellata* Linn.   Eyed Hawk-moth
Widespread and fairly common, flying from May to July.

Llandogo dist. (Nesbitt 1892). Usk 1931, one reared from larva and one from pupa (RBH). Bettws Newydd (Crowther 1933).

Usk, 1961, one reared from pupa found in garden (GANH). Abergavenny 1964 (AD). Usk, garden m.v.l. trap, recorded most years from 1966 to 1983. Ed. 16/5/69. Ld. 11/7/81. Llanbadoc (Church Wood) 24/6/58, one netted at dusk. Trelleck Common 9/7/68. Wentwood 1970, (svl.), 29/5 – 4/7 (JMD). Caerwent 4/6/71 (CT). Hael Wds 12/7/71. Magor Marsh 2/7/76. Llansoy 5/6/85, 1986 (3). (MEA) : Risca 25/6/86, Pontllanfraith 20/6/87.

Sq. **SO** 30,40,50,31. **ST** 48,19,29,49,59.

1981   *Laothoe populi* Linn.   Poplar Hawk-moth
Widespread and often numerous and undoubtedly the commonest Hawk-moth in Monmouthshire, flying from May to late July and in some years with a second generation in August and September.

Abertillery dist. 1912 (Rait-Smith 1913). Bettws Newydd 1930, larvae, four reared (RBH). Bettws Newydd, 1933 (Crowther).

Usk, 1952, several larvae on young poplars (RBH). Usk 12/6/58 (GANH). Abergavenny 1964 (AD). Usk, garden m.v.l. trap, recorded every year from 1966 to 82 and sometimes very plentiful eg. 29/5/69 (11) and in 1981 (65) 31/5 – 11/8. Ed. 11/5/82. Second generation moths in 1971, 23/8, and 1976, (9) 13/8 – 27/10. Rhyd-y-maen 23/5/70, 17/7/70. Slade Wds 1/6/71. Wyndcliff 24/6/71. Cwm-mawr 2/7/71. Hael Wds 3/7/71, 26/6/86. Redding's Inclosure 20/7/71. Magor Marsh 25/6/76. Wentwood 26/5/80. Hendre Wds 2/6/80. Llansoy 26/7/84. Clydach 18/7/89. (MEA) : Risca 7/7/83, Ynysddu 13/7/85, Cwmfelinfach 6/7/87, Nant Gwyddon 25/7/88.

Sq. **SO** 20,30,40,50,21,31,41,51. **ST** 48,19,29,39,49,59.

MACROGLOSSINAE

1982   *Hemaris tityus* Linn.   Narrow-bordered Bee Hawk-moth
No recent Monmouthshire records but a few old ones: Pontnewydd May 1833, one (Conway 1833). Pontllanfraith 1909, one (Rait-Smith 1912); fairly common May 17 1914 (Rait-Smith 1915). Bettws Newydd, 3/8/33, one taken at buddleia (Crowther 1933).

Sq. **SO** 30. **ST** 19,29.

1983   *Hemaris fuciformis* Linn.   Broad-bordered Bee Hawk-moth
Only one definite record for this county viz. Risca 1983, Aug 18th (one) (MEA).

Sq. **ST** 29.

1984 *Macroglossum stellatarum* Linn.   Humming-bird Hawk-moth
An immigrant moth of sporadic occurrence in Britain. I have the impression that it has appeared far less frequently in Monmouthshire since about 1950 than in the previous three decades. (GANH pers. obs.).

Monkswood (Rae, teste Tutt, *Br. Lep*. **4**: 33). Usk 1928, on red valerian (*Centranthus ruber*) at Usk Castle 29/7 (2), 31/7 (1); also 1933 (RBH).

1964: Caerwent dist. 7/6 (CT). 1970: Usk, one (GANH). 1976: Great Barnet's Wds, Mounton 4/7 (TGE); Caldicot 19/9 (DU); Usk 27/9 (1), 13/10 one on flowers of *Corydalis lutea* (GANH); Caerwent dist. 9/10 (CT).

1982: Usk 20/7 and 21/7 on red valerian. 1983: Usk 15/7 (GANH); Risca 22/8, 31/8 (MEA). 1986: Llanvaches 6/7 (CJ per MEA). 1989: Brynawel (Sirhowy Valley) 12/7 (AH).

Sq. **SO** 30,50. **ST** 48,29,49,59.

1987 *Hyles gallii* Rott.   Bedstraw Hawk-moth
An occasional migrant to the British Isles.

Skinner 1984, states "An erratic visitor; absent for many years and then appearing in good numbers as in 1870, 1888, 1955 and 1973, . . ." The only two Monmouthshire records correspond to the first and last of these incursions.

Caerwood, near Chepstow, Aug. 6 1870, one taken at petunia blossoms and another seen on the wing (Sellon 1870-71). Usk, garden m.v.l. trap, ♀ (1) 31/7/73 (GANH).

Sq. **SO** 30. **ST** 59.

1990 *Hyles lineata* Fabr.   Striped Hawk-moth
    ssp. *livornica* Esp.
An occasional immigrant to Britain. South 1961, states "The biggest migration on record was in 1943 when over six hundred were reported . . .". One was noted in Monmouthshire that year.

Monmouth 1904, ♀ "in coll. H. Green, Monmouth" (Thornewill 1904). Usk 1943, "one taken at dusk on red valerian at Usk Castle 31/5" (RBH).

Sq. **SO** 30,51.

1991 *Deilephila elpenor* Linn.   Elephant Hawk-moth
Widespread and common in Monmouthshire, in gardens, hedgerows woodland etc. flying from late May until the end of July with, in some years, a second brood in late August.

Wye Valley 1892 (Nesbitt 1892a). Tintern 1906, a larva (Bird 1907). Wye Valley, 1911, larvae (Bird 1912). Abertillery dist. 1912 (Rait-Smith 1913). Usk 1926, a larva (RBH).

Croesyceiliog 1928, imago 4/6 (GANH). Usk, 1956 a number at

honeysuckle flowers 5/7. 1961, Dinham 22/6, (CT). 1964, Abergavenny (AD). 1965, Usk, garden m.v.l. traps, recorded every year from 1966 to 1983, some years in large numbers eg. 1968: 15/6 (26), 16/6 (20) (RBH); 30/6 (17) (GANH). Ed. 20/5/82. Ld. 9/8/70 and in 1976, the year of a hot prolonged drought, 31/8, probably a second brood moth. Magor Marsh 2/6/70. Rhyd-y-maen 6/6/70. Hendre Wds 5/7/70, 11/7/86. Hael Wds 5/7/70, 26/6/86. Slade Wds 1/6/71. Cwm-mawr 2/7/71. Cwm Tyleri 30/7/72. Llansoy 22/6/86. Trelleck Bog 19/7/86. (MEA) : Black Vein 3/7/81, Risca 6/7/83, Ynysddu 12/7/85, Cwmfelinfach 10/6/87, Abercarn 27/6/87, Pontllanfraith 2/7/87, Cwmcarn 11/7/87, Nant Gwyddon 10/6/88, Ochrwyth 17/6/88, Dixton Bank 18/6/88.

1966, Pont-y-saeson, larvae on rosebay willow herb; 1974, Usk, larvae on fuschia 19/8, and on tomato plants 30/8 (GANH). 1976, Caldicot, larvae on fuschia 2/8 (CT).

Sq. **SO** 20,30,40,50,31,41,51. **ST** 28,38,48,19,29,39,49,59.

1992 *Deilephila porcellus* Linn. Small Elephant Hawk-moth
Widespread but rather scarce in Monmouthshire except in the west of the county where it occurs more frequently.

Wye Valley 1892 (Nesbitt 1892a). Abertillery dist. 1914 (Rait-Smith 1915).

Usk, garden m.v.l. trap, appeared sporadically: 1966, 9/6, 1969, 11/7; 1971, 2/7, 11/7; 1976, June 18, 22, 25, 28; 1977, 4/6; 1983, 6/7. Wyndcliff 23/6/67. Wentwood 22/6/69, 30/5/70 (JMD). Cwm Tyleri 1976, 29/6; 1977, 9/7(2), 12/7(3); 1986, 28/6. Trelleck Bog 28/5/87. (MEA) : Risca, to act.l., 1984, 16/6, 20/6, 4/7, 5/7 (2); 1985, 4/6, 8/7; 1986, 4/7, 16/7. Ynysddu 12/7/85. Ochrwyth 28/6/86 to m.v.l. (plfl). Nant Gwyddon 25/6/88.

Sq. **SO** 20,30. **ST** 19,29,49,59,28.

1993 *Hippotion celerio* Linn. Silver-striped Hawk-moth
A rare immigrant to Britain.

Chepstow, October 1888 (Mason, teste Tutt, *Br. Lep.* 4: 133).

The validity of this record, at least as a Monmouthshire one, would now appear to be in doubt in view of the fact that Mason was also said to have taken *Leucinodes vagans* Tutt at Chepstow in October 1888 but this latter moth has recently been shewn to have been captured at Tidenham in Gloucestershire just across the county boundary from Chepstow. vid. *L. vagans* (Lear 1986. *Ent. Gaz.* 37:197).

NOTODONTIDAE

1994 *Phalera bucephala* Linn. Buff-tip
Very common from mid-May to early August in hedgerows, gardens, woods,

and almost anywhere among trees and bushes.

Tintern, bred (Bird 1913). Usk 1927, larvae, reared (RBH).

Usk, to garden m.v.l. trap, every year from 1966 to 83 and usually plentiful, eg. 1977 (71) 25/5 to 5/8, inc. (9) on 23/7. Ed. 17/5/80, 18/5/76. Ld. 3/8/70, 5/877. Trostrey 1965, 1968. Rhyd-y-maen 6/6/70. Wyndcliff 24/6/71, 25/6/86. Cwm-mawr 2/7/71 (numerous). Hael Wds 3/7/71. Magor Marsh 6/7/71. Redding's Inclosure 20/7/71, 7/6/77. Cwm Tyleri 9/7/73. Wentwood 8/7/76 (CT). Trelleck Bog 16/7/79. Slade Wds 23/7/79. Llansoy 28/7/84. Hendre Wds 1/7/86. (MEA) : Risca 8/7/83, Cwm Merddog 8/7/85, Ochrwyth 27/6/86, Cwmfelinfach 15/7/86, Pontllanfraith 2/7/87.

Larvae: Trelleck Bog 28/8/87 on birch, Cwm Tyleri (1600 ft.) 22/8/76 on sessile oak (*Quercus petraea*) (GANH). Crick, 25/8/72 on hazel; Llantrisant 8/7/76, on sallow (CT).

Sq. **SO** 10,20,30,40,50,41,51. **ST** 28,48,19,29,39,49,59.

1995 *Cerura vinula* Linn.   Puss Moth

Flies in woods and gardens during May and June but rather scarce in Monmouthshire. Both sexes come to light.

Llandogo dist. 26/5/92, to light (Nesbitt 1892). Wye Valley, 1906, larva on poplar (Bird 1907). Tintern 1912 (Bird 1913).

Croesyceiliog 1923, ♀ resting by day on iron railings (GANH). Usk, garden m.v.l. trap, every year from 1966 – 1982 but in small numbers eg. 1966, 6/6 ♂ (1); 1969, 22/5 – 8/6 ♀ (1), ♂♂ (3); 1971 1/5 – 2/6 ♂♂ (10). Ed. 25/4/73. Ld. 23/6/77. Prescoed 31/5/68 ♀ (1) (GANH). Wentwood, to m.v.l. trap, 1970 May 6th (1), 7th (1), 24th (1) (JMD). Hael Wds 30/5/78 (1). Llansoy 1985 18/5 (1); 1986 22/6 ♀ (1), 23/6 ♀ (1), ♂ (1). (MEA) : Risca 1985 3/6 (1), 4/6 (1).

Sq. **SO** 30,40,50. **ST** 29,39,49.

1996 *Furcula bicuspis* Borkh.   Alder Kitten

Fairly common in Monmouthshire, flying during May and June in damp woods and river valleys.

Usk, garden m.v.l. trap (Plas Newydd), 1966, 31/5/66 (1) (1st v-c. rec.), 6/6 (1), 7/6 (2) (GANH) and recorded annually until 1982 except 1969, 72, 73 and 74; eg. 1982 24/5 (1), 29/5 (3), 31/5 (1), 5/6 (1). Ed. 14/5/80. Ld. 5/7/83. Prescoed, one to m.v.l. 31/5/68 (GANH), 8/6/69 (1) (CGMdW). Usk Castle 29/5/78 (2) (RBH). Slade Wds 1/6/71, 19/6/85. Hendre Wds 27/5/78 ♀ (1). Hael Wds 24/5/80 (1), 29/5/82 (3), 14/6/86 (1). Llansoy 8/5/88 (1), 14/5/88 (1), 28/5/89 (1). Dixton Bank 18/6/88. (MEA) : Risca 2/6/80, Cwmfelinfach 24/5/87, Abercarn 26/5/87.

Sq. **SO** 30,40,50,41,51. **ST** 19,29,39,48.

1997   *Furcula furcula* Cl.   Sallow Kitten
Occurs locally in moist woods and marshy places but not common in Monmouthshire.

Usk, garden m.v.l. trap, recorded sparingly most years from 1966 to 83 eg. 1966, 19/8 (1); 1968, (6) 29/5 – 22/8; 1971, (11) 1/7 – 25/8. Magor Marsh 23/8/68, 7/8/71, 25/6/76. Cwm Tyleri 12/7/77 (1). Hael Wds 31/7/82, 14/8/86. Llansoy 5/6/85. (MEA) : Risca 21/7/83, Upper Ochrwyth 27/6/86, Cwmfelinfach 24/5/87, Abercarn 26/5/87, Wyndcliff 11/7/88.

Sq. **SO** 20,30,40,50 **ST** 28,48,19,29,59.

1998   *Furcula bifida* Brahm.   Poplar Kitten
Occurs where there are poplars but rare in Monmouhshire.

Tintern, 1908, larva on poplar, reared (Bird 1910). Usk, Aug. 1931, larva on poplar, Usk Castle garden, reared (RBH).

Usk: m.v.l. traps 1968, 28/5 (1), 15/6 (1), 16/6 (1), (GANH, RBH). Usk, Plas Newydd, garden m.v.l. trap, 1974, 12/6 (1), 25/6 (1); 1975, 7/6 (1); 1978, 2/6 (1); 1981, 31/5 (2); 1982 1/8 (1). (MEA) : Risca 29/7/84 (1). No other records.

Sq. **SO** 30,50. **ST** 29.

1999   *Stauropus fagi* Linn.   Lobster Moth
This moth, flying from mid-May to mid-July, is common in eastern Monmouthshire in woods containing beeches. It comes readily to light.

Wye Valley, 1892 (Nesbitt 1892a). Tintern 1946 – 1948, several at light each year including "melanic specimens" (Blathwayt 1967). Tintern dist. 1965, to m.v.l. 19/6 (Mere 1965).

Wyndcliff 1966, 16/6/67 (abdt.), (GANH); 20/6/67 (CT); 22/6/71, 8/7/86 etc. Usk, garden m.v.l. trap, several every year from 1966 to 1984 except 1980 and 82. Ed. 25/5/76. Ld. 20/7/72. Pont-y-saeson 1966, 13/7/72. Wentwood 1969, 28/6, 11/7; 1972 29/6 (?) Hael Wds 3/7/71, 25/4/80, 5/7/86 (svl.) inc. f. obscura). Redding's Inclosure 7/6/71, 1/5/72, 23/6/81. Magor Marsh 9/7/74. Hendre Wds 10/5/82. Slade Wds 27/6/85, 13/6/87 (1). Dinham 4/6/87. Kilpale 30/6/87. (MEA) : Upper Ochrwyth 27/6/86, Risca 28/6/87, Abercarn 16/6/88.

Sq. **SO** 30,50,41,51. **ST** 28,48,29,49,59.

2000   *Notodonta dromedarius* Linn.   Iron Prominent
A double-brooded woodland moth common throughout Monmouthshire from May to September.

Wye Valley, 1906, larvae (Bird 1907).

Usk, garden m.v.l. trap, recorded most years from 1966 to 83 and sometimes plentiful eg. 1982 (47) 11/5 – 10/8 inc. (12) on 3/8. Ed 11/5/82. Ld. 26/8/71. Pont-y-saeson 13/6/67. Prescoed 31/5/68. Wentwood 28/6/69.

Slade Wds 1/6/71, 24/8/86. Hendre Wds 5/6/71. Cwm-mawr 2/7/71. Hael Wds 3/7/71, 31/8/80. Cwm Tyleri 17/6/73. Magor Marsh 2/7/76. Newport Docks 17/8/76. Wyndcliff 5/8/82. Llansoy 29/5/85. Trelleck Bog 30/8/86. Dinham 4/6/87. (MEA) : Risca 7/7/81, Ynysddu 12/7/85, Cwm Merddog 20/6/86, Ochrwyth 27/6/86, Cwmfelinfach 15/7/86, Abercarn 26/5/87, Pontllanfraith 20/6/87, Nant Gwyddon 10/6/88.

Sq. **SO** 10,20,30,40,50,21,41. **ST** 28,48,19,29,39,49,59.

2003 *Eligmodonta ziczac* Linn. Pebble Prominent
Widespread and common in Monmouthshire, flying from May to August in damp woods, river valleys and marshy places.

Llandogo dist. 13/5/92 (Nesbitt 1892). Bettws Newydd 1934, a larva, reared (RBH). Monmouth 13/5/39 (Crowther, *Entomologist* **72**: 186). Usk, garden m.v.l. trap, frequent most years from 1966 to 82. Ed. 19/4/76 (2). Ld. 2/9/70. Pont-y-saeson 13/6/67. Prescoed 31/5/68. Magor Marsh 5/5/70, 25/5/80. Rhyd-y-maen 23/5/70. Hendre Wds 26/5/70, 23/5/80. Wentwood 2/6/70, 2/6/81. Redding's Inclosure 7/6/71, 23/6/81. Cwm-mawr 2/7/71. Hael Wds 3/7/71, 5/7/86. Trelleck Bog 16/7/79, 18/7/86. Slade Wds 23/7/79. Wyndcliff 3/8/81. Llansoy 22/5/85. (MEA) : Risca 25/6/84, Abercarn 27/6/87, Cwmcarn 11/7/87, Nant Gwyddon 21/5/88.

Sq. **SO** 20,30,40,50,41,51. **ST** 48,29,39,49,59.

2005 *Peridea anceps* Goeze Great Prominent
Found in oakwoods from mid-April through May and June, this moth is locally common in the eastern half of Monmouthshire.

Llandogo dist., "to light" 26/5/92 (Nesbitt 1892). Tintern 1946 – 1948, "fairly common at light in all three years" (Blathwayt 1967).

Hael Wds, to m.v.l., 25/5/70 (5), 21/5/73 (3), 17/6/74, 27/5/75 (5), also 1978, 79, 82, 84, 86, 87 (inc. melanic form 1/6), 88. Redding's Inclosure 7/6/71, 17/4/72, 17/6/74, 30/5/77. Pont-y-saeson 22/5/72, 30/5/73 (5). Trelleck Bog 22/5/73, 25/5/87 (2). Wyndcliff 26/4/76, 13/6/86. Hendre Wds 28/5/72 (2), 27/5/78 (12), 4/6/79 (numerous), 19/5/80 (abdt., 25 on the sheet at the same time), 14/5/83 (4). Usk, garden m.v.l. trap, 9/5/76 (1). (GANH). Usk Castle, 27/5/76 (1) (RBH); also recorded at Usk in 1977, 1978 and 1984.

Sq. **SO** 30,50,41,51. **ST** 59.

2006 *Pheosia gnoma* Fabr. Lesser Swallow Prominent
Common and widespread in wooded areas of Monmouthshire. Double-brooded, flying from late April to early June and again in July and August.

Llandogo, one imago (Nesbitt 1892). Bettws Newydd (Crowther 1933). Usk 1934, one to light 9/8 (RBH).

Usk, 14/6/58 (GANH). Usk, garden m.v.l. trap, recorded annually 1966

– 84 and usually plentiful. Ed. 19/4/76, 16/4/82. Ld. 2/9/70, 6/9/78. Pont-y-saeson 13/6/67. Prescoed 31/5/68. Langstone 7/5/69 (CT). Wentwood 19/8/69. Hael Wds 25/5/70. Hendre Wds 26/5/70. Rhyd-y-maen 7/5/71. Wyndcliff 24/6/71. Cwm-mawr 2/7/71. Redding's Inclosure 27/8/71. St Pierre's Great Wds 15/5/73. Cwm Coed-y-cerrig 5/6/73. Trelleck Bog 22/5/73, 4/9/73. Cwm Tyleri 29/6/76. Slade Wds 23/7/79. Llansoy 17/9/86. (MEA) : Risca 27/4/82, Mescoed Mawr 18/5/85, Pant-yr-eos 7/7/85, Ynysddu 13/7/85, Abercarn 17/7/85, Ochrwyth 28/6/86, Cwm Merddog 9/8/86, Cwmfelinfach 13/8/86, Cwmcarn 11/7/97, Pontllanfraith 10/7/87, Lower Machen (Park Wood) 13/5/88, Nant Gwyddon 10/6/88, Clydach 15/6/90.

Sq. **SO** 10,20,30,40,50,21,41,51,22. **ST** 28,48,19,29,39,49,59.

2007 *Pheosia tremula* Cl. Swallow Prominent
A common and widespread woodland moth, double-brooded, flying in May and Jumne and again in August.
    Llandogo dist. 13/5/92 (Nesbitt 1892).
    Usk, garden m.v.l. trap, 1966 and most years to 1983, usually frequent. Ed. 4/5/77, 18/4/81. Ld. 31/8/70, 25/8/81. Pont-y-saeson 13/6/67, 1/8/70 (abdt.). Wyndcliff 24/6/71. Cwm-mawr 2/7/71. Hael Wds 3/7/71, 31/8/76. Magor Marsh 11/8/81. Redding's Inclosure 1/5/72. Hendre Wds 28/5/73, 11/7/86. Wentwood 29/6/73. Cwm Tyleri 29/6/74. Slade Wds 21/6/81. Llansoy 19/5/86. Trelleck Bog 30/8/86. (MEA) : Mescoed Mawr 15/8/81, Risca 28/8/81, Lower Machen (Park Wood) 3/5/88.

Sq. **SO** 20,30,40,50,41,51. **ST** 28,48,29,49,59.

2008 *Ptilodon capucina* Linn Coxcomb Prominent
Widespread and common in gardens and deciduous woods. Double-brooded, appearing in the second half of May and in June and again later, in July and August.
    Llandogo "one on tree-trunk" 9/5 1892 and "to light" 26/5/92, and also Llandogo dist. 13/5/92 (Nesbitt 1892). Wye Valley 1906 (Bird 1907).
    Devauden 5/7/61 (GANH). Usk, garden m.v.l. trap, recorded annually 1966 – 82 and fairly common eg. 1970, 14/5 to 25/6 and 28/7 to 29/8. Ed 14/5/70. Ld. 29/8/70. Wyndcliff 12/6/67, 3/8/81. Pont-y-saeson 13/6/67, 13/7/81. Prescoed 31/5/68. Hael Wds 25/5/70, 26/6/79 (plfl.). Wentwood 2/6/70, 26/5/80. Tintern 1/8/70. Cwm-mawr 31/5/71. Slade Wds 1/6/71, 24/8/86. Hendre Wds 5/6/71. Redding's Inlcosure 20/7/71. Cwm Tyleri 17/6/73. Magor Marsh 9/8/76. Llansoy 1/6/85. Trelleck Bog 30/8/86. Fedw Fach (Trelleck) 1/6/88, one on grass stem in daytime. (MEA) : Mescoed Mawr 15/8/81, Risca 2/7/83, Pant-yr-eos 7/7/85, Cwmfelinfach 16/7/86, Llanddewi 18/7/86, Abercarn 26/5/87, Pontllanfraith 20/6/87, Cwmcarn 11/7/87, Lower Machen (Park Wood) 13/5/88, Nant Gwyddon 7/6/88,

Clydach 26/7/90.

Sq. **SO** 20,30,40,50,21,41,51. **ST** 28,48,19,29,39,49,59.

2009   *Ptilodontella cucullina* D. & S.   Maple Prominent
Llandogo dist. (Nesbitt 1892). The only record.

Sq. **SO** 50.

2010   *Odontosia carmelita* Esp.   Scarce Prominent
Occurs during April and May in woods containing mature birches. Recorded from the eastern half of Monmouthshire but local and scarce.

Wentwood, to m.v.l. trap, 1970, 9/5 (2), 16/5 (1) (JMD) (1st v-c. rec.); 1971, 22/4 (1), 12/5 (1) (JMD). GANH: Trelleck Bog 22/5/73 ♀ (1) to m.v.l.; Pen y cae-mawr (Darren Wood) 4/5/76, one freshly-emerged ♀ to m.v.l.; Hael Wds 25/4/80 (2), 28/4/80 (1); Pont-y-saeson 27/4/80 (1).

Sq. **SO** 50. **ST** 49,59.

2011   *Pterostoma palpina* Cl.   Pale Prominent
Widespread and fairly common in damp woods, river valleys and marshy places. Double-brooded, it flies in May and June and during the second half of July and August.

Llandogo dist. 13/5/92 (Nesbitt 1892). Llandogo, "to light", 26/5/92 (idem). Wye Valley 1906 (Bird 1907). Bettws Newydd (Crowther 1935).

Usk, garden m.v.l. trap, frequent and recorded annually from 1966 to 1983. Ed. 8/5/69, 9/5/76. Ld. 30/8/71, 28/8/77. Pont-y-saeson 13/6/67. Prescoed 31/5/68. Rhyd-y-maen 23/5/70. Hendre Wds 26/5/70. Magor Marsh 2/6/70, 2/8/70 and recorded most years (GANH). Tintern 1/8/70. Slade Wds 1/6/71, 11/8/81. Redding's Inclosure 7/6/71. Cwm Coed-y-cerrig 27/5/73, 5/6/78. Trelleck Bog 16/6/73. Wentwood 29/6/73. Hael Wds 26/6/79. Wyndcliff 5/8/82. Llansoy 19/5/85. (MEA) : Risca 4/6/80, Pant-yr-eos 17/6/83, Mescoed Mawr 1/6/85, Abercarn 26/5/87, Cwmfelinfach 13/8/87, Lower Machen (Park Wood) 13/5/88, Nant Gwyddon 7/6/88.

Sq. **SO** 30,40,50,41,51,22. **ST** 28,48,19,29,39,49,59.

2014 *Drymonia dodonaea* D. & S.   Marbled Brown
Occurs locally and sparingly in oakwoods in the eastern half of the county during May and June.

Llandogo dist. 26/5/92, "to light" (Nesbitt 1892).

Wyndcliff: 1966, 7/6; 1967, 6/6, 16/6; 1971, 22/6; 1986, 25/6. Pont-y-saeson 13/6/67. Hael Wds 25/5/70, 27/6/78, 24/5/80. Hendre Wds, to m.v.l., 26/5/79 (pfl.), 27/5/72 (2); numerous on 4/6/79, 19/5/80 and 25/5/80. Redding's Inclosure 7/6/71. Usk (Plas Newydd), garden m.v.l. trap, 11/6/66 (1), 1/6/79 (1). Usk Castle 21/4/71, one to m.v.l. (RBH).

Trelleck Bog 28/5/87.

Sq. **SO** 30,50,41,51. **ST** 59.

2015   *Drymonia ruficornis* Hufn.   Lunar Marbled Brown
This moth, flying in April and May, occurs locally in some Gwent oakwoods but often abundant where found.

Abertillery dist. 1911 (Rait-Smith 1912). Tintern, larva on oak 1912 (Bird 1913). Bettws Newydd (Crowther 1933).

Usk (Park Wood), 29/4/67; Gwehelog (Camp Wood), 29/4/67, abundant at m.v.l. (GANH). Usk, garden m.v.l. trap, a few most years from 1968 to 1984. Ed. 5/4/82, Ld 22/5/78 and an exceptionally late one 16/6/81. Craig-y-Master (Llangwm), 7/5/71. Redding's Inclosure 24/4/72. Cwm Tyleri 26/5/73. Cwm Coed-y-cerrig 27/5/73. Hendre Wds 14/4/80 (abdt.). Llansoy 19/5/86 (2). Hael Woods 14/6/86, a late moth. (MEA) : Black Vein 8/5/81, Risca 25/4/82, Mescoed Mawr 19/4/85, Wyndcliff 13/6/86, Ochrwyth 24/4/87, Cwm Merddog 26/4/87, Abercarn 29/4/87, Cwmfelinfach 5/5/87, Pontllanfraith 6/5/88.

Sq. **SO** 10,20,30,40,50,41,51,22. **ST** 28,48,19,29,49,59.

2019   *Clostera curtula* Linn.   Chocolate-tip
Until recently the literature stated that in the British Isles this moth was to be found in England to the south of a line from the Severn to the Humber. However, it has been a local and uncommon resident in Monmouthshire for at least three decades, occurring where there are poplars. Double-brooded it flies in April and May and again in late July and August and comes readily to light.

Usk, 1961, one in gardens at Usk Castle May 17th (1st v-c. rec.) (RBH in litt.). Usk (Plas Newydd), garden m.v.l. trap, 13/5/67 (1) (GANH) and one or two noted most years until 1983 but occasionally seen more frequently eg. 1970, (9) from 14/5 to 5/6 and (4) from 2/8 to 10/8; 1979, one only on 29/7. Ed. 13/5/67, 5/5/71. Ld. 19/8/71, 18/8/81. Prescoed 31/5/68 (1). Hael Wds 30/5/78 (1). Hendre Wds 21/4/82 (1), 10/5/82 (1) (GANH). (MEA) : Mescoed Mawr, to m.v.l., single moths on 3/5/84, 17/5/85 and 31/5/85; Risca to act.l., 1985, 18/5 and in August (4) from 9/8 to 28/8, 1986, 16/8 (1); Lower Machen (Park Wood) 13/5/88.

Sq. **SO** 30,50,41. **ST** 28,29,39.

2020   *Diloba caeruleocephala* Linn.   Figure of Eight
Flying from late September to early November, this moth is locally common in the eastern half of Monmouthshire inhabiting woods, hedgerows and bushy places.

Abergavenny, T. Baxter 1889, one in BMNH (per JMC-H). Llandogo 1892 (Nesbitt 1892b). Tintern dist., larvae on *Cotoneaster* (Bird 1905).

147

Brockwells (Caerwent), 24/10/63, 1964 and 1965 (CT in litt.). Llantrisant 29/10/65, plentiful along hedgerow (GANH). Usk, garden m.v.l. trap, noted in fair numbers most years from 1966 to 1982, eg. 1978, (64) 23/9 – 25/10, inc. (14) on 16/10. Ed. 8/9/82. Ld. 2/11/81. Rhyd-y-maen 2/11/70. Hael Wds 22/10/71, 9/10/79. Slade Wds 24/10/71. Hendre Wds 19/9/78. Magor Marsh 13/10/78. Llansoy 13/10/86 (33) etc.

Llansoy, 5/6/88, larva on hawthorn (*Crataegus*).

Sq. **SO** 30,40,50,41. **ST** 48,39,49.

LYMANTRIIDAE

2026 *Orgyia antiqua* Linn.   The Vapourer

This moth, widespread in the eastern half of Gwent, does not appear to be very common as the day-flying males are seen only sporadically and sparingly. The females are wingless. The males occasionally come to light.

East Monmouthshire, larvae on bramble (Bird 1905). Tintern 1912 (Bird 1913).

Pont-y-saeson, 22/9/63, flying in abundance in a plantation of young oaks (GANH). Usk (Park Wood), 1966, one by day; 1969, one to m.v.l. 3/9. Llandenny Walks 1968, one. St Pierre's Great Wds 29/9/70 (1). Magor Marsh, 23/9/72 (1) to m.v.l. Usk (Plas Newydd), singletons flying in garden, 25/9/78, 9/9/79, 30/9/79; garden m.v.l. trap, 3/10/71 (1), 15/9/78 (1). Hendre Wds, to m.v.l., 4/9/78 (1). Llantrisant (Coed-y-prior), 30/9/79, (svl.). Trostrey Common 2/10/79 (1). Raglan dist. 2/10/79, one.

Wentwood, larva, 9/7/74 (CT). Usk 1979, larvae on *Wisteria sinensis* in Usk prison yard (reared), and again on 28/6/83 (GANH).

Sq. **SO** 30,40,41. **ST** 48,39,49.

2028 *Calliteara pudibunda* Linn.   Pale Tussock
     Dasychira auct.

Flying in May and June, this species is common in woodland, hedgerows and bushy places throughout the county.

Llandogo dist., to light, 26/5/92 (Nesbitt 1892). Llandogo, 1906 larva on birch, reared (Bird 1907). Usk Castle, 1931, larva on poplar, reared (RBH). Bettws Newydd (Crowther 1933). Bettws Newydd 29/5/35, at light (RBH). Caerwent 30/9/64, Wyndcliff 19/6/67 (CT).

Usk, garden m.v.l. trap, noted most years from 1966 to 1983. Ed. 10/5/71. Ld. 7/7/77. Prescoed 31/5/68. Hael Wds 26/5/70 (abdt.). Hendre Wds 26/5/70 (plfl.). Magor Marsh 2/6/70, 19/6/79. Wentwood 2/6/70. Rhyd-y-maen 6/6/70. Cwm-mawr 31/5/71. Slade Wds 1/6/71, 27/6/85. Redding's Inclosure 7/6/71 (abdt.). Pont-y-saeson 21/6/71. Cwm Coed-y-cerrig 24/5/76. Llansoy 4/6/85. Trelleck Bog 25/5/87. (MEA) : Mescoed Mawr 21/5/78, Risca 2/6/80, Cwm Merddog 20/6/86, Ochrwyth 27/6/86, Abercarn 16/5/87,

Lasgarn 20/5/87, Cwmfelinfach 23/5/87, Lower Machen (Park Wood) 13/5/88, Nant Gwyddon 10/6/88, Dixton Bank 18/6/88, Clydach 11/5/90.

Sq. **SO** 10,20,30,40,50,21,41,51,22 **ST** 28,48,19,29,39,49,59.

2030 *Euproctis similis* Fuess.   Yellow-tail
Common in woods, gardens and hedgerows in July and August.

Usk, garden m.v.l.trap, recorded annually from 1966 to 1983 and sometimes abundant as in 1969 with peak numbers on 20/7 and 27/7 and in 1970 with peak numbers on 31/7. Ed. 3/7/70. Ld. 1/9/79. Magor Marsh 11/7/67, 22/7/67 (abdt.), larva on sallow 12/6/79. Hendre Wds 5/7/70. Rhyd-y-maen 18/7/70. Tintern 1/8/70. Redding's Inclosure 20/7/71. Cwm Tyleri 8/7/76. Slade Wds 23/7/79. Llansoy 26/7/84. Trelleck Bog 30/8/86. (MEA) : Wentwood 10/7/77, Newport Docks 11/7/84, Ynysyfro 4/7/76, Mescoed Mawr 8/7/84, Risca 25/7/84, Llanddewi 17/7/86, Wyndcliff 27/7/78, Uskmouth 16/8/88.

Sq. **SO** 20,30,40,50,41,51. **ST** 28,38,48,29,39,49,59.

2031 *Leucoma salicis* Linn.   White Satin Moth
Flying from the end of June to early August, and coming readily to light, this moth frequents marshes, river valleys and places where poplars and willows are found. Here in Monmouthshire this species is at the western limits of its range and is scarce and local. It was first recorded at Magor in 1971 and at Usk in 1979. At that time these constituted the westernmost localities for this moth in the southern half of Wales. Since then it has become more frequent and widespread in the county and subsequent records suggest that it continues to extend its range westwards.

Magor Marsh: one to m.v.l. 7/7/71 (GANH) (1st v-c. rec.). 1976, 2/7 (10), 3/7 (4), 6/7 (6). Also 1977 (6), 1978 (8), 1979 (4), 1983 (2). Usk: 1979, one to electric light 3/7 and one to m.v.l. 5/7; 1983, 17/7 (1). Llansoy: 1985, (3) 2/7 – 7/7; 1986, 12/7 (1), 14/7 (1). Slade Wds 27/6/85 (1). Hael Wds 5/7/86, 19/6/88 (1). (MEA) : Risca 1984 (3) 20/6 – 7/7; 1986 9/7 (1). Wyndcliff 8/7/85. Strawberry Wood (Stanton) 14/8/91.

Sq. **SO** 30,40,50,32. **ST** 48,29,59.

2033 *Lymantria monacha* Linn.   Black Arches
Widespread but of sporadic occurrence in the eastern half of Gwent, flying from late July to early September in deciduous woodland. To date (1990) it has not been noted to the west of the River Usk.

Usk, garden m.v.l. trap, recorded annually from 1968 to 1983 eg. 1973, (12) 24/7 – 20/8; 1982, (18) 10/7 – 6/8. Ed. 10/7/82. Ld. 7/9/79. Tintern 1/8/70 (2). Hael Wds 6/8/74, 31/7/82 (13), 5/8/82 (5). Redding's Inclosure: 1971, 20/7 (10), 27/8 (1); 1983, 24/7 (12). Wyndcliff 3/8/81, 10/8/83 (2), 7/8/89 (2). Hendre Wds 29/8/78 (2), 4/9/78, 20/7/79. Slade Wds 28/7/82,

20/7/89. Llansoy 26/7/84.

Sq. **SO** 30,40,50,41.51. **ST** 48,59.

ARCTIIDAE
LITHOSIINAE

2035 *Thumata senex* Hb.   Round-winged Muslin

Frequents fens and marshy places in July and August. In Gwent it is found at Magor where it is sometimes abundant. Occasionally recorded in my garden m.v.l. trap at Usk but whence these moths came is uncertain. Magor Marsh 1969, abdt. on 1/7 and 11/7 to m.v.l. (GANH). (1st v-c. rec.). Also 1970, 10/7 and 29/8; 1971, 19/7; 1973, 3/7 (plfl.); 1975, 76, 77; 1983 numerous on 13/7. Usk, garden m.v.l. trap, 6/7/70 (1), 5/8/77 (1), 27/7/79 (1); 1983, July 15 (7), 16 (3), 20 (1).

Sq. **SO** 30. **ST** 48.

2037 *Miltochrista miniata* Forst.   Rosy Footman

In Monmouthshire this moth is very scarce and local being found in damp woods and moist situations during July and early August.

    Hael Wds: 1970, one to m.v.l. 5/7 (GANH), (1st v-c. rec.); 1973, 10/7 (3), 17/7(1); 1976, 22/6(1); 1982, 31/7 (2). Magor Marsh 20/7/76 (1). Slade Wds 28/7/81 (1), 21/7/83 (1), 6/8/84 (1); 1986 22/7 (3), 29/7 (1). (MEA) : Slade Wds 1983 17/7, 20/7; 1984 6/7.

Sq. **SO** 50. **ST** 48,59.

2038 *Nudaria mundana* Linn.   Muslin Footman

Of widespread but infrequent occurrence in Monmouthshire being found in late June and July associated with such petrous habitats as stone walls and old quarries.

    Wye Valley, 1906, larva (Bird 1907). Abertillery dist. 1911 (Rait-Smith 1912).

    Cwm Tyleri: to m.v.l. 30/7/72 (1), 29/6/76 (3) on a hillside at 1,400 ft in vicinity of dry stone walls and an old quarry (GANH). Usk, garden m.v.l. trap, 1983, 15/7(1), 17/7(1). Wyndcliff 1983, July 17, 20 and 25; 1984, July 6 and 26 (MEA & GANH). (MEA) : Mescoed Mawr 8/7/84, Risca 12/7/86, Ochrwyth 12/7/86, Cwmfelinfach 15/7/86, Cwmcarn 11/7/87.

Sq. **SO** 20,30. **ST** 28,19,29,59.

2039 *Atolmis rubricollis* Linn.   Red-necked Footman

This woodland moth flying in June and July is widespread but local and at low density in Monmouthshire.

    Tintern dist., 19/6/65 to m.v.l. (Mere 1965).

    Wyndcliff, to m.v.l., 1967 14/6, 16/6; 1969 13/7 (7); 1986 25/6.

Wentwood, 3/7/68, one flying by day in conifer plantation at Foresters' Oaks (GANH); 8/7/71 one to m.v.l. trap (JMD). Usk: to m.v.l. trap at Usk Castle 13/7/69 (2) (RBH). Hael Wds 3/7/71, 26/6/86. Slade Wds 8/7/83 (1). Llansoy 15/7/86 (1). (MEA) : Risca 7/7/83, 4/7/85; Ochrwyth 27/6/86 (plfl.); Cwmfelinfach 15/7/86, Abercarn 27/6/87.

Sq. **SO** 30,40,50. **ST** 28,48,19,29,49,59.

2040 *Cybosia mesomella* Linn. Four-dotted Footman

This moth, flying in June and July, occurs very locally in woods in the eastern half of Monmouthshire.

Pont-y-saeson, 13/6/67 (plfl.) to m v.l. (REMP, RBH, GANH); 24/6/69 one by day (GANH). Wentwood: to m.v.l. trap 1969, single moths on June 21, 23, 28 (JMD). Hael Wds 28/6/74(1), 7/7/87 (1). Usk Castle 21/6/76 one to m.v.l. (RBH). Trelleck Bog 16/7/79 (5), 18/7/86 (svl.). Llansoy 1987, 5/7 (2), 6/7(1); 19/6/88 (1); 28/5/89 (1).

Sq. **SO** 30,40,50. **ST** 49,59.

2043 *Eilema sororcula* Hufn. Orange Footman

Flies during May and June in deep woods containing oaks (Redding's Inclosure) or beech (Wyndcliff). Rare in Monmouthshire and noted only from these two localities in the extreme east of the county

Wyndcliff, to m.v.l.: 1966, 7/6 (1); 1967, 6/6 (2), 12/6 (1), 10/7 (GANH). 1971, 22/6, 24/6, (JMC-H, GANH); 1986, 13/6 (plfl.), 25/6; 1990, 6/5 (GANH). Redding's Inclosure: 7/6/71(1), 1/5/72(1) (GANH).

Sq. **SO** 51. **ST** 59.

2044 *Eilema griseola* Hb. Dingy Footman

Found in fens and damp woods in July and August, this moth is widespread in the eastern half of the county, usually in small numbers, except at a few favoured sites such as Magor Marsh and the Wyndcliff where it can be numerous. The pale yellow variant ab. *flava* Haw. (= *stramineola* Doubl.) occurs in a number of places.

Bica Common, 1968, to m.v.l. 23/8 f. *flava* (1) (GANH). Wentwood, to m.v.l. trap at Foresters' Oaks 1969, 2/8 (1) and f. *flava* (1) (JMD). Magor Marsh 1969, 22/7(1), 5/8(1); 1970; 1974 24/8 (abdt.); 1975 4/8 (plfl.); 1976, 77, 81, 83 and 1/8/88 (4) inc. f. *flava* (3). Usk, garden m.v.l. trap, 8/8/70 1983 (3). Piercefield Park 18/7/73 f. *flava*. Hael Wds 28/7/81, 31/7/82 inc. f. *flava*, 5/8/82. Slade Wds 11/8/81, 28/7/82 (f. *flava*). Wyndcliff 1982 28/7, 5/8, 11/8; 1983 25/7 (abdt.); 1984 8/8 (abdt.). Redding's Inclosure 24/7/83. Trelleck Bog 4/7/86 f. *flava*.

Sq. **SO** 30,50,51. **ST** 48,49,59.

2045   *Eilema caniola* Hb.   Hoary Footman

This Footman is found typically on sea cliffs and shingle beaches, and the nearest colony known to me flourished on sea cliffs in the vicinity of Woolacombe in North Devon. Skinner 1984, described its distribution as "locally common along the coasts of North and South Devon, Cornwall, Pembrokeshire and the other sea-board counties in West Wales. Nevertheless despite the absence in Monmouthshire of such typical habitats, on Aug. 3rd 1981 it surprisingly appeared at m.v.l. in some numbers, in the lower Wye Valley in a long-disused quarry, about 500 ft above the river and some 10 Km. upstream from the Wye's confluence with the Severn Estuary.

Although I had worked the area for moths many times over the preceeding fifteen years I had not previously encountered it there. The site was re-visited on August 5th and 11th but no more were seen, nor did further visits at the appropriate season in subsequent years produce it again until 1989.

Wye Valley, 1981, 3/8 (plfl.) (GANH); 1989, 7/8 (1) (GANH, MEA); 1991, 29/7 (1) (MEA).

Sq. **ST** 59.

2047   *Eilema complana* Linn.   Scarce Footman

Widespread and fairly frequent in Monmouthshire, flying in woods from late June to early August.

Hael Wds 5/7/70(1), 17/7/73, 22/6/76, 31/7/82 (6), 2/8/86. Redding's Inclosure 20/7/71 (2). Hendre Wds 29/8/78, 11/7/86. Trelleck Bog 16/7/79. Usk, garden m.v.l. trap, 29/7/82 (1). Slade Wds 2/8/82. Wyndcliff 5/8/82 (GANH); also 1983, 84, 85, 86 (MEA). (MEA) : Risca 7/8/81, 1983, 1985; Dinham 7/7/87, Magor Marsh 1/8/88, Nant Gwyddon 15/8/88.

Sq. **SO** 30,50,41,51. **ST** 48,29,49,59.

2049   *Eilema deplana* Esp.   Buff Footman

This moth occurs sporadically and at low density in woods in the eastern half of Monmouthshire during July and August but at the Wyndcliff it appears regularly and is often abundant.

Wyndcliff, to m.v.l., abundant on 10/7/67, 22/7/68, 13/7/69 and exceptionally so on 3/8/81 (GANH). Also noted 1983, 84, 85 and 86 (MEA). Tintern 1/8/70 (1). Redding's Inclosure 20/7/71. Usk, garden m.v.l. trap, 11/8/78 (1). Hael Wds 14/8/78. Slade Wds 2/8/80 (1), 17/7/87.

Sq. **SO** 30,50,51. **ST** 48,59.

2050   *Eilema lurideola* Zinck.   Common Footman

Widespread and common in woods, gardens and hedgerows during July and August.

Wye Valley, 1906 (Bird 1907).

Usk, garden m.v.l. trap, annually from 1966 to 1983 and sometimes abundant eg. 27/7/69 and 23/7/82. Ed. 27/6/70, 1/7/82. Ld. 13/8/79, 18/8/81. Magor Marsh 2/8/70. Hael Wds 5/7/70, 2/8/81 (abdt.). Rhyd-y-maen 18/7/70. Redding's Inclosure 20/7/71. Hendre Wds 18/7/71. Slade Wds 28/7/81. Wyndcliff 28/7/82 (abdt.). Llansoy 26/7/84. Trelleck Bog 18/7/86. MEA: Risca 12/7/82, also 1983, 84, 85 and 86; Mescoed Mawr 8/7/84; Cwm Merddog 8/8/86.

Sq. **SO** 10,20,30,40,50,41,51. **ST** 48,29,39,49,59.

ARCTIINAE

2054 *Utetheisa pulchella* Linn.   Crimson Speckled
A scarce migrant from the Mediterranean region, seen only rarely and very sporadically in the British Isles and recorded once (1871) in Monmouthshire. South 1907, states that 1871 was a record year for this species in Britain when more than thirty were recorded.

Monmouthshire, "flying in the sunshine", Sept. 1871 (S.H. Gustard. *Entomologist* 5: 414). (S.H. Gustard had already taken one at Swanage a little earlier loc. cit.).

2056 *Parasemia plantaginis* Linn.   Wood Tiger
ssp. *plantaginis* Linn.
Flies by day in open heathy woods and on moorland but is local and very scarce in Monmouthshire.

Wye Valley 1906 (Bird 1907). Llandogo, several, 1909 (Bird 1910). Aber-tillery dist. (Rait-Smith 1912, 1913). Monmouth dist. (Charles c. 1937).

Pont-y-saeson, 1958, one flying by day along woodland path 8/6 (GANH). Wentwood, 1/6/68 (JMD). Wentwood: 1974 13/6, 20/6; 1975 9/6, 10/6; 1976 11/6 (2) (CT in litt.). Trelleck Bog, 22/6/69 (MWH in litt.). Cwm Tyleri: 1976 27/6, 29/6 (GANH); 1977 9/7 (CGMdW), 12/7 (4) (GANH). Blorenge (WEK).

Sq. **SO** 20,50,21. **ST** 49,59.

2057 *Arctia caja* Linn.   Garden Tiger
Widespread and common in gardens, woods, waste ground etc., flying from late June to mid-August.

Usk, garden m.v.l. trap, noted every year from 1966 to 1982 and often plentiful eg. 24/7/81 (23). Ed. 27/6/70, 25/6/76. Ld. 8/8/77, 11/8/81. Caerwent 19/7/64 (CT). Magor Marsh 10/7/70 (svl.); larva feeding on burdock (*Arctium lappa*) (GANH). Rhyd-y-maen 18/7/70 (4) to m.v.l. Cwm-mawr 2/7/71. Redding's Inclosure 20/7/71. Hendre Wds 17/7/79. Llansoy 26/7/84. (MEA) : Risca 7/7/81, Newport Docks 11/7/84, Cwmfelinfach 15/8/86.

Sq. **SO** 20,30,40,41,51. **ST** 38,48,19,29,39,49,59.

2058   *Arctia villica* Linn   Cream-spot Tiger
         ssp. *britannica* Ob.

This moth, flying during May and June, appears locally and sporadically in the eastern half of the county mainly in woodland.

Tintern 1911, one male. (Bird 1912). Tintern, "several, mainly by day." (Blathwayt 1967). 1955, Usk, Park Wood, ♀ (1) 15/6 and another on 22/6, also one ♀ at Usk 9/6/56 (RBH). 1961, Highmoor Hill 7/6 (CT in litt.). 1964, Tintern (MJL in litt.). 1965, Whitebrook 18/6 to m.v.l. (Mere 1965). 1966, Pont-y-saeson, one on woodland path (REMP). Magor Marsh ♂ (1) to m.v.l. 18/6/79 (MJL in litt.).

Usk, to garden m.v.l. trap 14/6/66, 1/7/71, 8/7/72, also 1973, 76 (2), and 82. Hendre Wds 5/6/71. Redding's Inclosure 7/6/71 (4), 17/6/73. Hael Wds 14/6/86. Dinham 4/6/87. (MEA) : Magor Pill (sea-wall) 1/6/90 (1).

Sq. **SO** 30,50,41,51. **ST** 48,49,59.

2059   *Diacrisia sannio* Linn.   Clouded Buff

Local and very scarce in Monmouthshire, flying in open woodland in June and early July.

Abertillery dist. (Rait-Smith 1912, 1913).

Wentwood, 1968, one to m.v.l. 8/7 (GANH). Wentwood (Foresters' Oaks), 1969, single males on 21/6 and 23/6 to m.v.l. and a ♀ flying by day on 24/6, also several in 1971 (JMD). Wentwood, 1974, 25/6 (2), 28/6 (CT). Trelleck Bog 22/6/69 (MWH in litt.). Slade Wds 1970 (JMD).

Sq. **SO** 20,50. **ST** 48,49.

2060   *Spilosoma lubricipeda* Linn.   White Ermine

Very common and widespread in woods, gardens and waysides from mid-May through June and July.

Llandogo dist. "to light" 26/5/92 (Nesbitt 1892). Llandogo, 24/4/93 (Nesbitt 1893). Abertillery dist. (Rait-Smith 1912, 1913).

Croesyceiliog 15/6/28 (GANH). Bettws Newydd "at light" 29/5/35 (RBH). Usk, garden m.v.l. trap, every year from 1966 to 1982 and often plentiful. Ed. 11/5/71, 8/5/75. Ld. 1/7/71, 18/7/78. Prescoed 5/7/68. Rhyd-y-maen 23/5/69. Hendre Wds 26/5/70. Magor Marsh 2/6/70. Wentwood 2/6/70. Redding's Inclosure 7/6/71, 23/6/81. Wyndcliff 22/6/71. Cwm-mawr 2/7/71. Llanvair Discoed, larvae 24/9/74 (CT). Pont-y-saeson, ♀ visiting foxglove flowers in daytime 23/6/71 (GANH). Cwm Tyleri 5/8/74. Trelleck Bog 11/7/79. Slade Wds 21/6/81. Llansoy 31/5/86. (MEA) : Risca 16/6/83, Blorenge 7/7/84, Abercarn 26/5/87, Dinham 4/6/87, Pontllanfraith 22/6/87, Ochrwyth 17/6/88, Dixton Bank 18/6/88, Nant Gwyddon 25/6/88.

Sq. **SO** 20,30,40,50,21,41,51. **ST** 28,48,19,29,39,49,59.

2061   *Spilosoma luteum* Hufn.   Buff Ermine

Very common and widespread inhabiting gardens, woods and waysides from mid-May to mid-July.

Llandogo dist. "to light" 26/5/92 (Nesbitt 1892). Abertillery dist. (Rait-Smith 1912, 1913).

Croesyceiliog, 14/6/28 (GANH). Bettws Newydd, "at light", 29/5/35 (RBH). Usk, garden m.v.l. trap, annually 1966 to 83 and often abundant eg. (29) on 20/6/79. Ed. 18/5/70, 20/4/82 Ld. 15/7/71, 21/7/77. Wyndcliff 20/6/67 (CT). Prescoed 5/7/68. Hendre Wds 26/5/70. Wentwood 2/6/70. Rhyd-y-maen 6/6/70. Redding's Inclosure 7/6/71, 23/6/81. Pont-y-saeson 21/6/71. Hael Wds 3/7/71. Magor Marsh 19/7/71. Cwm Coed-y-cerrig 24/6/76. Llansoy 12/6/85. Slade Wds 27/6/85. (MEA) : Allt-yr-yn 20/6/76, Mescoed Mawr 25/6/76, Risca 7/7/83, Cwm Merddog 7/7/85, Ynysyddu 12/7/85, Pant-yr-eos 6/7/85, Dinham 4/6/87, Cwmcarn 11/7/87, Dixton Bank 18/6/88, Nant Gwyddon 25/6/88.

Sq. **SO** 10,20,30,40,50,41,51,22. **ST** 28,38,48,19,29,39,49,59.

2062   *Spilosoma urticae* Esp.   Water Ermine

This very local moth inhabits fens and marshes in S.E. England mainly in counties to the south and east of a line from the Wash to the Severn Estuary. It does however, occur in Monmouthshire at Magor Marsh where it is very scarce. There are only one or two other known stations for this species in Wales. There is an old record (1854) for South Pembrokeshire (Barrett 1895, **2**: 291) and it was found by Demuth at Borth Bog, Cardiganshire in 1960 (RPD 1981).

Magor Marsh, 1969, one to m.v.l. 11/7 (GANH) (1st v-c. rec.). Also 1976 21/6 (3), 25/6 (1); 1979 19/6 (2); 1981 19/6 (1) (GANH). 18/6/79, one to m.v.l. (MJL in litt.).

Sq. **ST** 48.

2063   *Diaphora mendica* Cl.   Muslin Moth

Flies from late April until the beginning of June in gardens and open woodland. The female is only occasionally seen, either resting by day or flying in sunshine, while the male comes readily to light. Although it occurs regularly and often plentifully at Usk, this species appears to be scarce and local in the county as a whole.

Llandogo (Nesbitt 1893). Llanock Wood, Crumlin, ♀ 1911 (Rait-Smith 1912); [= Llanerch Wood. GANH]. Tintern 1912. (Bird 1913).

Usk, garden m.v.l. trap 22/4/67 and every year until 1983, often in fair numbers eg. 1978 (150) 8/5 to 3/6, 1977 (109). Max.c. 18/5/76 (17), 26/5/77 (15). Ed. 21/4/76, 15/4/81. Ld. 8/6/71, 2/7/79. Usk, 21/5/71 ♀ on lighted door panel; 26/5/78 ♀ in house porch; 26/5/80 ♀ in garden, flying in bright sunshine with Green-veined White butterflies. Wentwood, 1968 ♂ to m.v.l.

(JMD). Rhyd-y-maen 26/5/80, ♀ flying at noon. Llansoy 1/6/85 (1), 13/6/86 (1). Wolvesnewton 29/5/89 ♀ flying in sunshine. (MEA) : Risca 8/5/85.

Sq. **SO** 30.40.50. **ST** 29,49.

2064   *Phragmatobia fuliginosa* Linn.   Ruby Tiger
       ssp. *fuliginosa* Linn.

Widespread and locally common in Gwent, inhabiting marshes, open woods and moorland. Double-brooded and noted from May to September but most frequently in late July and August.

Wye Valley, 1906, a larva (Bird 1907). Abertillery dist. one larva, 1910, reared (Rait-Smith 1912).

Usk, garden m.v.l. trap 15/8/66 (2) and recorded most years until 1982 when it was plentiful (40) 31/5 – 18/9. Ed. 10/5/71, 6/5/76. Ld. 28/8/71, 18/9/82. Magor Marsh 23/8/68 and often common as on 22/7/69 and 2/8/70. Prescoed 5/7/68. Undy (Collister Pill) 30/8/70, 18/8/87. Redding's Inclosure 7/6/71, 20/7/71. Caerwent 10/4/74 (2) (CT). Cwm Tyleri 27/6/76. Llansoy 27/6/86. (MEA) : Wentwood 25/5/77, Risca 21/7/83, Ochrwyth 27/6/86, Pontllanfraith 22/6/87, Wyndcliff 27/7/88, Magor Pill (sea-wall), plentiful 19/7/89.

Sq. **SO** 20,30,40,51. **ST** 28,48,19,29,39,49,59.

2068   *Callimorpha dominula* Linn.   Scarlet Tiger

Rare in Monmouthshire.

Tintern, 1887, "in numbers" (Goss 1887); one only in 1890 "where it was formerly very common" (Goss 1890).

Magor Marsh, 1976, two flying at dusk 2/7 (GANH). Slade Wds, 1984, one to m.v.l. 30/6 (GANH). Wyndcliff: 1984, one to m.v.l. 7/7; 2/7/92, (1) (MEA).

Sq. **SO** 50. **ST** 48,59.

2069   *Tyria jacobaeae* Linn.   The Cinnabar

Flying from late May through June and July this moth is widespread and common in rough fields, gardens, waste ground and waysides where its larval food-plants, groundsel and ragwort (*Senecio* spp.) grow, but it is not as plentiful as it was sixty-five years ago when it was known to the local children (myself included) as the "Soldier Moth". The imago occasionally comes to light but is readily disturbed from herbage by day. The conspicuous yellow and black larvae often completely defoliate their food-plants.

Monmouthshire, 1889, larvae (Patten 1890). Llandogo (Nesbitt 1893). Wye Valley (Bird 1907). Abertillery dist. (Rait-Smith 1912).

Croesyceiliog: 1923 (abdt.), 1928 "one flying by day" 31/5 (GANH). Pont-y-saeson 1966. Usk, garden m.v.l. trap, recorded most years from 1967 to 1982. Ed. 14/5/71. Ld. 2/8/72. Prescoed 31/5/68. Wentwood 19/6/68 (JMD). St Pierre's Great Wds, larvae 23/7/68. Hadnock Quarry, larvae 20/7/71. Slade Wds, imago 15/6/73. Hendre Wds, larvae on hoary ragwort (*S. erucifolius*) 8/8/73. Redding's Inclosure 30/5/77. (MEA) : Risca 29/6/84, Gaer Hill (larvae) 17/7/86, Dinham 22/5/87.

Sq. **SO** 30,40,41,51. **ST** 28,38,48,29,39,49,59.

NOLIDAE

2075 *Meganola strigula* D. & S.   Small Black Arches
Skinner 1984, described this species as local and declining, inhabiting mature oakwoods in a few southern counties including Monmouthshire. *MBGBI* Vol. **9** (1979) indicates square **SO** 50 which spans two counties. However, BS (in litt. 1992) now considers the record dubious, a view endorsed by P.T. Harding of the Biological Records Centre.

Sq. **SO** 50. (?).

2077 *Nola cucullatella* Linn.   Short-cloaked Moth
Widespread and frequent in woods and hedgerows in June and July.

Wye Valley 1911 (Bird 1912). Usk, garden m.v.l. trap, 10/7/69 and most years to 1983. Ed. 1/7/70. Ld. 30/7/78. Magor Marsh 11/7/69 to m.v.l., also 1971,75,76,77 etc. Hendre Wds 5/7/70, 11/7/86. Redding's Inclosure 18/5/71. Piercefield Park 18/7/73. Hael Wds 28/6/74. Wyndcliff 3/8/81. Trelleck Bog 16/7/79. Llansoy 11/7/86. Kilpale 30/6/87. (MEA) : Mescoed Mawr 11/7/77, Risca 14/7/80, Ynysddu 12/7/85, Ochrwyth 9/8/86, Cwmfelinfach 6/7/87, Pontllanfraith 10/7/87.

Sq. **SO** 30,40,50,41,51. **ST** 28,48,19,29,49,59.

2078 *Nola confusalis* H.-S.   Least Black Arches
This moth is common and widespread in woodland in the eastern half of the county. It flies in May and June and at night comes to light.

Wye Valley 1906 (Bird 1907).

Usk, garden m.v.l. trap 24/5/69 and most years until 1983 but, in general, singly or in small numbers. Ed. 11/5/80. Ld. 15/7/71(2). Hendre Wds 26/5/70 (a few), 21/4/80. Rhyd-y-maen 6/6/70. Hael Wds 25/5/70 (abdt.), 21/5/73 (plfl.), 14/6/86. Wyndcliff 22/6/71. Redding's Inclosure 1/5/72 (plfl.), 21/4/81. Slade Wds 28/5/78. Wentwood 26/5/80. Llansoy 31/5/85. Trelleck Bog 28/5/87. Clydach 11/5/90.

Sq. **SO** 30,40,50,21,41,51. **ST** 48,49,59.

NOCTUIDAE
NOCTUINAE

2080 *Euxoa obelisca* D. & S.   Square-spot Dart
   ssp. *grisea* Tutt

This moth occurs locally on coastal cliffs chiefly in south-west England and in south and west Wales. As a probable vagrant, it has once been recorded in Monmouthshire.

   Usk 1974, one to garden m.v.l. trap 18/8. (GANH). (1st v-c. rec.). (Detn. confirmed by WHTT and JDB at BMNH).

Sq. **SO** 30.

2082 *Euxoa nigricans* Linn.   Garden Dart
Very scarce in Monmouthshire.

   Usk, garden m.v.l. trap, 21/8/72, 24/8/75, 11/8/78 ♀, 30/8/78 ♂ (GANH). (MEA) : Risca, to act.l. 6/8/84, 6/8/88.

Sq. **SO** 30. **ST** 29.

2084 *Agrotis cinerea* D. & S.   Light Feathered Rustic
North-west Gwent, in old quarries on the Carboniferous Limestone, 4/5/90, six to m.v.l. (MEA); 11/5/90 (15) (MEA, GANH). Males fairly pale in colour but the females all dark blackish-brown.

Sq. **SO** 21.

2087 *Agrotis segetum* D. & S.   Turnip Moth
This species, coming readily to light, frequents gardens and cultivated land, flying in May and June and as a second generation in August and September. Although plentiful some years, its numbers have very noticeably declined in the last two decades.

   Llandogo, "black variety" Oct 1892 (Nesbitt 1892b). Abertillery dist. 1911 (Rait-Smith 1912).

   Usk, m.v.l. trap, most years from 1966 to 1983. Ed. 23/5/71. Ld. 4/11/83 (4). Prescoed 8/6/69. Wentwood 19/9/69. Magor Marsh 23/9/69. Cwm Tyleri July 1977 (BG in litt.). Tal-y-coed 9/9/83. (MEA) : Risca 7/7/83, Coed Mawr 1/6/85, Ochrwyth 27/6/86, Cwm Tyleri 7/7/86, Cwm Merddog 14/7/86.

Sq. **SO** 10,20,30,31. **ST** 28,48,29,39,49.

2088 *Agrotis clavis* Hufn.   Heart and Club
Occurs at low density in Monmouthshire, flying during June and July and coming to light.

   Usk, garden m.v.l. trap, recorded most years from 1967 to 1984 but not more than three appeared in any one year eg. 1970, 26/6(1), 8/7(1), 28/7 (1).

1984, 3/6(1). Ed. 3/6/84. Ld. 28/7/70. Prescoed 5/7/68. Magor Marsh 8/8/76. Slade Wds 27/6/85. Llansoy 3/7/85. (All the foregoing records were of single moths). (MEA) : Risca 18/7/86, Cwm Tyleri 5/7/87.

Sq. **SO** 20,30,40. **ST** 29,39,48.

2089 *Agrotis exclamationis* Linn.   Heart and Dart
Widespread and often abundant, especially in gardens and on arable land. It flies from mid-May to August and comes readily to light.

East Monmouthshire (Bird 1905). Wye Valley 1906 (Bird 1907). Abertillery dist. (Rait-Smith 1912).

Usk, garden m.v.l. trap, every year from 1966 to 1982 and exceedingly numerous some years, especially 1968, 70, 76, 77 and 82; eg. 24/6/70 (1,700), 25/6/70 (390). Ed. 15/5/76, 12/5/82. Ld. 9/8/70, 10/10/71, 10/9/77.

Gwernesney, 23/6/71, one visiting flowers of hemlock water dropwort (*Oenanthe crocata*) by day (GANH). Also noted: 1970: Rhyd-y-maen, Magor Marsh, Hendre Wds, Hael Wds. 1971: Redding's Inclosure, Prescoed, Wyndcliff, Cwm-mawr. 1974: Cwm Tyleri, Mynyddislwyn. 1979: Trelleck Bog, Slade Wds. 1980: Wentwood. 1985: Llansoy. 1987: Kilpale. (MEA) : 1976: Allt-yr-yn, Llantarnam, Mescoed Mawr, Ynysyfro. 1977: Sugar Loaf. 1978: Tintern. 1980: Risca. 1984: Newport Docks, The Blorenge. 1985: Ynysddu, Cwm Merddog. 1986: Cwmfelinfach, Ochrwyth. 1987: Pontllanfraith, Trefil. 1988: Dixton Bank, Nant Gwyddon. 1990: Clydach.

Sq. **SO** 10,20,30,40,50,11,21,41,51. **ST** 28,38,48,19,29,39,49,59.

2091 *Agrotis ipsilon* Hufn.   Dark Sword Grass
An immigrant and, to quote Skinner, "transitory resident" coming readily to light. Its numbers, which are usually greater in September and October than in earlier months, vary a good deal from year to year.

Llandogo (Nesbitt 1892). Abertillery dist. (Rait-Smith 1912).

Llanbadoc, two on ivy blossom at night 7/10/68 (GANH). Usk, garden m.v.l. trap, recorded most years from 1966 to 83 and noted in every month from March to November. Very plentiful in the years 1966 to 1970 eg. 5/10/69 (25), 20/10/69 (18). Ed. 4/5/69, 23/3/77. Ld. 10/11/82, 9/11/83 (4).

Magor Marsh 2/8/70, 29/8/70. Gwehelog (Camp Wood) 28/8/70. Wentwood 29/8/70. Hael Wds 18/8/75 (2). Tal-y-coed 9/9/83. Llansoy 7/7/85, 17/10/85. Undy (Collister Pill) 18/8/87. Slade Wds 20/7/89. (MEA) : Wyndcliff 2/9/77, Risca 28/8/85, Trelleck Bog 29/8/86.

Sq. **SO** 30,40,50,31. **ST** 48,29,49,59.

2092 *Agrotis puta* Hb.   Shuttle-shaped Dart
        ssp. *puta* Hb.
Widespread and often abundant in Monmouthshire in cultivated areas, marshes and woodland. There are several broods during the year, flying from

late April to October and coming to light.

Usk, garden m.v.l. trap, noted annually from 1967 to 1982 being very numerous in 1975, 76, 77 and 82. eg. 1976: 29/4 – 1/11 (800+), with peak numbers on 13/8 (200+) and 14/8 (144). Ed. 1/5/72, 24/4/82. Ld. 26/10/75, 3/11/82 (3). Usk, 20/8/77, one ab. *nigra* Tutt to m.v.l. (GANH). Magor Marsh 2/6/70, 20/8/77. Rhyd-y-maen 6/6/70. Mynyddislwyn 10/6/74. Newport Docks 17/8/76. Slade Wds 28/8/79. Hendre Wds 16/9/79. Hael Wds 31/7/82. Wynd-cliff 5/8/82. Tal-y-coed 9/9/83. Llansoy 26/7/84. (MEA) : Risca 12/8/80, Mescoed Mawr 31/5/85, Ochrwyth 9/8/86, Cwm Tyleri 29/9/86, Abercarn 1987, Clydach 11/5/90.

Sq. **SO** 20,30,40,50,21,31,41. **ST** 28,38,48,19,29,49,59.

2098 *Axylia putris* Linn.   The Flame
Widespread and common in woodland, gardens and cultivated land, flying from late May until early August.

Usk, garden m.v.l. trap, frequent most years from 1966 to 82 but rarely more than one or two in any one night, (5) on 12/6/81 was unusual. Ed. 31/5/81, 21/5/82. Ld. 7/8/81, 13/8/79. Magor Marsh, seen most years eg. 2/6/70, plentiful on 8/8/76 and 3/7/79. Rhyd-y-maen 6/6/70. Slade Wds 1/6/71. Redding's Inclosure 7/6/71. Prescoed 5/7/68. Cwm-mawr 2/7/71. Hael Wds 3/7/71. Newport Docks 17/8/76. Hendre Wds 17/7/79. Llansoy 2/7/85. Kilpale 30/6/87. (MEA) : Ynysyfro 28/6/76, Tintern 30/6/78, Risca 7/7/81, Wyndcliff 17/7/83, Trelleck 19/7/83, Pant-yr-eos 6/7/85, Cwm Merddog 7/7/85, Ynysddu 12/7/85, Abercarn 16/7/85, Ochrwyth 12/7/86, Cwmfelinfach 15/7/86, Pontllanfraith 22/6/87, Cwmcarn 11/7/87, Dixton Bank 18/6/88, Nant Gwyddon 25/6/88.

Sq. **SO** 10,20,30,40,50,21,41,51. **ST** 28,38,48,19,29,39,49,59.

2102 *Ochropleura plecta* Linn.   Flame Shoulder
Widespread and common throughout Gwent. Double-brooded, flying from late April to mid-September and frequenting woodland, marshes, cultivated land and gardens.

Llandogo dist. "to light", 26/5/92 (Nesbitt 1892). Wye Valley 1906 (Bird 1907). Abertillery dist. (Rait-Smith 1912, 1913). Tintern 31/5/68 (JMC-H 1969).

Usk: common at m.v.l. most years from 1966 to 1982. Ed. 18/4/76. Ld. 28/9/70. Rhyd-y-maen 23/5/70. Magor Marsh 2/6/70. Undy 30/8/70. Gwehelog (Camp Wood) 28/8/70. Wentwood 3/6/70. Hael Wds 5/7/70. Hendre Wds 5/7/70. Slade Wds 1/6/71. Redding's Inclosure 7/6/71. Prescoed 21/6/71. Cwm-mawr 2/7/71. Mynyddislwyn 10/6/74. Cwm Tyleri 9/7/77. Cwm Coed-y-cerrig 5/6/78. Trelleck Bog 8/7/78. Tal-y-coed 9/9/83. Llansoy 26/7/84. Kilpale 30/6/87. (MEA) : Allt-yr-yn 20/6/76, Ynysyfro 22/6/76, Mescoed Mawr 25/6/76, Tintern 30/6/78, Risca 27/8/81, Black Vein 12/6/81,

Wyndcliff 20/7/83, Cwm Merddog 7/7/85, Ynysddu 12/7/85, Ochrwyth 12/7/86, Cwmfelinfach 5/7/86, Pontllanfraith 20/6/87, Trefil 3/7/87, Cwmcarn 11/7/87, Dixton Bank 18/6/88, Nant Gwyddon 25/6/88, Clydach 15/6/90.

Sq. **SO** 10,20,30,40,50,11,21,31,41,51,22. **ST** 28,38,48,19,29,39,49,59.

2104 *Standfussiana lucernea* Linn.   Northern Rustic
Scarce in Gwent and occurring mainly in the northern hills.
   Usk, to m.v.l. trap 14/7/69, ♂ one (GANH) (1st v-c. rec.). Cwm Tyleri to m v.l., 1972, 22/8 (1); 1973, 9/7 ♂ (1); 1974, 5/8 (2) (GANH): 1976, 8/7 (1) to m.v.l. (AR), 9/7 three by day from a clump of wood sage growing in a quarry (BG in litt.). (MEA) : Clydach 9/7/89, Trefil 31/8/90, Risca 21/8/91.

Sq. **SO** 20,30,11,21. **ST** 29.

2107 *Noctua pronuba* Linn.   Large Yellow Underwing
Virtually ubiquitous and often abundant, flying from June to October.
   Wye Valley 1906 (Bird 1907). Abertillery dist. (Rait-Smith 1912, 1913).
   Croesyceiliog 13/7/28 (GANH). Usk, garden m.v.l. trap, every year from 1966 to 1982, often numerous eg. 9/8/73 (many hundreds), 10/9/77 (121). Ed. 8/6/71. Ld. 30/10/78. 1970: Hendre Wds 5/7, Hael Wds 5/7, Rhyd-y-maen 18/7, Tintern 1/8, Magor Marsh 2/8, Wentwood 29/8, Undy 30/8. 1971: Redding's Inclosure 15/6, Cwm-mawr 2/7. 1974: Cwm Tyleri 29/7. 1979: Trelleck Bog 16/7, Slade Wds 23/7, Pont-y-saeson 18/9. 1981: Wyndcliff 6/7. 1983: Tal-y-coed 9/9. 1984: Llansoy 26/7. 1987: Kilpale 30/6. St Mary's Vale 8/7. (MEA) : 1976: Ynysyfro 22/6, Llantarnam 24/6, Mescoed Mawr 1/7. 1977: Sugar Loaf 6/7. 1979: Risca 9/7. 1982: Black Vein 15/7. 1985: Pant-yr-eos 6/7, Cwm Merddog 7/7, Ynysddu 12/7, Abercarn 16/7. 1986: Ochrwyth 27/6, Llanddewi (Sor Brook) 17/7, Cwmfelinfach 12/8. 1987: Pontllanfraith 20/6, Trefil 3/7, Cwmcarn 11/7.

Sq. **SO** 10,20,30,40,50,11,21,31,41,51. **ST** 28,38,48,19,29,39,49,59.

2109 *Noctua comes* Hb.   Lesser Yellow Underwing
Widespread and common, flying from mid-July to early October.
   Wye Valley 1906 (Bird 1907). Abertillery dist. (Rait-Smith 1912).
   Croesyceiliog, Aug. 1928, (plfl.) (GANH). Usk, m.v.l. trap, plentiful every year from 1966 to 1982, eg. 9/9/77(19), 10/9/77(31), 27/8/78(28). Ed. 22/7/70, 7/7/72. Ld. 25/9/70, 10/10/78, 14/10/79. Rhyd-y-maen 18/7/70, Magor Marsh 2/8/70. Undy 30/8/70. Wentwood 29/8/70. Redding's Inclosure 27/8/71. Cwm Tyleri 5/8/74. Hael Wds 30/8/77. Trelleck Bog 22/7/78. Hendre Wds 29/8/78. Slade Wds 28/7/82. Wyndcliff 5/8/82. Llansoy 22/9/84. Dixton Bank 18/6/88. (MEA) . Mescoed Mawr 29/8/79, Risca 7/9/79, Newport Docks 11/7/84, Ynysddu 12/7/85, Ochrwyth 12/7/86, Pontllanfraith

20/7/87, Cwmfelinfach 13/8/87, Trefil 15/8/87, Uskmouth 16/8/88, Nant Gwyddon 8/9/88, Clydach 26/7/90.

Sq. **SO** 20,30,40,50,11,21,41,51. **ST** 28,38,48,19,29,39,49,59.

2110   *Noctua fimbriata* Schreb.   Broad-bordered Yellow Underwing
This species, flying from July to September, is at low density and of sporadic appearance in Monmouthshire.

Wye Valley, one, 1906 (Bird 1907).

Usk, garden m.v.l. trap, a total of (24) was recorded in the seventeen years 1966 – 82, the highest number for a single year being (5) as in 1973 and 1978. 1966, July (2), 24/9 (1). 1971, single moths on July 29, 30 and Aug. 1, 10, 30. Wentwood, to m.v.l. 1969, several 2/8 – 27/8 (JMD); 28/7/80 (1) (CT). Hendre Wds 4/9/78 (2). Slade Wds 28/8/79. Tal-y-coed 9/9/83. Llansoy 11/8/84. Magor Marsh 24/7/90. Hael Wds 3/8/90. (MEA): Wyndcliff 12/7/84, Pant-yr-eos 6/7/85, Risca 18/9/85 and 4/9/86, Ochrwyth 9/8/86, Cwmfelinfach 5/9/86, Cwm Tyleri 5/7/87, Cwmcarn 11/7/87, Pontllanfraith 20/7/87.

Sq. **SO** 20,30,40,50,31,41 **ST** 28,48,19,29,49,59.

2111   *Noctua janthina* D. & S.   Lesser Broad-bordered Yellow
                                        Underwing
Common and widespread, flying in gardens, woodland etc. from late July to mid-September.

Wye Valley 1906, imagines, also larvae on honeysuckle (Bird 1907).

Usk, m.v.l. trap, recorded most years from 1966 to 82 and often abundant. Ed. 16/7/81. Ld. 17/9/79. Magor Marsh 2/8/70. Gwehelog (Camp Wood) 28/8/70. Wentwood 29/8/70. Redding's Inclosure 27/8/71. Cwm Tyleri 8/7/76. Hendre Wds 23/8/77. Slade Wds 21/8/79. Hael Wds 2/8/81. Wyndcliff 15/9/81. Tal-y-coed 9/9/83. Llansoy 11/8/84. Undy (Collister Pill) 18/8/87. (MEA) : Mescoed Mawr 29/8/79, Risca 7/9/79, Cwmfelinfach 12/8/86, Pontllanfraith 14/8/87, Cwm Coed-y-cerrig 17/8/87, Uskmouth 16/8/88, Nant Gwyddon 8/9/88.

Sq. **SO** 20,30,40,50,31,41,51,22. **ST** 38,48,19,29,49,59.

2112   *Noctua interjecta* Hb.   Least Yellow Underwing
        ssp. *caliginosa* Schaw.
Scarce and local in Monmouthshire.

Usk, garden m.v.l. trap, (39) recorded from 1966 to 1983 inc. (15) in 1973 (18/7 – 15/8). Ed. 2/7/71. Ld. 29/8/74. Magor Marsh: 1969 22/7 (1), 1970 (1), 1974 (1), 1976 (2), 1977 (2), 1978 (1). Newport Docks 17/8/76 (1). Magor Pill 1989 19/7, 6/8. (MEA) : Risca 17/8/85 and 1/9/86, Ochrwyth 9/8/86, Pontllanfraith 16/8/87.

Sq. **SO** 30. **ST** 28,38,48,19,29.

2113 *Spaelotis ravida* D. & S.   Stout Dart
Rare in Monmouthshire
    Risca, 1984, one to act.l. 12/8 (MEA) (1st v-c. rec.) and 1992 10/8 (1). 1990 Clydach, one to m.v.l. 1/10 (MEA). The only records.

Sq. **SO** 21. **ST** 29.

2114 *Graphiphora augur* Fabr.   Double Dart
This moth, local and scarce in Gwent, flies during June and July and comes to light.
    Abertillery dist. (Rait-Smith 1912, 1913).
    Usk, recorded sparingly at garden m.v.l. trap with a total of (20) over the eighteen years from 1966 to 1983 eg. 1967, 1/7 (1) and 4/7 (1); 1969, (7), 2/7 – 31/7. Wentwood 1968 (2) (JMD). Magor Marsh 1969 11/7(1), 22/7(1); 1970 10/7; 1971 (1); 1973 (4); 1975 24/6(1); 1981 (1). Llansoy 8/7/86 (1). (MEA) : Ynysyfro 4/7/76, Risca 8/7/83, Abercarn 16/7/85, Cwmfelinfach 15/7/86.

Sq. **SO** 20,30,40. **ST** 28,48,19,29,49.

2117 *Paradiarsia glareosa* Esp.   Autumnal Rustic
       ssp. *glareosa* Esp.
Widespread and common in Monmouthshire, flying in late August and September and coming readily to light. The pink ab. *rosea* Tutt occurs commonly and tends to be more frequent in the earlier part of the flight period.
    Llandogo 1892 (Nesbitt 1892b). Tintern dist. (Bird 1905). Abertillery dist. 1911 (Rait-Smith 1912).
    Usk, m.v.l. trap, noted every year from 1966 to 1983 and in some years plentiful. Ed. 28/7/71. Ld. 26/9/70. 1966, 4/9 (ab. *rosea*); 1968 12/9 (d°); 1969 31/8 (2) (inc. one ab. *rosea*), 4/9 (3) (inc. two ab. *rosea*), 6/9 (7) (inc. three ab. *rosea*). Wentwood 1968 (JMD). Usk (Park Wood) 3/9/70, numerous. Cwm-mawr 30/8/71. Cwm Tyleri 22/8/72. Trelleck Bog 18/9/72. Pont-y-saeson 7/9/73. Hael Wds 30/8/77. Hendre Wds 29/8/78. Magor Marsh 24/8/81. Slade Wds 9/9/82. Tal-y-coed 9/9/83. Llansoy 12/9/86 (ab. *rosea*). (MEA) : Mescoed Mawr 5/9/77 (2) (inc. ab. *rosea*), Wyndcliff 1/9/79, Risca 14/9/81, Cwmfelinfach 15/9/86, Ochrwyth 28/9/86, Pontllanfraith 29/8/87.

Sq. **SO** 20,30,40,50,31,41. **ST** 28,48,19,29,49,59.

2118 *Lycophotia porphyrea* D. & S.   True Lover's Knot
Common and widespread in Monmouthshire, flying from mid-June to mid-August and coming freely to light. In the central and eastern parts of the county it inhabits open heathy woods but is much more plentiful on the moors and hills of the north and west.

Llandogo 1892 (Nesbitt 1892b).

Usk, garden m.v.l. trap, occurred very sporadically eg. 1966, one only 17/7; 1969, frequent from 28/6 (1) to 20/7 (numerous); 1971 (5); 1972 to 1979 (none); 1978 (4); 1979 and 1981 frequent. Cwm-mawr 2/7/71. Cwm Tyleri abundant on 28/7/72, 58/8/74 etc. Trelleck Bog 16/6/73 (abdt.). Magor Marsh 3/7/73. Hael Wds 28/6/74. Redding's Inclosure 14/7/74. Hendre Wds 23/7/79. Slade Wds 28/7/81. Llansoy 13/7/86. Clydach 18/7/89. (MEA) : Sugar Loaf 7/7/77, Risca 13/7/83, Blorenge 18/7/83, Mescoed Mawr 28/7/84, Cwm Merddog 7/7/85, Ochrwyth 27/6/86, Cwmfelinfach 15/7/86, Pontllanfraith 22/6/87, Trefil 3/7/87, Cwmcarn 11/7/87, Nant Gwyddon 25/6/88.

Sq. **SO** 10,20,30,40,50,11,21,41,51. **ST** 28,48,19,29,49.

2119   *Peridroma saucia* Hb.   Pearly Underwing
An immigrant species with numbers varying widely from year to year.

Llandogo 1892 (Nesbitt 1892b). Tintern dist. (Bird 1905). Brockwells near Caerwent, 21/10/62 (CT in litt.). Cwmyoy Mar/Apr 1965 (SHR 1966).

Usk, garden m.v.l. trap, fairly common from 1966 to 1970, but none in the years 1975 to 81. Sometimes seen in May and June, but more plentiful from late August to October eg. 1968, frequent 4/10 – 24/10; 1969, 27/5 and later from 6/9 to 24/10; 1970, 7/9 – 31/10; 1973, 22/5(1), 7/9(1). Magor Marsh 29/8/70. Llansoy 13/10/86, 12/10/90.

Sq. **SO** 30,40,50,32. **ST** 48.

2120   *Diarsia mendica* Fabr.   Ingrailed Clay
         ssp. *mendica* Fabr
A very variable species, common and widespread throughout Gwent, flying from late May to mid-July with peak numbers usually in late June. It inhabits deciduous woodland and also the hills and moorlands in the north-west of the county.

Wye Valley, 1906, larvae (Bird 1907). Abertillery dist. 1911 and 1912 (Rait-Smith 1912, 1913). Bettws Newydd, 1935, "in abundance at light" 29/5 (RBH).

Usk, garden m.v.l. trap, frequent every year from 1966 to 1982. Ed. 17/5/76, 19/5/82. Ld. 8/7/71, 13/7/79. 1968, Usk (Park Wood) 4/6, Prescoed 9/6, Wentwood 8/7. 1970, Hael Wds 25/5, Hendre Wds 5/7. 1971, Slade Wds 1/6 (abdt.), Redding's Inclosure 7/6, Wyndcliff 22/6, Cwm-mawr 2/7. 1973, Cwm Coed-y-cerrig 5/6 (plfl.), Trelleck Bog 16/6, Magor Marsh 3/7. 1974, Cwm Tyleri 5/8 (abdt.). 1985, Llansoy 2/7. 1987, Kilpale 30/6. (MEA) : Llantarnam 10/6/76, Mescoed Mawr 25/6/76, Ynysyfro 25/6/76, Sugar Loaf 7/7/77, Tintern 30/6/78, Black Vein 3/7/81, Risca 7/7/81, Blorenge 7/7/84, Cwm Merddog 7/7/85, Ynysddu 12/7/85, Ochrwyth 12/7/86, Cwmfelinfach

15/7/86, Pontllanfraith 20/6/87, Abercarn 27/6/87, Trefil 3/7/87, St. Mary's Vale 8/7/87, Nant Gwyddon 25/6/88.

Sq. **SO** 10,20,30,40,50,11,21,41,51,22. **ST** 28,38,48,19,29,39,49,59.

2121 *Diarsia dahlii* Hb.   Barred Chestnut
Rare in Monmouthshire.

Wye Valley, one, 1906 (Bird 1907).

1972: Trelleck Bog, one to m.v.l. 18/9 (GANH). 1986: Trelleck Bog, 30/8 (1), 5/9 (2) (GANH, MEA).

Sq. **SO** 50.

2122 *Diarsia brunnea* D. & S.   Purple Clay
Widespread and fairly common in Monmouthshire flying in deciduous woods in June and July.

Wye Valley 1906 (Bird 1907) Abertillery dist. (Rait-Smith 1912, 1913).

Usk, garden m.v.l. trap, recorded annually from 1966 to 1982 but generally in small numbers, eg. 1982, 3/6 – 24/7. Prescoed 5/7/68. Wentwood 8/7/68 (plfl.), 26/7/74. Trelleck Common 9/7/68. Hendre Wds 5/7/70. Hael Wds 5/7/70, 27/6/78. Cwm Tyleri 9/7/73. Trelleck Bog 8/7/78. Pont-y-saeson 13/7/81. Slade Wds 19/6/85. Clydach 18/7/89. Llansoy 28/6/93 (MEA) : Ynysyfro 22/6/76, Black Vein 3/7/81, Risca 7/7/81, Blorenge 7/7/84, Mescoed Mawr 8/7/84, Cwmfelinfach 12/8/86, Abercarn 27/6/87, St. Mary's Vale 8/7/87, Pontllanfraith 10/7/87, Wyndcliff 12/7/87, Ochrwyth 17/6/88, Nant Gwyddon 25/6/88.

Sq. **SO** 20,30,40,50,21,41. **ST** 28,48,19,29,39,49,59.

2123 *Diarsia rubi* View.   Small Square-spot
Double-brooded, and noted from May to September, this moth is widespread and common in Monmouthshire in both wooded and cultivated areas. It was very plentiful in the county during the late 1960s but my records suggest its numbers have greatly declined since then.

Llandogo dist. "to light" 1892 (Nesbitt 1892). Wye Valley 1906 (Bird 1907). Abertillery dist. (Rait-Smith 1912, 1913).

Usk: garden m.v.l. trap, recorded every year from 1966 to 1982 but over the years its numbers have diminished. Ed. 11/5/82. Ld. 18/9/71, 11/10/82. Prescoed 9/6/68. Llangwm 24/6/68. Magor Marsh 23/8/68, a late moth on 13/10/78. Ravensnest Wood 25/8/68. Bica Common 27/8/68. Wyndcliff 2/6/69. Wentwood 21/6/69. Hael Wds 26/5/70. Rhyd-y-maen 6/6/70. Gwehelog (Camp Wood) 28/8/70. Undy (Collister Pill) 30/8/70. Trelleck Bog 18/9/72. Cwm Tyleri 14/6/74. Mynyddislwyn 10/6/74 (abdt.). Redding's Inclosure 17/6/74. Newport Docks 17/8/76. Hendre Wds 27/5/79. Slade Wds 21/8/79. Llansoy 3/7/85. (MEA) : Risca 3/6/80, Mescoed Mawr

14/8/81, Ochrwyth 1/6/85, Pontllanfraith 15/7/88, Uskmouth 16/8/88.

Sq. **SO** 20,30,40,50,41,51. **ST** 28,38,48,19,29,39,49,59.

2126   *Xestia c-nigrum* Linn.   Setaceous Hebrew Character
Flies in gardens, woodland and marshy areas from May to October, being especially plentiful from August onwards. Although it still remains fairly common, this is another species which in Monmouthshire has suffered a great reduction in numbers since its extreme abundance in the late 1960s.

Llandogo, 1892. (Nesbitt 1892b). Usk Castle 1934, "at sugar", 23/8 (1), 25/8 (1). (RBH). Cwmyoy Oct/Nov 1965 (SHR 1966).

Usk: garden m.v.l. trap, every year from 1966 to 1983. eg. 1970, 27/5 to 27/6 and later, from 5/8 to 26/10 with huge numbers on 27/8. 1971, 28/5 to 8/7, then later, apart from a single moth on 29/7, from 17/8 to 24/10 with peak numbers in late August and mid-September eg. 15/9 (103), and finally, one on 1/11. Ed. 27/5/70, 24/5/82. Ld. 26/10/70, 6/11/78. Prescoed 8/6/69. Rhyd-y-maen 6/6/70. Gwehelog (Camp Wood) 28/8/70. Magor Marsh 29/8/70, 24/8/81. Wentwood 29/8/70. Undy 30/8/70. Redding's Inclosure 15/6/71. Wyndcliff 24/6/71. Cwm Tyleri 29/6/76. Newport Docks 17/8/76. Slade Wds 28/7/82, 16/9/82. Tal-y-coed 9/9/83. Llansoy 2/7/85, 19/10/85. (MEA) : Risca 12/8/80, Cwmfelinfach 26/9/86, Ochrwyth 28/9/86, Uskmouth 16/8/88.

Sq. **SO** 20,30,40,50,31,51,32. **ST** 28,38,48,19,29,39,49,59.

2127   *Xestia ditrapezium* D. & S.   Triple-spotted Clay
Thinly distributed but not uncommon in Monmouthshire, inhabiting deciduous woodland, parkland and also the scantily-wooded hillsides in the north-west of the county, where it occurs more frequently.

Abertillery dist. one, 1910 (Rait-Smith 1912).

Usk: garden m.v.l. trap, recorded in small numbers most years from 1967 to 1983, the highest count in one year being (10) in 1983 eg. 1967, (1) 24/6; 1969, (6) 2/7-19/7; 1970, (8) 9/7-3/8; 1975, (3) 10/7; 1981, (9) 1/7 – 4/8. Ed. 24/6/67. Ld. 18/8/79. Usk (Park Wood) 7/7/69. Magor Marsh 22/7/69 (1), 4/8/75. Cwm-mawr 2/7/71. Redding's Inclosure 20/7/71. Wentwood 1971 (JMD), 26/7/74. Cwm Tyleri: 1972 7/8, 28/8; 1973 8/8 (4); 1974 5/8 melanic f. (det. W.T.Tams); 1975 13/7; 1976 29/6, 8/7. Piercefield Park 18/7/73. Hael Wds 12/7/74, 2/8/81 (2). Trelleck Bog 22/7/78. Wyndcliff 3/8/81. Llansoy 26/7/84. (MEA) : Risca 7/7/83, Ynysddu 12/7/85, Cwm Merddog 14/7/86, Cwmfelinfach 12/8/86, Abercarn 27/6/87, Pontllanfraith 2/7/87, St Mary's Vale 8/7/87, Cwmcarn 11/7/87.

Sq. **SO** 10,20,30,40,50,51. **ST** 38,48,19,29,49,59.

2128   *Xestia triangulum* Hufn.   Double Square-spot
Widespread and common in Monmouthshire, flying in deciduous woods and

parkland from late June to mid-August.

    Tintern 1912, larva, bred. (Bird 1913).

    Usk, garden m.v.l. trap, 12/7/66 and recorded annually until 1982 and often plentiful eg. 1969, 24/6 – 3/8 with (10) on 10/7 and 12/7; 1978, 29/6 – 5/8. Ed. 21/6/76, 9/6/82. Ld. 13/8/77, 15/8/79. Wentwood 8/7/68. Rhyd-y-maen 18/7/70. Tintern 1/8/70. Hael Wds 3/7/71. St Pierre's Great Wds 13/7/73. Piercefield Park 18/7/73. Wyndcliff 2/7/74. Magor Marsh 16/7/76. Cwm Tyleri 27/6/76. Trelleck Bog 16/7/79. Slade Wds 28/7/82. Llansoy 9/7/86. Kilpale 30/6/87. Clydach 18/7/89. (MEA) : Mescoed Mawr 1/7/76, Risca 7/7/81, Ynysddu 12/7/85, Abercarn 16/7/85, Ochrwyth 12/7/86, Cwm Merddog 14/7/86, Cwmfelinfach 12/8/86, St Mary's Vale 8/7/87, Pontllanfraith 16/8/87, Dixton Bank 18/6/88.

Sq. **SO** 10,20,30,40,50,21,51. **ST** 28,38,48.19,29,49,59.

2130   *Xestia baja* D. & S.   Dotted Clay
Local and infrequent in Monmouthshire, flying in woodland during the second half of July and in August. Recent records suggest that it may be more plentiful and widespread in the hills and moorland of the western half of the county.

    Abertillery dist. 1912 (Rait-Smith 1913).

    Usk, garden m.v.l. trap, 1967 (8) 5/8 – 18/8, and occurring irregularly and in small numbers to 1983, eg. 1968, (14) 8/8 – 28/8; 1970, (8) 8/7 – 3/8. Ed. 8/7/70. Ld. 28/8/78. Wentwood 10/6/68 (1). Bica Common 23/8/68 (3). Cwm Tyleri, to m.v.l., 1972 28/7; 1983 16/7, 7/8, 8/8; 1974 5/8 (plentiful to m.v.l.) and several settled in the moonlight on the flowers of the moorland rushes (*Juncus* spp.). Trelleck Bog 16/7/86. (MEA) : Mescoed Mawr 3/7/76, Risca 25/7/84 (2), Ochrwyth 9/8/86, Cwmfelinfach 12/8/86, Pontllanfraith 10/7/87, Nant Gwyddon 15/8/88.

Sq. **SO** 20,30,40,50. **ST** 28,19,29,49.

2132   *Xestia castanea* Esp.   Neglected Rustic
Inhabits moorlands and heaths. Scarce in Monmouthshire.

    1973, Cwm Tyleri, one to m.v.l. 20/8 (GANH) (1st v-c. rec.); Blaenavon Mountain 25/8 (1) (GANH). Cwm Tyleri 15/9/86 (10) to m.v.l., 29/9/86 (GANH, MEA). (MEA) : Trelleck Bog 6/9/86 (1), Trefil 15/8/87 and 31/8/90, Pontllanfraith 1/9/87 (1).

Sq. **SO** 20,50,11. **ST** 19.

2133   *Xestia sexstrigata* Haw.   Six-striped Rustic
Local in Monmouthshire, occurring sparingly in gardens and damp situations from late June to September.

    Usk, garden m.v.l. trap, noted in small numbers every year from 1966 to 81 eg. 1966, (2); 1969, (9) 23/6 – 29/8; 1970, (12) 6/8 – 27/8 inc. (5) on

11/8; 1980, (1) 25/8. Wentwood. Magor Marsh 23/8/68, 24/8/81. Newport Docks 17/8/76. Undy (Collister Pill) 18/8/87. Llansoy 22/8/90, 26/8/90. (MEA) : Risca 25/9/86, Cwmfelinfach 13/8/87, Pontllanfraith 14/8/87, Uskmouth 16/8/88.

Sq. **SO** 30,40. **ST** 38,48,19,29,49.

2134 *Xestia xanthographa* D. & S. Square-spot Rustic
Widespread and common, flying from mid-August to early October in open woods, gardens and weedy places.

Llandogo 1892 (Nesbitt 1892b).

Usk, garden m.v.l. trap, recorded annually from 1967 to 1982. eg. 1971, plentiful, 17/8 – 22/9; 1978, (66) 20/8 – 23/9; 1982, (34) 23/8 – 19/9. Ed. 12/8/76. Ld. 24/9/70. Gwehelog (Camp Wood) 28/8/70. Magor Marsh 29/8/70. Undy (Collister Pill), sea-wall, 30/8/70, 18/8/87. Redding's Inclosure 27/8/71. Cwm-mawr 30/8/71. Cwm Tyleri 14/8/73. Trelleck Bog 4/9/73. Wyndcliff 6/9/75. Newport Docks 17/8/76. Hendre Wds 29/8/78. Slade Wds 9/9/82. Hael Wds 18/9/82. Tal-y-coed 9/9/83. Llansoy 1986, 28/8 – 10/10. (MEA) : Mescoed Mawr 5/9/77, Wentwood 8/9/77, Risca 7/9/79, Ochrwyth 28/9/86, Cwmfelinfach 12/8/86, Pontllanfraith 14/8/87, Uskmouth 16/8/88.

Sq. **SO** 20,30,40,50,31,41,51. **ST** 28,38,48,19,29,39,49,59.

2135 *Xestia agathina* Dup. Heath Rustic
ssp. *agathina* Dup.
Several recent Monmouthshire records viz. Llansoy, 4/9/86 one to u.v.l. (GANH) (1st v-c. rec.). Risca, to act.l. 28/9/86 (1), 28/8/87 (2) (MEA).

Sq. **SO** 40. **ST** 29.

2136 *Naenia typica* Linn. The Gothic
Occurs during July and August but local and scarce in Gwent.

Abertillery dist., "series bred 1911 from ova collected with dock leaves in 1910" (Rait-Smith 1912).

Usk 1970, one in outhouse 21/7. Usk, garden m.v.l. trap, 1966 – 83 (all records): 1967 4/7; 1971 20/8; 1972 22/7; 1973 18/7; 1977 9/8; 1982 10/7, 17/7, 2/8; 1983 26/7, 9/8. (all singletons). Magor Marsh 8/8/76 (1), 11/7/78 (1). Magor Pill 19/7/89 (1). (MEA) : Risca, to act.l. 14/7/79, 4/7/84, 5/7/86.

Sq. **SO** 20,30. **ST** 29,48.

2137 *Eurois occulta* Linn. Great Brocade
An immigrant species twice recorded in Monmouthshire.

Cwm Tyleri, 15/7/77, a "large pale" specimen to m.v.l. (BG in litt.).

Usk, one ♂ to garden m.v.l. trap 3/8/82 (GANH).

Sq. **SO** 20,30.

2138 *Anaplectoides prasina* D. & S.   Green Arches
Widespread and fairly frequent in deciduous woods throughout the county, flying from late June to early August.

Wye Valley 1892 (Nesbitt 1892a). Tintern, "one rather dark specimen at light in 1948" (Blathwayt 1967).

Usk, garden m.v.l. trap, a few most years from 1966 to 1983, eg. 1969 (6), 23/6 – 23/7; 1970 (6), 24/6 – 3/7; 1977 (1), 3/6; 1979 (9), 14/7 – 28/7. Usk (Park Wood) 15/6/67. Pont-y-saeson 13/6/67. Wentwood 22/7/68. Hael Wds 5/7/70. Wyndcliff 12/6/71. Cwm-mawr 2/7/71. Redding's Inclosure 20/7/71. Magor Marsh 1/7/75. Slade Wds 30/6/84. Hendre Wds 11/7/86. Trelleck Bog 18/7/86. Kilpale 30/6/87. (MEA) : Ynysddu 12/7/85, Abercarn 16/7/85, Risca 11/7/86, Ochrwyth 12/7/86, Cwm Merddog 14/7/86, Cwmfelinfach 12/8/86, Cwmcarn 11/7/87, Nant Gwyddon 25/6/88.

Sq. **SO** 10,20,30,50,41,51. **ST** 28,48,19,29,49,59.

2139 *Cerastis rubricosa* D. & S.   Red Chestnut
Common and widespread in deciduous woodland, flying from early March to mid-May and readily attracted to light.

Abertillery dist. 1911 and 1912 (Rait-Smith 1912, 1913).

Usk, garden m.v.l. trap, recorded annually from 1967 – 83 eg. 1971 (22), 12/3 – 19/5; 1977 (12), 4/3 – 24/4. Ed.1/3/80. Ld. 29/5/68. Usk (Park Wood) 26/4/68. Redding's Inclosure 23/3/73. Pont-y-saeson 3/4/72. St Pierre's Great Wds 23/3/73. Slade Wds 24/3/73 (abdt.). Cwm Tyleri 26/5/73. Wyndcliff 22/3/74. Hael Wds 6/4/74. Hendre Wds 23/4/79. Llansoy 10/5/85. (MEA) : Wentwood 1/3/77, Llangybi 25/3/77, Risca 17/4/85, Mescoed Mawr 19/4/85, Cwmfelinfach 11/4/87, Ochrwyth 24/4/87, Cwm Merddog 26/4/87, Abercarn 3/5/87, Pontllanfraith 6/5/88, Clydach 11/5/90.

Sq. **SO** 10,20,30,40,50,21,41,51. **ST** 28,48,19,29,39,49,59.

2140 *Cerastis leucographa* D. & S.   White-marked
This moth, flying in deciduous woods from late March until the end of May, is local and occurs at low density mainly in the eastern half of the county.

Wye Valley (near Bigsweir), "on sallows" 31/3 1907 (Barraud 1907b).

Usk, garden m.v.l. trap, a few most years from 1967 to 1982, eg. 1967 30/3 (1); 1968 1/5 (1); 1969 9/4 (1), 10/4 (1), 12/5 (1); 1972 (6) 30/3 – 23/4; 1977 21/5 (1), 2/6 (3), 3/6 (2). Usk (Park Wood) 28/4/67. Wentwood 4/5/70 (1), 1/4/71 (1) (JMD). Redding's Inclosure 11/4/72, 6/4/81. Pont-y-saeson 16/4/73, 24/4/79. St Pierre's Great Wds 28/3/73. Hendre Wds 23/4/79, 9/5/87

(6). Llansoy 7/5/85. Wyndcliff 27/4/87. (MEA) : Ochrwyth 24/4/87.

Sq. **SO** 30,40,41,51. **ST** 28,49,59.

HADENINAE

2142 *Anarta myrtilli* Linn.   Beautiful Yellow Underwing
To date there are only two Monmouthshire records for this heath and moorland species viz. Trelleck Bog, larvae (12), 12/7/70 (MWH in litt.). Trefil, imago (1), 7/8/88 (MEA).

Sq. **SO** 50,11.

2145 *Dicestra trifolii* Hufn.   The Nutmeg
Scarce in Monmouthshire.

   Abertillery dist., one, 1911 (Rait-Smith 1912).

   Magor Marsh, 1976, one to m.v.l. 8/8 and Newport Docks 17/8 (2) to m.v.l. (GANH). Risca, to act.l. (1) 29/7/84 (MEA). 1989, Magor Pill 19/7 (svl.) and Slade Woods 8/8 (1) (MEA, GANH).

   Sq. **SO** 20. **ST** 29,38,48.

HADENINAE

2147 *Hada nana* Hufn.   The Shears
This species, flying from mid-May to early July, occurs at low density in Monmouthshire.

   Llandogo dist. 13/5 1892 (Nesbitt 1892). East Monmouthshire (Bird 1905). Wye Valley 1906 (Bird 1907). Abertillery dist. (Rait-Smith 1912, 1913). Bettws Newydd 31/5/35 (RBH). Tintern 1946-48, "fairly common in all three years". (Blathwayt 1967).

   Usk, garden m.v.l. trap, 7/6/66 (1) and a few most years to 1983 inc. 1977 (10) 30/5 – 6/7. Ed. 24/5/75, 19/5/82. Ld. 6/7/72 (2), 9/7/79. Cwm Tyleri 1974, 31/5, 14/6; 1976, 26/6. Hael Wds 27/6/78. Llansoy 2/7/85. Wyndcliff 23/6/86. Dinham 4/6/87 (3) to m.v.l. Clydach 11/5/90. (MEA) : Risca, 17/6/79, 2/6/80, 1982, 84, 85, 86; Pant-yr-eos 6/7/85; Cwm Merddog 7/7/85; Ochrwyth 27/6/86; Cwmfelinfach 10/6/87.

Sq. **SO** 10,20,30,40,50,21. **ST** 28,19,29,49,59.

2149 *Polia trimaculosa* Esp.   Silvery Arches
   = *hepatica* auctt.
Abertillery dist., three, 1911 (Rait-Smith 1912). Abertillery dist., 1912, scarce (Rait-Smith 1913). The only Monmouthshire records.

Sq. **SO** 20.

2150   *Polia nebulosa* Hufn.   Grey Arches

Fairly common in Monmouthshire, flying in deciduous woods in the second half of June and throughout July.

Wye Valley 1906, (Bird 1907). Tintern 1912, bred (Bird 1913). Abertillery dist. 1912 (Rait-Smith 1913).

Usk, garden m.v.l. trap, 26/6/67 and occurring in small numbers most years until 1983. eg. 1968 (8) 29/6 – 13/7. Ed. 20/6/69 (2), 15/6/76. Ld. 28/7/79, 7/8/81. Wyndcliff 14/6/67. Wentwood 8/7/68 (plfl.). Prescoed 5/7/68 (5) to m.v.l. Usk (Park Wood) 7/7/69. Hael Wds 5/7/70, 3/7/83. Cwm-mawr 2/7/71. Pont-y-saeson 18/7/72. Magor Marsh 3/7/73. St Pierre's Great Wds 13/7/73. Redding's Inclosure 17/6/74. Cwm Tyleri 8/7/76. Hendre Wds 17/7/79. Slade Wds 28/7/81. Llansoy 11/7/86. Trelleck Dog 24/7/86. St Mary's Vale 8/7/87. (MEA) : Mescoed Mawr 13/7/82, Risca 7/7/83, Pant-yr-eos 6/7/85, Ynysddu 12/7/85, Abercarn 16/7/85, Sor Brook 17/7/86, Cwmfelinfach 15/7/86, Ochrwyth 17/6/88.

Sq. **SO** 20,30,40,50,31,41,51. **ST** 28,48,19,29,39,49,59.

2153   *Heliophobus reticulata* Goeze   Bordered Gothic
         ssp. *marginosa* Haw.

Llandogo, 1892 (Nesbitt 1892b). The only Monmouthshire record.

Sq. **SO** 50.

2154   *Mamestra brassicae* Linn.   Cabbage Moth

Common in Monmouthshire, especially in gardens and cultivated ground, flying in June and July and again in September.

Abertillery dist. 1911, 1912 (Rait-Smith 1912,1913).

Usk, garden m.v.l. trap, recorded most years from 1966 to 84 and some years plentiful eg. 1979 when it was noted from June 17 to Sept. 17 being most frequent in July and with a single very late specimen on Oct. 2nd. Ed. 16/5/78, 26/5/82. Ld. 2/10/79, 22/9/84. Wentwood 3/7/68 (JMD). Newport Docks 17/8/76. Magor Marsh 20/8/77. (MEA) : Risca 6/7/83, Wyndcliff 17/7/83, Cwm Tyleri 28/6/86, Nant Gwyddon 10/6/88.

Sq. **SO** 20,30. **ST** 38,48,29,49,59

2155   *Melanchra persicariae* Linn.   Dot Moth

Common in gardens, waste ground and woodland during June and July.

Usk, to light, 1958 and 1965. Usk, garden m.v.l. trap, recorded most years from 1966 to 1983 and sometimes abundant as on 20/7/69. Ed. 6/6/71, 2/6/80. Ld. 6/8/77, 7/8/81. Pont-y-saeson, larvae on sallow Sept. 1966 and 1/10/69 (reared). Prescoed 5/7/68. Tintern 1/8/70, imago visiting bramble flowers in afternoon sunshine. Wentwood 8/7/68 to m.v.l. Usk (Park Wood) 9/6/69. Hael Wds 5/7/70. Magor Marsh 10/7/70, 25/9/77. Cwm-mawr 2/7/71. Redding's Inclosure 20/7/71. Trelleck Bog 16/7/79. Hendre Wds 17/7/79.

Slade Wds 23/7/79. Wyndcliff 3/8/81. Llansoy 24/7/86. (MEA) : Risca 7/7/81, Ynysddu 12/7/85, Cwmfelinfach 12/8/86, Abercarn 27/6/87, Dinham 7/7/87, Pontllanfraith 10/7/87, Cwmcarn 11/7/87, Ochrwyth 17/6/88.

Sq. SO 20,30,40,50,41,51. ST 28,48,19,29,39,49,59.

2156 *Lacanobia contigua* D. & S.   Beautiful Brocade
Flying from late May to July in deciduous woodland, this species occurs only sparingly in the eastern half of the county but is more plentiful in the hilly districts of western Monmouthshire.

Abertillery dist. 1911, 1912 (Rait-Smith 1912, 1913).

Usk Castle, to m.v.l., June 1966 (1), 26/5/70 (1) (RBH). Pont-y-saeson, to m.v.l., 13/6/67 (2). Hael Woods 10/7/73 (1). Trelleck Bog 29/6/88. Dixton Bank 18/6/88 (2). Llansoy 26/8/90 (3). (MEA) : Ochrwyth 27/6/86 (10+) to m.v.l., Cwmfelinfach 23/5/87 (1), Pontllanfraith 22/6/87 (plfl.), Risca 25/6/87, Nant Gwyddon 25/6/88.

Sq. SO 20,30,40,50,51. ST 28,19,29,59.

2157 *Lacanobia w-latinum* Hufn.   Light Brocade
This moth, flying from mid-May through June, is found mainly in the central and eastern parts of the county, but is local and at low density, except in the vicinity of Usk, where it occurs with greater frequency.

Abertillery dist. 1912 (Rait-Smith 1913). Bettws Newydd, 31/5/35 "at light" (RBH in litt.).

Usk, garden m.v.l. trap, 7/6/66 and recorded most years until 1982, sometimes in fair numbers eg. 1977 (24) 26/5 – 19/6; 1979 (21) 29/5 – 26/6. Ed. 12/5/71, 17/5/76. Ld. 1/7/71, 26/6/79. Pont-y-saeson 13/6/67 (GANH). Wyndcliff 14/6/67 (REMP). Prescoed 31/5/68. Usk (Park Wood) 1968. Wentwood 1971 (JMD). Llansoy 1985 4/6(1), 5/6(9). Dinham 4/6/87, several to m.v.l. (MEA) : Risca 29/6/87, to act.l., Cwmfelinfach 8/5/88, Clydach 11/5/90 (5).

Sq. SO 20,30,40,21. ST 19,29,39,49,59.

2158 *Lacanobia thalassina* Hufn.   Pale-shouldered Brocade
A woodland species widespread and common in Monmouthshire, flying from mid-May to mid-July.

Wye Valley 1906 (Bird 1907). Abertillery dist. 1911, 1912 (Rait-Smith 1912,1913).

Usk, garden m.v.l. trap, recorded annually 1966 – 1983. Ed. 13/5/76, 19/5/80. Ld. 11/7/71, 15/7/79, and a very late record 29/8/82, probably representing a second generation. Pont-y-saeson 13/6/67. Wyndcliff 14/6/67 (REMP). Usk (Park Wood) 15/6/67. Wentwood 19/6/68. Prescoed 8/6/69. Hael Wds 25/5/70. Hendre Wds 26/5/70. Redding's Inclosure 7/6/71. Magor Marsh 6/7/71. Mynyddislwyn 10/6/74. Trelleck Bog 16/7/79. Llansoy 4/7/85.

Dinham 4/6/87. (MEA) : Llantarnam 10/6/76, Risca 16/6/83, Ochrwyth 27/6/86, Abercarn 16/5/87, Cwmfelinfach 23/5/87, Nant Gwyddon 25/6/88.

Sq. **SO** 20,30,40,50,41,51. **ST** 28,48,19,29,39,49,59.

2159 *Lacanobia suasa* D. & S.   Dog's Tooth
A mainly coastal species recorded sparingly in Monmouthshire from small areas of salt-marsh on the Severn Estuary littoral and from two inland localities.

Usk, one, to garden m.v.l. trap 25/6/73 (GANH). (1st v-c. rec.). Undy (Collister Pill), sea-wall, 13/7/73 (1) to m.v.l., 18/8/87 (1) (GANH). Uskmouth 16/8/88 (4), Magor Pill (sea wall), 19/7/89 (plfl); Slade Wds 20/7/89 (1); Magor Marsh 12/5/90 (1) (MEA, GANH). (MEA) : Risca 5/7/85 (1) to act.l.

Sq. **SO** 30. **ST** 38,48,29.

2160 *Lacanobia oleracea* Linn.   Bright-line Brown-eye
This moth, flying from late May through June and July and sometimes in August, is widespread and common in Monmouthshire though its numbers have markedly declined over the last twenty years.

Abertillery dist. 1912 (Rait-Smith 1913).

Usk, garden m.v.l. trap, recorded most years from 1966 to 1982. Ed. 21/5/70, 8/5/71. Ld. 21/8/71, 4/8/81. Prescoed 5/7/68. Wentwood 8/7/68. Usk (Park Wood) 23/7/68. Rhyd-y-maen 23/5/70. Magor Marsh 10/7/70. Tintern 1/8/70. Cwm-mawr 2/7/71. Cwm Tyleri 28/8/72 (abdt.). Cwm Coed-y-cerrig 5/6/73. Mynyddislwyn 10/6/74. Newport (Coldra) 27/6/75. Redding's Inclosure 23/6/81. Llansoy 6/7/85. (MEA) : Mescoed Mawr 1/7/76, Sugar Loaf 7/7/77, Risca 7/7/81, Wyndcliff 20/7/83, Blorenge 7/7/84, Pant-yr-eos 6/7/85, Cwm Merddog 7/7/85, Ynysddu 12/7/85, Abercarn 16/7/85, Ochrwyth 27/6/86, Cwmfelinfach 12/8/86, Pontllanfraith 10/7/87, Dixton Bank 18/6/88.

Sq. **SO** 10,20,30,40,50,21,51,22. **ST** 28,38,18,19,29,39,49,59

2162 *Papestra biren* Goeze   Glaucous Shears
Flying in May, June and early July, this species occurs locally in central and southern Monmouthshire but is more frequent in the hilly areas and moorland of the north and west of the county.

Usk, garden m.v.l. trap, 7/6/66 (2). (GANH) (1st v-c. rec.), and sporadically from 1966 to 1982 during which period at Usk a total of (34) was recorded at m.v.l. (GANH, RBH), inc. (21) in 1982 viz. May 14 (1), 15 (11), 18 (1), 19 (5), 20 (3). Ed 24/5/70, 14/5/82. Ld. 7/6/66, 4/6/77 (3). Magor Marsh 2/6/70. Wentwood 1970, 27/5(2), 30/5(1), 2/6(1) (JMD). Cwm Tyleri 26/5/73 (11) to m.v.l., also 1974, 76, 77 and 86. Llansoy 5/6/85, 5/5/90 (2). Clydach 11/5/90. (MEA) : Ochrwyth 27/6/86, Cwmfelinfach 29/5/87,

Nant Gwyddon 7/6/88.

Sq. **SO** 20,30,40,21. **ST** 28,48,19,29,49.

2163   *Ceramica pisi* Linn.   Broom Moth

This moth, flying in June and July, is widespread and fairly common in Gwent. It comes to light.

   Llandogo dist., "to light" 26/5 1892 (Nesbitt 1892). Abertillery dist. 1912 (Rait-Smith 1913).

   Usk, garden m.v.l. trap, 7/6/66 and a few most years until 1983. Ed. 2/6/70, 31/5/82. Ld. 26/7/79, 10/7/82. Wentwood 8/7/68. Rhyd-y-maen 6/6/70. Magor Marsh 10/7/70. Cwm-mawr 2/7/71. Cwm Tyleri 28/7/72 (also 1973, 74, 76, 77, 85, 86). Gray Hill, larvae on broom 17/8/73 (CT). Hael Wds 28/6/74. Llansoy 9/7/86. Trelleck Bog 25/5/87. (MEA) : Pant-yr-eos 6/7/85, Ynysddu 12/7/85, Ochrwyth 27/6/86, Cwm Merddog 20/6/86, Cwmfelinfach 15/7/86, Trefil 3/7/87, Pontllanfraith 10/7/87, Nant Gwyddon 25/6/88.

Sq. **SO** 10,20,30,40,50,11. **ST** 28,48,19,29,49,59.

2164   *Hecatera bicolorata* Hufn.   Broad-barred White

Flying from late May to August, this species is very scarce in the eastern half of Monmouthshire but occurs more frequently in the west.

   Bettws Newydd (Crowther 1935).

   Usk, garden m.v.l. trap, 12/7/81, 29/6/83. Slade Wds, 8/7/83 (GANH). (MEA) : Mescoed Mawr 9/7/79 to m.v.l.: Risca (to act.l.) 1983 16/6, 11/8; 1984 6/7, 5/8; 1985 3/7, 4/7; 1986 6/6, 18/6, 26/6, 11/7: Cwmfelinfach 24/5/87 (3) to m.v.l., 8/5/88: Pontllanfraith 10/7/87.

Sq. **SO** 30. **ST** 48,19,29.

2166   *Hadena rivularis* Fabr.   The Campion

Very scarce in Monmouthshire occurring sporadically from late May to mid-August.

   Usk, garden m.v.l. trap, one or two most years from 1966 to 1983 with three recorded in 1982. Ed. 27/5/70, 5/6/82. Ld. 14/8/67, 11/8/83. Magor Marsh 11/8/71, 3/8/76, 6/8/76. Wentwood 1971 (JMD). (MEA) : Risca, to act.l. 18/6/84, 5/6/85, 12/6/87.

Sq. **SO** 30. **ST** 48,29,49.

2173   *Hadena bicruris* Hufn.   The Lychnis

Widespread and fairly common in Monmouthshire flying from the middle of May to late August.

   Wye Valley, larvae on sweet William 1906 (Bird 1907). Abertillery dist. 1911 (Rait-Smith 1912).

Usk, garden m.v.l. trap, recorded most years from 1966 to 1983. Ed. 12/5/70, 14/5/80. Ld. 28/8/70, 20/8/79. Pont-y-saeson 13/6/67 (REMP). Trelleck Common 9/7/68. Magor Marsh 2/6/70. Rhyd-y-maen 6/6/70. Redding's Inclosure 1971, 7/6, 20/7. Wentwood 1971 (JMD). Hael Woods 26/6/79. Slade Woods 21/6/80. Llansoy 3/7/85. (MEA) : Llantarnam 10/6/76, Wyndcliff 26/5/77, Risca 19/5/82, Pant-yr-eos 16/6/83.

Sq. **SO** 20,30,40,50,51. **ST** 48,29,39,49,59.

2175 *Eriopygodes imbecilla* Fabr. The Silurian

A Palaearctic species, widely distributed as an alpine moth in the mountains of Europe, but unknown in the British Isles until 1972 when I took one at m.v.l. in the hills of north-west Monmouthshire at an altitude of 1,400 ft. Mr. D.S. Fletcher of the BMNH kindly identified the moth and I named it "The Silurian" as it is undoubtedly a relict species found in the region once occupied by the Silures. Despite many subsequent searches I did not see this elusive moth again until 1976 when I was able to confirm its status as a resident British species. It flies in June and July. There is a diurnal flight on hot sunny afternoons and at night the males especially come readily to light. To date (1993) it has not been found in other localities and its larval foodplant in the wild in this country is as yet unknown.

North-west Monmouthshire 1972, one male to m.v.l. 29/7 (GANH) (1st British rec.). This specimen is now in the National Collection at the BMNH. 1976: 26/6 ♂♂ (2) to m.v.l.; 27/6, ♂♂ (6) and ♀♀ (3) to m.v.l.; 29/6 ♂♂ (7) and ♀ (1) flying in hot (80 deg. F.) afternoon sunshine (GANH). 1977: 9/7 (afternoon), two freshly-emerged males taken in flight (CGMdW, JLM, GANH); 12/7, ♂ (1) by day and ♂♂ (4) to m.v.l. (DSF, JDB, ECP-C, GANH).

Over the years, many lepidopterists have visited the site to collect the moth but, sadly, do not seem to search for it elsewhere. It continues to be recorded annually in the original locality eg. 7/7/86 (8) to m.v.l. (MEA).

Refs. (vide appendix): Horton & Heath 1973, Horton 1976, Bretherton 1977, de Worms 1978, Goater 1978, Haggett 1981.

2176 *Cerapteryx graminis* Linn. Antler Moth

This moth, flying from mid-July to mid-September, is more common in the hills of the western half of the county than elsewhere in Gwent.

Llandogo 1892 (Nesbitt 1892b). Tintern "several at light" 1909 (Bird 1910). Abertillery dist. 1911, "swarmed on the hills" (Rait-Smith 1912).

Usk, garden m.v.l. trap, 1967, 14/8 (2), 26/8 (1) and recorded sporadically until 1983 but in small numbers. Ed. 7/7/76. Ld. 16/9/82. Wentwood 28/6/69. Piercefield Park 18/7/73. Cwm Tyleri 6/8/73, 14/8/73 (abdt.). Magor Marsh 4/8/75. Newport Docks 17/8/76. Clydach 26/7/90. (MEA) : Ynysyfro 22/6/76, Mescoed Mawr 14/8/81, Risca 27/8/81, Cwm Merddog 8/8/86, Ochrwyth 9/8/86, Cwmfelinfach 12/8/86, Trelleck Bog

6/9/86, Pontllanfraith 20/7/87, Trefil 15/8/87, Nant Gwyddon 8/9/88.

Sq. **SO** 10,20,30,50,11,21. **ST** 28,38,48,19,29,49,59.

2177   ***Tholera cespitis*** D. & S.   Hedge Rustic
Flying from mid-August through the first half of September, this moth is widespread in Monmouthshire but at low density.

Llandogo 1892 (Nesbitt 1892b). Tintern, Aug. 20 1912 (Bird 1913).

Usk, garden m.v.l. trap, 28/8/66 (1), 12/9/67 (1) and recorded most years in small numbers eg. 1977 7/9 (3), 8/9 (2). Ed. 24/8/68, 11/8/70. Ld. 12/9/71, 19/9/79. Wentwood 1970. Cwm-mawr 30/8/71. Magor Marsh 21/8/76. Llansoy 5/9/86. Cwm Tyleri 15/9/86 (10) to m.v.l. (MEA) : Risca 4/9/80 and recorded in 1984, 85 and 86; Cwmfelinfach 26/9/86; Ochrwyth 28/9/86.

Sq. **SO** 20,30,40,50. **ST** 28,48,19,29,49,59.

2178   ***Tholera decimalis*** Poda   Feathered Gothic
Widespread and common in Monmouthshire flying in late August and the first half of September.

Llandogo 1892 (Nesbitt 1892b). Wye Valley 1906 (Bird 1907). Bettws Newydd (Crowther 1935).

Usk, garden m.v.l. trap, fairly common most years from 1966 to 1982 eg. 1969, plentiful from 9/8 to 4/9 (10). Ed. 9/8/69, 23/8/82. Ld. 15/9/71, 12/9/81. Usk (Park Wood) 3/9/69. Gwehelog (Camp Wood) 28/8/70. Magor Marsh 29/8/70, 31/8/76. Undy (Collister Pill) 30/8/70. Wentwood 1970. Slade Wds 29/8/72. Rhyd-y-maen 1/9/73, plentiful to m.v.l. Wyndcliff 6/9/75. Hendre Wds 29/8/78. Hael Wds 18/9/82. Llansoy 1/9/86. (MEA) : Risca 30/8/82, Mescoed Mawr 29/8/79, Cwmfelinfach 5/9/86.

Sq. **SO** 30,40,50,41. **ST** 48,19,29,49,59.

2179   ***Panolis flammea*** D. & S.   Pine Beauty
Flies in April and May but very scarce in Monmouthshire.

Usk, garden m.v.l. trap, 3/5/68 (1) (GANH). (1st v-c. rec.) and (12) further specimens recorded up to 1984, usually not more than one per year but (2) singletons noted in 1980 and (3) in 1982. Ed. 15/4/80, 3/4/81. Ld. 13/5/71, 3/5/78. Usk Castle, to m.v.l. 20/4/84 (1), 28/4/87 (1) (RBH). Hael Woods 1/5/86 (1). Llansoy 27/4/87 (1). (MEA) : Ochrwyth 13/4/87 (1).

Sq. **SO** 30,40,50. **ST** 28.

2181   ***Egira conspicillaris*** Linn.   Silver Cloud
Rare in Monmouthshire with few records.

Monmouth 1939, 13/5 (1), 19/5 (1), "at lamp in garden" (Crowther) (1st v-c. rec.). (Crowther Entomologist **72**:186).

Usk, to garden m.v.l. trap, ab. *melaleuca* 18/5/73 (1), 28/5/77 (1) (GANH).

Sq. **SO** 30,51.

2182   *Orthosia cruda* D. & S.   Small Quaker
A widespread and often abundant woodland moth flying from late February to the middle of May. Plentiful at sallow blossom and readily comes to light.

Wye Valley, near Bigsweir, "on sallows" 31/3 1907 (Barraud 1907b). Wye Valley 1906 (Bird 1907). Tintern 1912 (Bird 1913). Abertillery dist. 1911, 1912 (Rait-Smith 1912, 1913). Cwmyoy 1965 (SHR).

Usk, garden m.v.l. trap, recorded annually 1967   84, usually in abundance. Ed. 1/3/77, 25/2/80. Ld. 7/5/71, 25/5/79. Very abundant in 1979 from 30/3/79 to 25/5/79 especially in April when the following counts were recorded . . . 16/4 (36), 17/4 (151), 18/4 (172), 19/4 (96), 20/4 (66), 21/4 (45) . . . Usk (Park Wood) 23/4/67. Llangybi (Cae Cnap) 23/4/68. Wentwood 1970 (JMD). Craig-y-Master 9/5/70. Hadnock 30/3/71 (abdt.). Cwm-mawr 6/4/71. Redding's Inclosure 20/3/72 (abdt.). Pont-y-saeson 20/3/72. Hendre Wds 25/3/72, 15/4/79 (abdt.). Prescoed 16/4/72. Slade Wds 24/3/73. Hael Wds 26/3/73. Trelleck Bog 27/3/73. Llantrisant 11/4/80. Llansoy 2/5/85. Wyndcliff 27/4/87. (MEA) : Llantarnam 18/3/77, Risca 8/4/83, Mescoed Mawr 19/4/85, Cwmfelinfach 11/4/87, Ochrwyth 13/4/87.

Sq. **SO** 20,30,40,50,41,51,32. **ST** 28,48,19,29,39,49,59.

2183   *Orthosia miniosa* D. & S.   Blossom Underwing
Flying during March and April, this moth occurs sparingly in oak-woods in the eastern half of Monmouthshire.

Wye Valley, about 40 larvae on oak-twig, 3/6 1906 (Barraud 1907a). Tintern, larvae on oak 1912 (Bird 1913).

Usk, garden m.v.l. trap, 8/4/67 (1), 1969 (1), 71 (1), 77 (1), 80 (4), 81 (2), 83 (1), 84 (1). Ed. 18/3/84. Usk (Park Wood) 28/4/67. Redding's Inclosure 1972 4/4; 1981 6/4, 21/4 Hendre Wds 1979 15/4 (1), 1983 30/4 (1), 1987 23/4 (4). Llansoy 25/4/87 (2), 30/4/90 (1).

Sq. **SO** 30,40,41,51. **ST** 59.

2184   *Orthosia opima* Hb.   Northern Drab
This species flying in March and April is scarce in Gwent.

Tintern, 1912 (Bird 1913), Abertillery dist. 1912 (Rait-Smith 1913).
Pont-y-saeson, 2/4/72, one to m.v.l. (CGMdW, JLM). Usk, garden m.v.l. trap, 3/4/72 (1), 31/3/73 (1), 1977 25/3 (1) and 20/4 (1). (MEA) : Risca, 21/4/87 (1) to act.l; Ochrwyth 24/4/87 (1) to m.v.l.; Clydach 4/5/90.

Sq. **SO** 20,30,21. **ST** 28,29,59.

2185   *Orthosia populeti* Fabr.   Lead-coloured Drab

Flies from late March to early May but uncommon in Monmouthshire.

Wye Valley, larvae on poplar 1906 (Bird 1907).

Usk, garden m.v.l. trap, 1979 21/4 (1), 28/4 (1), 15/5 (2). Also 1980 (12) 13/4 – 11/5, 1981, 82, 83, 84. Hendre Wds 14/4/80 (6), 21/4/80 (2), 28/4/82 (1). Redding's Inclosure 6/4/81, 21/4/81. Hael Wds 30/3/81, 28/4/82. Llansoy 18/5/85, 28/4/87.

Sq. **SO** 30,40,50,41,51.

2186   *Orthosia gracilis* D. & S.   Powdered Quaker

Flying from late March to mid-May, this moth occurs in Monmouthshire but at low density.

Wye Valley 1906 (Bird 1907).

Usk, garden m.v.l. trap, 30/3/67 and noted every year until 1984, usually in small numbers, but in 1977 a total of (54) was recorded from 25/3 to 19/5 with a maximum count of (9) on 22/4. Ed. 25/3/77 (4). Ld. 30/5/69. Gwehelog (Camp Wood) 29/4/67. Wentwood 1970 (JMD). Magor Marsh 5/5/70 (2). Craig-y-Master 7/5/71. Pont-y-saeson 17/4/72. Hendre Wds 23/4/78. Llansoy 1985 (6). Clydach 11/5/90. (MEA) : Risca 14/4/85, Cwmfelinfach 21/4/87, Abercarn 3/5/87.

Sq. **SO** 30,40,41,21. **ST** 48,19,29,39,49,59.

2187   *Orthosia cerasi* Fabr.   Common Quaker
       = *stabilis* D. & S.

Widespread and often abundant in Monmouthshire, flying from early March well into May.

Wye Valley, Bigsweir, on sallows 31/3/07 (Barraud 1907b). Wye Valley 1906 (Bird 1907). Abertillery dist. 1911 and 1912 (Rait-Smith 1912, 1913). Cwmyoy 1965 (SHR). Wyndcliff 1971, larva on sycamore 22/6 (JMC-H).

Usk, garden m.v.l. trap, recorded annually 1967 – 84 and very common. Ed. 26/2/76, 29/2/80. Ld. 30/5/77, 29/5/78 and a very late moth on 8/6/73. Usk (Park Wood) 28/4/67. Llangybi (Cae Cnap) 23/4/68. Craig-y-Master 21/5/68. Prescoed 31/5/68. Wentwood 25/4/69 (JMD). Magor Marsh 5/5/70 (abdt.). Rhyd-y-maen 23/5/70. Hael Wds 25/5/70. Hendre Wds 26/5/70. Hadnock Quarry 30/3/71 (abdt.). Cwm-mawr 6/4/71. Pont-y-saeson 3/4/72. Redding's Inclosure 1/5/72, 6/4/81 (plfl.). Slade Wds 24/3/73. Trelleck Bog 27/3/73. Cwm Coedycerrig 27/5/73. Wyndcliff 26/4/76. Llantrisant 11/4/80. Llansoy 2/5/85. (MEA) : Mescoed Mawr 10/5/78, Risca 3/4/82, Cwmfelinfach 11/4/87, Ochrwyth 13/4/87, Cwm Merddog 26/4/87, Abercarn 3/5/87, Pontllanfraith 6/5/88, Lower Machen (Park Wood) 13/5/88.

Sq. **SO** 10,20,30,40,50,21,41,51,22,32. **ST** 28,48,19,29,39,49,59.

2188   *Orthosia incerta* Hufn.   Clouded Drab

This very variable moth is widespread and abundant in Gwent flying from mid-March to late May. On 23/10 1970, a most unusual date, a male (freshly-emerged) came to my garden m.v.l. trap at Usk.

Wye Valley 1906 (Bird 1907). Wye Valley, near Bigsweir, "on sallows" 31/3/07 (Barraud 1907b). Abertillery dist., "scarce" 1911, 1912 (Rait-Smith 1912, 1913). Cwmyoy 1965 (SHR).

Usk, garden m,v,l, trap, recorded annually 1967 – 1984 and often abundant eg. 1977 (98) 4/3 – 5/6. Ed. 4/3/77, 6/3/84. Ld. 5/6/77, 4/6/79. Usk (Park Wood) 28/4/67. Llangybi (Cae Cnap) 23/4/68. Craig-y-Master 21/5/68. Prescoed 31/5/68. Magor Marsh 5/5/70 (abdt.). Rhyd-y-maen 23/5/70. Hael Wds 25/5/70. Hendre Wds 26/5/70. Wentwood 1970 (JMD). Hadnock Quarry 30/3/71 (abdt.). Cwm-mawr 6/4/71, 27/3/73 (abdt.). Redding's Inclosure 20/3/72. Pont-y-saeson 3/4/72. St Pierre's Great Wds 23/3/73. Trelleck Bog 22/5/73. Llantrisant 11/4/80. Llansoy 2/5/85. Wyndcliff 28/4/87. (MEA) : Mescoed Mawr 3/5/77, Risca 8/4/83, Cwmfclinfach 11/4/87, Cwm Merddog 26/4/87, Abercarn 3/5/87, Pontllanfraith 6/5/88.

Sq. **SO** 10,20,30,40,50,31,41,51,32. **ST** 48,19,29,39,49,59.

2189   *Orthosia munda* D. & S.   Twin-spotted Quaker

Widespread and fairly common in Monmouthshire, flying in woodland during March and April.

Wye Valley 1906 (Bird 1907). Cwmyoy 1965 (SHR).

Usk, garden m.v.l. trap, recorded annually 1966 – 1984 and sometimes fairly plentiful. eg. 15/4/70 (17), 1979 (46) 31/3 to 27/4. Ed. 3/3/77. Ld. 30/4/71. Llangybi (Cae Cnap) 23/4/68. Wentwood 25/4/69 (JMD). Usk (Park Wood) 3/5/69 (plfl.). Hadnock Quarry 30/3/71 (abdt.). Cwm-mawr 6/4/71. Pont-y-saeson 20/3/72. Redding's Inclosure 20/3/72 (plfl.). Hendre Wds 25/3/72, 15/4/79 (plfl.). St Pierre's Great Wds 23/3/73. Slade Wds 24/3/73. Hael Wds 26/3/73. Wyndcliff 22/3/74. Llansoy 7/5/86. (MEA) : Llantarnam 18/3/77, Mescoed Mawr 3/5/77, Risca 12/4/83, Cwmfelinfach 11/4/87, Ochrwyth 13/4/87.

Sq. **SO** 30,40,50,21,41,51,32. **ST** 28,48,19,29,39,49,59.

2190   *Orthosia gothica* Linn.   Hebrew Character

Widespread and generally abundant in Monmouthshire from early March until late May in woodland, gardens and bushy places.

Wye Valley, near Bigsweir, "on sallows" 31/3/07 (Barraud 1907b). Wye Valley 1906 (Bird 1907). Tintern 1912 (Bird 1913). Abertillery dist. 1911, 1912 (Rait-Smith 1912, 1913). Cwmyoy 1965 (SHR).

Usk, garden m.v.l. trap, recorded annually from 1967 to 1984 and often in abundance, eg. 25/4/69 (41), 3/5/70 (43), 18/4/79 (36). 1980, a total of (348) from 25/2 to 24/5 thus: Feb. (1), Mar. (45), Apr. (216), May (86). Ed.

20/2/71, 25/2/76. Ld. 8/6/77, 12/6/79. Usk (Park Wood) 28/4/67. Llangybi (Cae Cnap) 23/4/68. Craig-y-Master 21/5/68. Prescoed 31/5/68. Rhyd-y-maen 23/5/70. Hael Wds 25/5/70. Hendre Wds 26/5/70. Hadnock Quarry 30/3/71. Cwm-mawr 6/4/71. Redding's Inclosure 7/3/72. Pont-y-saeson 20/3/72. Slade Wds 24/3/73. Trelleck Bog 27/3/73. Cwm Coed-y-cerrig 27/5/73. Cwm Tyleri 31/5/74. Mynyddislwyn 10/6/74. Wyndcliff 26/4/76. Wentwood 4/5/76. Llantrisant 11/4/80. Llansoy 2/5/85. (MEA) : Llantarnam 18/3/77, Mescoed Mawr 19/5/77, Risca 21/3/82, Ochrwyth 1/6/85, Cwmfelinfach 11/4/87, Cwm Merddog 26/4/87, Abercarn 29/4/87, Pontllanfraith 6/5/88, Lower Machen (Park Wood) 13/5/88, Clydach 11/5/90.

Larvae on beech and oak, Pont-y-saeson 25/5/76 (GANH). Larvae on broom, Hael Wds 30/6/76 (MWH). Larvae on dogwood and hazel, Pont-y-saeson 5/6/78 (REMP).

Sq. **SO** 10,20,30,40,50,21,41,51,22,32. **ST** 28,48,19,29,39,49,59.

2191 *Mythimna turca* Linn.   Double Line
Usk, one to m.v.l. trap at Usk Castle 13/7/69 (RBH). The only Monmouthshire record.

Sq. **SO** 30.

2192 *Mythimna conigera* D. & S.   Brown-line Bright-eye
This moth, flying during July and August, occurs at low density in Monmouthshire.

Usk, garden m.v.l. trap, 13/7/67 and a few recorded most years eg. 1968, 14/7 (1); 1969, (6) 3/7 – 27/7; 1970, (5) 8/7 – 10/8; 1982, (4) 14/7 – 4/8. Prescoed 5/7/68. (MEA) : Risca 12/7/86, Cwmfelinfach 12/8/86.

Sq. **SO** 30. **ST** 19,29,39.

2193 *Mythimna ferrago* Fabr.   The Clay
This moth, flying from late June to mid-August, is widespread and fairly common in Gwent, especially in the western half of the county.

Abertillery dist. 1911, 1914 (Rait-Smith 1912, 1915).

Usk, garden m.v.l. trap, recorded annually from 1966 to 1982 eg. 1978, (11) 14/7 – 5/8; 1982, (23) 4/7 – 7/8. Ed. 24/6/69. Ld. 15/8/79. Glascoed 1965. Wentwood 12/7/68. Magor Marsh 23/8/68, 11/7/78. Wyndcliff 13/7/69. Rhyd-y-maen 18/7/70. Redding's Inclosure 20/7/71. Hael Wds 10/7/73, 2/8/81. St Pierre's Great Wds 13/7/73. Piercefield Park 18/7/73. Cwm Tyleri 5/8/74. Newport (Coldra) 27/6/75. Hendre Wds 20/7/79. Slade Wds 30/6/84. Llansoy 16/7/86. Kilpale 30/6/87. (MEA) : Tintern 30/6/78, Risca 7/7/81, Trelleck 19/7/83, Newport Docks 11/7/84, Abercarn 16/7/85, Ynysddu 12/7/85, Ochrwyth 9/8/86, Cwmfelinfach 12/8/86, Pontllanfraith 10/7/87.

Sq. **SO** 20,30,40,50,41,51. **ST** 28,38,48,19,29,49,59.

2196 *Mythimna pudorina* D. & S.   Striped Wainscot

In Monmouthshire known from two localities only. One, an acid bog in the east of the county and the other a boggy heath in western Gwent, both at an elevation of 700 – 750 ft.

Trelleck Bog 8/7/78 (12) to m.v.l. (GANH) (1st v-c. rec.). Also 16/7/79 (3), 1986 19/7 (plfl.) and 24/7. Pontllanfraith 22/6/87 (1), 2/7/87 (plfl.) (MEA).

Sq. **SO** 50. **ST** 19.

2197 *Mythimna straminea* Treit.   Southern Wainscot

Occurs in marshes in the south of the county but scarce.

Magor Marsh, one to m.v.l., 10/7/70 (GANH) (1st v-c. rec.); 1974 (3) to m.v.l. 3/8 (MJL); 1/7/75 (1), 3/8/76 (1), 1/8/88 (5). Newport Docks 17/8/76 (2) (GANH). Magor Pill 19/7/89. (MEA) : Pontllanfraith 16/8/87 (1), Uskmouth 16/8/88.

Sq. **SO** 19. **ST** 38,48.

2198 *Mythimna impura* Hb.   Smoky Wainscot
        ssp. *impura* Hb.

Common throughout Gwent, flying in grassy places during July and August with a second generation in late September and early October.

Abertillery dist. 1911, 1912 (Rait-Smith 1912, 1913).

Usk, garden m.v.l. trap, recorded annually 1966 – 1983 and usually plentiful eg. 1970, 27/6 to 24/8 and 26/9 to 30/9; 1978, 12/7 to 14/8 and 4/10 to 9/10. Ed. 27/6/70. Ld. 11/10/79. Prescoed 5/7/68 (abdt.). Wentwood 8/7/68. Magor Marsh 23/8/68, 10/7/70 (abdt.). Hendre Wds 5/7/70. Rhyd-y-maen 18/7/70. Undy (Collister Pill) 13/7/73. Newport Docks 17/8/76. Trelleck Bog 16/7/79. Wyndcliff 28/7/82. Usk (Park Wood) 29/7/82. Hael Wds 31/7/82. Llansoy 14/7/86. Slade Wds 29/7/86. Kilpale 30/6/87. (MEA) : Mescoed Mawr 3/7/76, Ynysyfro 4/7/76, Risca 7/7/81, Black Vein 15/7/82, Ynysddu 12/7/85, Cwm Tyleri 10/7/85, Pant-yr-eos 6/7/85, Ochrwyth 9/8/76, Cwmfelinfach 15/7/86, Cwm Merddog 8/8/86, Pontllanfraith 10/7/87, Tretil 15/8/87, Uskmouth 16/8/88.

Sq **SO** 10,20,30,40,50,11,41. **ST** 28,38,48,19,29,39,49,59.

2199 *Mythimna pallens* Linn.   Common Wainscot

Common throughout the county, flying in grassy places from mid-June to early August and again from early September to mid-October.

Abertillery dist. 1911, 1912 (Rait-Smith 1912, 1913).

Usk, recorded at garden m,v,l, trap every year from 1966 to 1983 and usually plentiful eg. 1970, 28/6 – 9/8 and 26/8 – 3/10; 1971, 26/6 – 5/8 and 28/8 – 23/10. Extremely abundant in the hot summer of 1976 when large numbers appeared in the trap eg. 24/8 (57), 26/8 (39). Usk (Park Wood)

15/6/67. Magor Marsh 23/8/68, 10/7/70 (abdt.). Rhyd-y-maen 6/6/70. Redding's Inclosure 20/7/71. St Pierre's Great Wds 13/7/73. Cwm Tyleri 8/7/76. Newport Docks 17/8/76. Hael Wds 17/7/79. Slade Wds 9/9/82. Llansoy 6/10/86. (MEA) : Wyndcliff 3/7/77, Risca 7/7/81, Wentwood 13/8/81, Mescoed Mawr 28/7/84, Cwmfelinfach 15/7/86, Pontllanfraith 20/7/87, Uskmouth 16/8/88.

Sq. **SO** 20,30,40,50,41,51. **ST** 38,48,19,29,49,59.

2203 *Mythimna unipuncta* Haw.   White-speck
One Monmouthshire record for this migrant viz. Usk, 1978 a male to garden m.v.l. trap 12/10 (GANH).

Sq. **SO** 30.

2204 *Mythimna obsoleta* Hb.   Obscure Wainscot
Scarce in Monmouthshire, frequenting reed-beds in the south of the county.
    Usk, garden m.v.l.trap, 25/6/74 ♀ (1) (GANH) (1st v-c. rec.), probably a vagrant. Magor Marsh, to m.v.l., 1975 24/6(1), 1977 8/7(2), 1978 11/7(1), 1979 19/6(1).

Sq. **SO** 30. **ST** 48.

2205 *Mythimna comma* Linn.   Shoulder-striped Wainscot
Common, frequenting grassy places from late May to mid-August.
    Abertillery dist. 1914 (Rait-Smith 1915).
    Usk, garden m.v.l. trap, recorded annually from 1966 to 1983 and often plentiful eg. 1979 20/6 (15), 22/6 (15). Ed. 28/5/70, 25/5/80. Ld. 13/8/77, 20/8/81. Very late record 4/11/78. Usk (Park Wood) 15/6/67. Hael Woods 3/7/71. Magor Marsh 6/7/71. Newport (Coldra) 27/6/75. Cwm Tyleri 9/7/77. Hendre Wds 9/7/79. Slade Wds 21/6/81. Llansoy 13/6/85. Wyndcliff 23/6/86. Dinham 5/6/87. (MEA) : Wentwood 6/7/77, Sugar Loaf 7/7/77, Mescoed Mawr 16/7/78, Risca 16/6/84, Pant-yr-eos 6/7/85, Ynysddu 12/7/85, Ocluwyth 27/6/86, Cwmfelinfach 15/7/86, Pontllanfraith 22/6/87, Dixton Bank 18/6/88, Nant Gwyddon 25/6/88.

Sq. **SO** 20,30,40,50,21,41,51. **ST** 28,38,48,19,29,49,59.

CUCULLIINAE

2214 *Cucullia chamomillae* D. & S.   Chamomile Shark
Scarce in Monmouthshire, the only records being: Usk, garden m.v.l. trap, 12/5/70 (1), 10/5/72 (1), 9/5/77 (1), 6/5/83 (1) (GANH).

Sq. **SO** 30.

2216 *Cucullia umbratica* Linn.   The Shark

Occurs in Monmouthshire but at low density, flying from mid-June to mid-August in gardens, waste ground etc.

Wye Valley 1906 (Bird 1907). Abertillery dist. 1911 (Rait-Smith 1912).

Usk, garden m.v.l. traps 1967 13/6 (1), 11/7 (1), and a few recorded most years to 1983 (GANH, RBH), eg. 1977, 9/5 (1), 19/6 (1), 21/7 (4). Ed. 9/5/77. Ld. 17/8/72. Wentwood 1970, 29/5, 3/6 (JMD). Magor Marsh 6/7/71. Undy (Collister Pill) 13/7/73. St Pierre's Great Wds 18/7/73. Llansoy, 1984, 28/7; 1985, 4/7, 5/7; 1989, 9/7. Slade Wds 29/7/86. Clydach 15/6/90. (MEA) : Risca 13/7/82, also 1983, 84, 85, 86; Newport Docks 11/7/84. Wyndcliff 4/7/87 (RFMcC in litt.); 31/7/91.

Sq. **SO** 20,30,40,21. **ST** 28,48,29,49,59.

2221 *Cucullia verbasci* Linn.   The Mullein

Flies in May and June but local and uncommon in Monmouthshire.

Usk, garden m.v.l. trap, recorded sporadically eg. 7/6/66 (1), 13/5/67 (1), 1970 (8) 6/5 – 15/5, 1977 (3) 10/5 – 15/6. Usk Castle, Apr 1987 (1) (RBH). Caerwent 25/6/74 (CT). Magor Marsh 1969, larvae on water figwort (PC-W in litt.). Wentwood (Foresters' Oaks) 9/7/69, three larvae on *Buddleia globosa* and one on knotted figwort (*Scrophularia nodosa*) (GANH). Usk 10/7/69, plants of Mullein (*Verbascum thapsus*) completely defoliated by numerous larvae. Llantrisant 22/7/77, larvae on water figwort (*S. auriculata*).

Llansoy 13/5/91 (1).

Sq. **SO** 30,40. **ST** 48,39,49.

2225 *Brachylomia viminalis* Fabr.   Minor Shoulder-knot

Widespread and fairly common in Monmouthshire flying during July and early August.

Abertillery dist. 1912 (Rait-Smith 1913).

Usk, garden m.v.l. trap, 9/7/67 and recorded most years until 1982 eg. 1970 (8) 6/7 – 3/8, 1982 (13) 30/6 (3) – 29/7. Ed. 27/6/72. Ld. 7/8/82. Wentwood 22/7/68. Magor Marsh 22/7/69. Hael Wds 5/7/70, 31/7/82 etc. Rhyd-y-maen 18/7/70. Tintern 1/8/70. Redding's Inclosure 20/7/71. Wyndcliff 3/8/81. Slade Woods 28/7/82. Usk (Cwm Cayo) 29/7/82. Kilpale 30/6/87. (MEA) : Risca 9/7/84, Ochrwyth 12/7/86, Cwmfelinfach 12/8/86.

Sq. **SO** 20,30,40,50,51. **ST** 28,48,19,29,49,59.

2227 *Brachionycha sphinx* Hufn.   The Sprawler

Occurs locally in the eastern half of Monmouthshire, flying in late October and November and coming readily to light.

Tintern 1909, at light, (Bird 1910). Cwmyoy 1965 (SHR).

Usk, garden m.v.l. trap, 1966 (10) and noted most years until 1982 eg. 1970 (15) 23/10 – 31/10; 1971 (29) 12/10 – 25/11 inc. (8) on 13/11 and (6) on 15/11. Ed. 12/10/82. Pont-y-saeson, one to light 30/10/67. Rhyd-y-maen 2/11/70. Also recorded from Wentwood, Hael Wds, Hendre Wds and Magor Marsh. Llansoy 27/10/86.

Sq. **SO** 30,40,50,41,32. **ST** 48,49.

2232 *Aporophyla nigra* Haw.   Black Rustic

This moth, widespread in Monmouthshire, and locally common as at Usk, flies during the latter half of September and throughout October and comes readily to light.

Tintern dist. (Bird 1905). Cwmyoy 1965 (SHR).

Usk, garden m.v.l. trap, recorded every year from 1966 to 1983 eg. 1970 (61) from 8/9 to 31/10, with a maximum count of (11) on 24/9, (though the trap was not operated from 5/10 to 22/10). Magor Marsh 23/9/69, 16/10/79. Hendre Wds 29/8/78. Hael Wds 3/10/78. Slade Wds 9/9/82. Llansoy 1985 (9) 1/10 – 14/10. (MEA) : Risca 25/9/79 and annually 1980 to 1986, Cwmfelinfach 9/11/86, Ochrwyth 17/9/88.

Sq. **SO** 30,40,50,41,32. **ST** 28,48,19,29,59.

2235 *Lithophane semibrunnea* Haw.   Tawny Pinion

This moth, rare in Monmouthshire, has been recorded in October and again in April and May following hibernation.

Tintern dist. (Bird 1905).

Usk Castle 25/10/69 one to m.v.l. (RBH). Usk, garden m.v.l. trap, 15/5/79 (1), 24/4/82 (1), 26/4/82 (1) (GANH).

Sq. **SO** 30,50.

2236 *Lithophane socia* Hufn.   Pale Pinion

Not uncommon in some Monmouthshire localities, flying in October and early November but more often seen in April and May after hibernation.

Llandogo 1892 (Nesbitt 1892b). Tintern dist. (Bird 1905). Wye Valley 1906 (Bird 1907). Usk Castle, at ivy blossom, 20/10/34 (RBH); one to m.v.l. 30/4/86 (RBH). Cwmyoy Mar/Apr 1965 (SHR).

Usk, garden m.v.l. trap, a few recorded most years from 1967 to 1982 mainly in the spring after hibernation eg. 8/4/67, 1968 (4) 27/3 to 25/4 and a freshly emerged specimen on 25/9, 1971 (5) 4/4 – 2/6 and 10/10 (1). Hendre Wds 26/5/70 (1). Llansoy 1987 (3) 30/3 – 22/4. (MEA) : Wyndcliff 21/5/77, 6/11/92; Risca 1984 (7) 5/10 – 28/10, 1985, 86, 87; Cwmfelinfach 9/11/86.

Sq. **SO** 30,40,50,41,32. **ST** 19,29,59.

2237 *Lithophane ornitopus* Hufn.   Grey Shoulder-knot
       ssp. *lactipennis* Dadd

This moth, frequenting parkland and deciduous woods, occurs at low density in the eastern half of Monmouthshire. It flies in late September and October and again in March and April after hibernation.

Llandogo 1892 (Nesbitt 1892b). Tintern dist., (Bird 1905). Wye Valley, 1906 (Bird 1907). Cwmyoy, Oct/Nov 1965 (SHR).

Usk, garden m.v.l. trap, a few recorded most years from 1966 to 1982 eg. 1966 5/10 (1), 1968 (4) 25/9 to 29/10, 1982 5/4 (1). Usk (Park Wood) 22/4/68 (3) to m.v.l. Wentwood 30/4/70 (JMD), 5/4/71 (GANH). Hael Wds 22/10/71, 19/3/83. Wyndcliff 22/3/74, 19/3/83. Hendre Wds 21/4/80, 19/5/80, 1982. Llanbadoc (Cilfeigan Wds) 14/4/81 one on fir trunk. Llansoy 1985 (3) 12/10 – 20/10; 17/4/87 (2).

Sq. **SO** 30,40,50,41,32. **ST** 49,59.

2238 *Lithophane furcifera* Hufn. The Conformist
ssp. *suffusa* Tutt

This rare resident species was first found at Llantrisant, in Glamorgan in 1859 by Evan John and appears to have been more or less confined to Glamorgan and Monmouthshire, but there have been few definite records for many years. G.A. Birkenhead (1891. 177) wrote "*conformis* was first found at Llantrisant by Mr Evan John who gives the information that it has since been caught in Monmouthshire". Meyrick (1895. 52) gives "Glamorgan, Monmouth, scarce and local".

Monmouth, taken by T. Philipson at sugar Oct. 2 1869 and another taken by his friend a few days previously (Buckler 1869). "Near Bigsweir, male at sallow blossom Mar. 31 1907" (Barraud 1907b). This specimen was taken on the Gloucestershire bank of the River Wye, the county boundary with Monmouthshire.

There has been one more recent Glamorgan record viz. "at light near Cardiff on 10/10/1959" (Skinner 1984. 111).

2240 *Lithophane leautieri* Boisd. Blair's Shoulder-knot
ssp. *hesperica* Bours.

First reported from the Isle of Wight in 1951, this moth is now steadily extending its range in Britain, its larvae feeding on Monterey and Leyland cypresses. First seen in Gwent in 1979, within a few days of its Welsh debut at Cowbridge, S. Glamorgan in early October (DRS pers. comm.) it is now (1990) well established in some gardens in Monmouthshire.

Usk, garden m.v.l. trap, 1979 17/10 (1) (GANH) (1st v-c. rec.); 1982 10/9 (1); 1983 1/10 (1), 5/11 (1), 8/11 (1). Llansoy 1985 11/10 (1), 16/10 (2), 20/10 (?); 1986 9/10 (1); 1990 12/10 (3), 14/10 (1), 19/10 (1). (MEA) : Risca 1988, Oct. 12, 16, 21; 1989, 25/9 etc. (frqt.); 1990, (plfl.) inc. 9/10 (14); 1991, (plfl.).

Sq. **SO** 30,40. **ST** 29.

2241   *Xylena vetusta* Hb.   Red Sword-grass

Very scarce in Monmouthshire. Has been noted in October and also in April and May after hibernation.

Llandogo 1892 (Nesbitt 1892b). Abertillery dist. 1911 (Rait-Smith 1912).

Usk, garden m.v.l. trap, 1968 27/5 (1), 1971 31/10 (1) (freshly emerged), 1973 27/4 (1), 1976 23/5 (1), 1982 31/5 (1).

Sq. **SO** 20,30,50.

2242   *Xylena exsoleta* Linn.   Sword-grass

Wye Valley 1892 (Nesbitt 1892a). Llandogo 1892 (Nesbitt 1892b). No recent records.

Sq. **SO** 50.

2243   *Xylocampa areola* Esp.   Early Grey

Widespread and common in Monmouthshire, inhabiting woods, gardens and hedgerows. It flies from early March, sometimes earlier, to late April. The pinkish form ab. *rosea* Tutt is frequent, especially in the earlier days of the emergence period.

Tintern, 1906, larvae on honeysuckle, reared (Bird 1907). Cwmyoy, Mar. 1965 (SHR).

Usk, garden trap, 5/3/67 and noted annually until 1984 eg. 1969 (29) 29/3 to 6/5; 1980 (30) 18/2 to 18/4. Ed. 5/3/67, 24/2/76 (ab. *rosea*), 2/3/77 (ab. *rosea*), 18/2/80. Ld. 15/5/70, 5/5/71. Wentwood 25/4/69. Hadnock Quarries 30/3/71. Cwm-mawr 6/4/71. Hendre Wds 25/3/72. Redding's Inclosure 29/3/72, 6/4/81. Pont-y-saeson 3/4/72. St Pierre's Great Wds 23/3/73 (abdt.). Slade Wds 24/3/73. Trelleck Bog 27/3/73. Wyndcliff 22/3/74. Hael Wds 6/4/74. Llansoy 7/5/85 (7). (MEA) : Llantarnam 18/3/77, Risca (to act.l.) 8/4/83, 30/3/87 (14) etc., Cwmfelinfach 8/4/88.

Sq. **SO** 20,30,40,50,21,41,51,32. **ST** 48,19,29,39,49,59.

2245   *Allophyes oxyacanthae* Linn.   Green-brindled Crescent

Widespread and locally common in Gwent, flying throughout October and early November, in gardens, hedgerows and open woodland. The melanic form ab. *capucina* Mill. is frequent wherever this species occurs.

Llandogo, 1892, inc. ab. *capucina* (Nesbitt 1892b). Tintern dist., larvae on cotoneaster, ab. *capucina* (Bird 1905). Wye Valley 1906, inc. ab. *capucina* (Bird 1907). Abertillery dist. 1911, inc. ab. *capucina* (Rait-Smith 1912). Brockwells near Caerwent 21/10/62 (CT in litt.). Cwmyoy 1965 (SHR).

Usk, garden m.v.l. trap, recorded every year from 1966 to 1982 eg. 1970, 4/10 to 1/11 inc. 23/10 (13) and 28/10 (10). Ed. 2/10/76, 30/9/81. Ld. 2/11/81, 10/11/82. Usk 25/10/69 at ivy blossom. Rhyd-y-maen 2/10/70.

Redding's Inclosure 19/10/71. Slade Wds 24/10/71. Hendre Wds 25/10/71. Hael Wds 9/10/79. Magor Marsh 16/10/79. Wentwood. Llansoy, garden trap, 1985 9/10 – 30/10 total of (93) inc. (30) ab. *capucina*. (MEA) : Risca 6/10/79.

Sq. **SO** 30,40,50,41,51. **ST** 48,29,49.

2247 *Dichonia aprilina* Linn.   Merveille du Jour
This moth, frequenting oakwoods, occurs locally and at low density in Gwent. It flies in the second half of September and in October and comes to light.

    Llandogo 1892 (Nesbitt 1892b). Brockwells (Caerwent) 24/10/63 (CT). Cwmyoy 1965 (SHR).

    Usk (Plas Newydd), garden m.v.l. trap, noted most years from 1966 to 1983 but in small numbers eg. 1968 4/10 (1), 8/10 (1); 1972 18/9 (1), 8/10 (1), 26/10 (6); 1976 (13) 23/10 – 28/10. Redding's Inclosure 19/10/71. Hael Wds 22/10/71, 9/10/79 ♀ Hendre Wds 10/10/78. Pont-y-saeson (3) to m.v.l. 30/9/79. Magor Marsh 16/10/79. Llansoy 1985 (8) 9/10 – 20/10. (MEA) : Pontypool dist. 10/10/86, Risca 1/10/87, Clydach 1/10/90.

Sq. **SO** 30,40,50,21,41,51,32. **ST** 48,29.

2248 *Dryobotodes eremita* Fabr   Brindled Green
Widespread but infrequent in Monmouthshire, flying in deciduous woodland during September and the first half of October.

    Usk, garden m.v.l. trap, occurred most years from 1966 to 1982 but rarely more than two or three in a season. Ed. 3/9/66. Ld. 14/10/62. Usk (Park Wood) 17/9/66 one to m.v.l. (REMP). Hendre Wds 4/9/78, 10/10/78. Slade Wds 9/9/82. Hael Wds 19/9/82 (3). Tal-y-coed 9/9/83. Llansoy 1/10/85. (MEA) : Risca, to act.l. 6/9/84, 1986; Cwmfelinfach 9/10/86.

Sq. **SO** 30,40,50,31,41. **ST** 48,19,29.

2250 *Blepharita adusta* Esp.   Dark Brocade
Flying during June and July this moth is scarce and of sporadic occurrence in Monmouthshire.

    Wentwood, to m.v.l. trap 28/5/70 (2) (JMD); 2/6/70 (1) (GANH). Usk, garden m.v.l. trap, 1970, single moths on June 24, 25, 26. (MEA) : Trelleck Common 19/7/83, Cwm Tyleri 7/7/86.

Sq. **SO** 20,30,50. **ST** 49

2252 *Polymixis flavicincta* D. & S.   Large Ranunculus
This species, flying in September and October, is very scarce in Monmouthshire.

    Llandogo 1892 (Nesbitt 1892b). Tintern, larvae (Bird 1905). Wye Valley 1906 (Bird 1907). Monmouth dist. (Charles c. 1937). Usk 1931,

"one taken on wing" at Usk Castle (RBH).
Usk, garden m.v.l. trap, 25/9/70 (1), 3/10/83 (1) (GANH).

Sq. **SO** 30,50,51.

2254 *Antitype chi* Linn.   Grey Chi
Not uncommon in Monmouthshire, frequenting open woods and moorland during August and September, by day it is to be found at rest on walls, gateposts, tree trunks etc. and at night it comes to light.

Tintern, larvae (Bird 1905). Wye Valley 1906 (Bird 1907). Abertillery dist. 1912 (Rait-Smith 1913).

Llanbadoc 23/9/63, 1965 (GANH). Usk, garden m.v.l. trap, a few most years from 1966 to 1982. Ed. 9/8/70, 8/8/81. Ld. 10/9/82. Wentwood 23/8/68, 24/8/74. Slade Wds 9/9/81. (MEA) : Risca 30/8/82, 1984; Pontllanfraith 1987 29/8, 1/9; Ochrwyth 17/9/88.

Sq. **SO** 20,30,50. **ST** 28,48,19,29,39,49.

2255 *Eumichtis lichenea* Hb.   Feathered Ranunculus
   ssp. *lichenea* Hb.
Rare in Monmouthshire.

Usk, garden m.v.l. trap, 30/9/83 (1); Llansoy 12/10/90 (1) (GANH). Risca 13/10/90, (1) to act. l. (MEA). *M.B.G.B.I.* Vol. **10** Map 34 gives Sq. **ST** 28.

Sq. **SO** 30,40. **ST** 28,29.

2256 *Eupsilia transversa* Hufn.   The Satellite
A woodland species, widespread and common in Monmouthshire, especially in the central and eastern parts of the county. It flies from mid-September to November and is seen again from February to April after hibernation.

Wye Valley near Bigsweir, on sallows, 31/3 1907 (Barraud 1907b). Usk Castle 1934, "one to sugar 29/9 and one on ivy blossom 20/10" (RBH). Brockwells, Caerwent 15/3/63 (CT in litt.). Cwmyoy 1965 (SHR)

Usk, garden m.v.l. trap, 14/10/66 and recorded most years until 1984 eg. 1970 (4) 13/3 to 28/3 and (4) 24/10 to 1/11. Ed. 16/9/69, 3/10/76. Ld. after hibernation 28/3/70, 5/4/74. Usk (Park Wood) 22/4/68. Slade Wds 24/10/71. Pont-y-saeson imago 29/2/72; larva on oak 25/5/76, reared (GANH). Redding's Inclosure 24/3/72. Hendre Wds 25/3/72. Hael Wds 6/4/74, 1/5/86. Llansoy 9/10/85. (MEA) : Cwmfelinfach 9/11/86, Ochrwyth 14/4/87.

Sq. **SO** 30,40,50,41,51,32. **ST** 28,48,19,49,59.

2258 *Conistra vaccinii* Linn.   The Chestnut
A woodland species common and widespread in Monmouthshire, flying in

October and November and reappearing in the spring after hibernation and continuing to fly through March and April. However, on mild nights it can be seen at any time during the winter.

Llandogo 1892 (Nesbitt 1892b). Wye Valley 1906 (Bird 1907). Wye Valley, near Bigsweir, on sallows 31/3/07 (Barraud 1907b). Tintern 1912 (Bird 1913). Abertillery dist. 1912 (Rait-Smith 1913). Usk 1934, "at ivy blossom", Usk Castle 20/10 (RBH). Brockwells near Caerwent 21/10/62 (CT in litt.). Cwmyoy 1965 (SHR).

Usk, garden m.v.l. trap, plentiful most years from 1967 to 1984. Ed. 23/9/71, 18/10/72. Winter records 5/12/71, 25/12/71. Earliest appearances after hibernation 4/3/67, 27/2/82. Ld. 4/5/83 Trostrey, plentiful at ivy blossom 25/10/69. Hael Wds 22/10/71. Hendre Wds 25/10/71. Redding's Inclosure 27/2/72. Pont-y-saeson 20/3/72. Cwm-mawr 27/3/73 (10) to m.v.l. Llansoy 13/10/85. Also noted at Wentwood, Prescoed and The Blorenge. (MEA) : Llantarnam 18/3/77, Wyndcliff 9/3/78, Risca 2/10/83, Cwmfelinfach 9/10/86.

Sq. **SO** 20,30,40,50,21,41,51,32. **ST** 19,29,39,49,59.

2259 *Conistra ligula* Esp. Dark Chestnut

Of widespread occurrence in Monmouthshire but at low density, flying from mid-October into the winter and sometimes noted as late as early March.

Llandogo 1892 (Nesbitt 1892b). Wye Valley 1906 (Bird 1907). Abertillery dist. 1911 (Rait-Smith 1912). Usk, 1934, "in abundance on ivy blossom at Usk Castle" 20/10 (RBH).

Usk, garden m.v.l. trap, a few most years from 1966 to 1982. Ed. 17/10/77, 2/10/82. Ld. 27/1/73, 12/3/83. Usk, at ivy blossom 25/10/69. Slade Wds 24/10/71. Hendre Wds 10/10/78. Hael Wds 18/9/82. Llansoy 1985, (24) 9/10 – 30/10 inc. (6) on 14/10. (MEA) : Risca 26/10/86.

Sq. **SO** 30,40,50,41. **ST** 48,29.

2260 *Conistra rubiginea* D. & S. Dotted Chestnut

Monmouthshire, T. Philipson of Newport, Mon on November 17 1869 "took *Dasycampa rubiginea* settled on the trunk of a tree". (Buckler 1969). Tintern, one at ivy bloom Oct. 14 1876 (A.H. Jones 1876).

No recent records.

2262 *Agrochola circellaris* Hufn. The Brick

Flies in the second half of September and in October but is not common in Monmouthshire.

Wye Valley 1906 (Bird 1907). Abertillery dist. 1911 (Rait-Smith 1912). Tintern 1912 (Bird 1913). Usk 1934, "three on ivy blossom at Usk Castle" 20/10 (RBH). Cwmyoy 1965 (SHR).

Usk, garden m.v.l. trap, occurred sporadically and sparingly from 1966 to 1982 eg. 1970, 4/10 (1), 23/10 (1); 1977 18/10 (2), 19/10 (2). Ed. 22/9/71. Ld. 26/10/71 and two very late records in 1969 viz. 1/11 & 9/12. Magor Marsh 23/9/69. Trostrey Church, plentiful at ivy bloom 25/10/69. Llansoy 1984 (10) Oct. 12 to 27. Noted also at Wyndcliff and Wentwood. (MEA) : Hilston Park, Skenfrith 4/10/86.

Sq. **SO** 20,30,40,41,32. **ST** 48,49,59.

2263 *Agrochola lota* Cl.   Red-line Quaker
Fairly common in Monmouthshire, flying in woodland and marshy places in late September and October.
  Llandogo 1892 (Nesbitt 1892). Wye Valley 1906 (Bird 1907). Abertillery dist. 1911 (Rait-Smith 1912). Tintern 1912 (Bird 1913). Cwmyoy 1965 (SHR).
  Usk, garden m.v.l. trap, frequent most years from 1966 to 1983. Ed. 25/9/70, 22/9/71. Ld. 31/10/70, 27/11/71. Pont-y-saeson 1968, larvae on leaf-buds of sallow, reared. Slade Wds 24/10/71 (plfl.). Magor Marsh 16/10/79. Llansoy 1985, (32) to garden trap 10/10 to 30/10 inc. (7) on 23/10. Also noted at Prescoed, Wentwood and Trelleck Bog. (MEA) : Risca 28/10/84, 1985, 1986; Cwmfelinfach 9/11/86.

Sq. **SO** 30,40,50,32. **ST** 48,19,29,39,49,59.

2264 *Agrochola macilenta* Hb.   Yellow-line Quaker
Widespread and fairly common, flying from late September until late November, in woodland and bushy places generally.
  Llandogo 1892 (Nesbitt 1892). Wye Valley 1906 (Bird 1907). Abertillery dist. 1911 (Rait-Smith 1912). Tintern 1912 (Bird 1913).
  Usk, garden m.v.l. trap, seen most years from 1966 to 1982 and sometimes very plentiful as in 1970 when the counts on Oct. 23rd and 31st were respectively (20) and (16). Ed. 4/10/69, 29/9/81. Ld. 28/11/70, 10/11/81. Llanbadoc 7/10/68 on ivy blossom. Magor Marsh 23/9/69, 16/10/79. Trostrey 25/10/69 plentiful on ivy bloom. Rhyd-y-maen 2/11/70. Redding's Inclosure 19/10 and 2/11/71. Hael Wds 22/10/71, 9/10/79. Llansoy garden trap, 1985 (40) 9/10 – 23/10. Also noted at Wentwood and Hendre Wds. (MEA) : Risca 16/10/84, Hilston Park 4/10/86, Cwmfelinfach 9/10/86.

Sq. **SO** 20,30,40,50.41,51. **ST** 48,19,29,49.

2265 *Agrochola helvola* Linn.   Flounced Chestnut
This mainly woodland species flies from mid-September to mid-October and though widespread in Monmouthshire its occurrence is sporadic and infrequent.
  Llandogo 1892 (Nesbitt 1892b). Abertillery dist. 1911 (Rait-Smith 1912). Cwmyoy 1965 (SHR).
  Usk, garden m.v.l. trap, recorded infrequently from 1966 to 1983 eg.

1967 (1) 23/9; 1968 (5) 1/10 – 10/10; 1981 30/9 (1). Wentwood, to m.v.l. trap, 1969 (8) 18/9 – 18/10 (JMD). Wyndcliff 28/9/70 (3) to m.v.l. Redding's Inclosure 19/10/71. Slade Wds 24/10/71. Trelleck Bog 5/9/86. Cwm Tyleri 15/9/86. Llansoy 12/10/86. (MEA) : Risca 12/10/84, Ochrwyth 28/9/86, Cwmfelinfach 9/11/86, Pontllanfraith 15/9/87.

Sq. SO 20,30,40,50,51,32. ST 28,48,19,29,49,59.

2266 *Agrochola litura* Linn.   Brown-spot Pinion
At low density in Monmouthshire frequenting gardens and woodland during the second half of September and in October.

Llandogo 1892 (Nesbitt 1892b). Wye Valley 1906 (Bird 1907). Abertillery dist. 1911 (Rait-Smith 1912).

Usk, garden m.v.l. trap, recorded sporadically from 1966 to 1983 eg. 1966 14/9 (1), 17/9 (2); 1971 22/9 (1); 1982 16/9 (5), 17/9 (3), 18/9 (3); 1983 30/9 (1). Wentwood 15/9/69 (?) (JMD). Hael Wds 21/9/73(1), 20/9/74(1). Llansoy 3/10/86. (MEA) : Slade Wds 20/9/85, Risca 7/10/86.

Sq. SO 20,30,40,50. ST 48,29,49.

2267 *Agrochola lychnidis* D. & S.   Beaded Chestnut
Very common, especially in the eastern half of the county, flying from mid-September to the middle of November in gardens, open woodland, waysides and bushy places.

Llandogo 1892 (Nesbitt 1892b). Wye Valley 1906 (Bird 1907). Abertillery dist. 1911 (Rait-Smith 1912). Usk Castle 1934, (3) at ivy blossom, 20/10 (RBH). Cwmyoy 1965 (SHR).

Usk, garden m.v.l. trap, recorded annually 1966 – 1983 and often extremely plentiful eg. 23/10/68 (117), 12/10/69 (123), 3/10/70 (118), 12/10/71 (104). Its abundance declined somewhat after 1976 but it still remains plentiful. Ed. 15/9/71, 16/9/82. Ld. 16/11/71, 5/11/82. Coed-y-paen and Llanbadoc, numerous on ivy bloom 7/10/68. Wentwood 19/9/69. Magor Marsh 23/9/69, 16/9/79. Wyndcliff 28/9/70. Rhyd-y-maen 2/11/70 (11) to m.v.l. Redding's Inclosure 19/10/71. Slade Wds 24/10/71. Hendre Wds 25/10/71. Trelleck Bog. Llansoy, 1985 plentiful 1/10 to 30/10. (MEA) : Risca 14/9/84.

Sq. SO 20,30,40,50,41,51,32. ST 48,29,39,49,59.

2268 *Parastichtis suspecta* Hb.   The Suspected
Rare in Monmouthshire and to date recorded from one locality only.

Trelleck Bog 16/7/75 (12) to m.v.l. (MJL in litt.) (1st v-c. rec.). Trelleck Bog 30/8/86 (2), 17/8/88 (2), 25/7/90 (GANH, MEA).

Sq. SO 50.

2269   *Atethmia centrago* Haw.   Centre-barred Sallow
Widespread and not uncommon, frequenting hedgerows and open woodland, flying in the latter half of August and throughout September.

Bettws Newydd (Crowther 1935). Usk, garden m.v.l. trap, 21/8/66 and a few most years until 1982 eg. 1970 (13) 26/8 – 24/9; 1982 (2) 1/9 – 12/9. Ed. 18/8/69, 15/8/76. Ld. 15/9/68, 24/9/70. Usk (Park Wood) 3/9/69 plentiful to m.v.l. Wyndcliff 2/9/69. Gwehelog (Camp Wood) 28/8/70. Hael Wds 15/9/80. Slade Wds 16/8/82. Wentwood. Llansoy 1986 (6) 10/9 – 12/9. (MEA) : Risca 22/8/83, Cwmfelinfach 26/9/86, Ochrwyth.

Sq. **SO** 30,40,50. **ST** 28,48,19,29,49,59.

2270   *Omphaloscelis lunosa* Haw.   Lunar Underwing
Locally common in Monmouthshire and often abundant, flying from mid-September to the middle of October and coming readily to light.

Llandogo 1892 (Nesbitt 1892b). Cwmyoy 1965 (SHR).

Usk, garden m.v.l. trap, noted annually from 1966 to 1984 with numbers varying widely from year to year eg. 1966 (5) only, 1976 (abdt.), 19/9 (56), 20/9 (77), 30/9 (39), 1/10 (43). Ed. 15/9/66, 8/9/77. Ld. 3/10/70, 18/10/82. Coed-y-paen, to ivy blossom, 7/10/68. Magor Marsh 23/9/72. Slade Wds 7/10/82. Wentwood. Llansoy 1986 (82) 10/9 – 14/10. (MEA) : Risca 8/9/77, Ochrwyth 17/9/88.

Sq. **SO** 30,40,50,32. **ST** 28,48,29,39,49.

2271   *Xanthia citrago* Linn.   Orange Sallow
This moth, local and scarce in the eastern half of Monmouthshire, frequents woods where lime trees occur, it flies during September and the first half of October and at night comes to light.

Llandogo 1892 (Nesbitt 1892b). Near Bigsweir bridge, one, Sept. 17 1906 (Bird 1907).

Usk, garden m.v.l. trap, 1967 23/9 (2); 1971 15/9(1), 22/9(1); 1975 2/9 (1); 1977 16/10 (1). Wyndcliff 28/9/70 (2). Hael Wds 9/10/79 (2), 15/9/80 (1). Hendre Wds. Slade Wds 9/9/82(1), 16/9/82(1). Llansoy 7/10/86 (1).

Sq. **SO** 30,40,50,41. **ST** 48.

2272   *Xanthia aurago* D. & S.   Barred Sallow
Widespread and not uncommon in Monmouthshire, flying in deciduous woods from mid-September to late October.

Llandogo 1892 (Nesbitt 1892b).

Usk, garden m.v.l. trap, 10/10/66 (1) and several most years until 1983 eg. 1968 25/9 (5) and singletons on Oct. 1, 2, 4 and 6; 1977 (13) 26/9 – 27/10. Ed. 18/9/71, 19/9/82. Ld. 28/10/70, 27/10/77. Llangybi Park 5/11/66, one on old tree-stump. Wyndcliff 28/9/70 (6) to m.v.l. Hael Wds 22/10/71, 9/10/79 (3). Pont-y-saeson 30/9/79. Hendre Wds 8/10/79. Llansoy 6/10/86.

(MEA) : Risca 29/9/83, 1984, 85; Slade Wds 20/9/85; Ochrwyth 28/9/86; Cwmfelinfach 26/9/86.

Sq. **SO** 30,40,50,41. **ST** 28,48,19,29,39,59.

2273   *Xanthia togata* Esp.   Pink-barred Sallow

Widespread in Monmouthshire, but at low density, this species flies during September and the first half of October in marshy places and damp deciduous woodland and comes readily to light.

Llandogo 1892 (Nesbitt 1892a).

Usk, garden m.v.l. trap, 10/9/66 and recorded sparingly most years until 1983 eg. 1970 (5) 19/9 – 28/9; 1976 23/8(1), 26/8(1). Ed. 10/9/66. Ld. 20/10/77. Magor Marsh 23/9/69 (2), 16/10/79 (GANH). Magor Marsh 28/9/75 (MJL). Wentwood 16/9/69 (JMD). Trelleck Bog 4/9/73 (2). Newport Docks 17/8/76. Llantrisant 21/9/76. Pont-y-saeson 30/9/79 (2). Llansoy 21/9/86. Hendre Wds. (MEA) : Mescoed Mawr 5/9/77, Risca 17/9/85, Cwmfelinfach 9/10/86, Pontllanfraith 15/9/87 (plfl. to m.v.l.), Ochrwyth 17/9/88.

Sq. **SO** 30,40,50,41. **ST** 28,38,48,19,29,39,49,59.

2274   *Xanthia icteritia* Hufn.   The Sallow

Widespread but sparsely distributed in Monmouthshire, this moth flies in damp woods, marshes and river valleys from late August to mid-October and comes to light. The form ab. *flavescens* Esp. is common.

Llandogo 1892 (Nesbitt 1892b).

Usk, garden m.v.l. trap, 15/9/66 and a small number recorded most years until 1983 inc. ab. *flavescens* eg. 1977 (10) 8/9 – 18/10; 1982 (15) 6/9 – 19/9. Ed. 19/8/68, 31/8/69. Ld. 2/10/76, 18/10/77. Usk (Park Wood) 3/9/69. Magor Marsh 23/9/69, 16/10/79. Slade Wds 24/10/71. Newport Docks 17/8/76 (2) inc. one ab. *flavescens*). Pont-y-saeson 18/9/79. Hael Wds 9/10/79. Llansoy 1986 (7) 11/9-13/10. Also noted at Hendre Wds, Trelleck Bog and Wentwood. (MEA) : Mescoed Mawr 4/9/77, Risca 9/9/85, Pontllanfraith 29/8/87.

Sq. **SO** 30,40,50,41. **ST** 38,48,19,29,49,59.

2275   *Xanthia gilvago* D. & S.   Dusky-lemon Sallow

Rare in Monmouthshire.

Usk (Castle garden), 15/9/66, one to m.v.l. (GANH) (1st v-c. rec.). Usk (Plas Newydd), garden m.v.l. trap, 1967, 27/9 (1); 1971, (4) 15/9 – 23/9. Wentwood, one to m.v.l. 19/9/69 (JMD).

Sq. **SO** 30. **ST** 49.

ACRONICTINAE

2278 *Acronicta megacephala* D. & S.   Poplar Grey
Widespread and fairly common in Monmouthshire, frequenting damp woods and river valleys and flying during the second half of May and throughout June and July.

Wye Valley, larva on poplar 1906 (Bird 1907).

Usk, garden m.v.l. trap, 7/6/66 and several most years until 1983 eg. 1971 (7) 22/5 – 23/7, 1983 (11) 3/7 – 18/7. Ed. 9/5/76, 19/5/80. Ld. 23/7/70, 18/7/83. Hael Wds 22/6/81. Hendre Wds 20/7/79, 11/7/86. Usk (Park Wood) 5/6/67. Magor Marsh 11/7/69 (plfl.), 24/6/75. Wentwood 11/7/69. Llansoy 1986 (4) 6/7 – 15/7. (MEA) : Tintern 30/6/78, Risca 6/6/83, Mescoed Mawr 17/5/85, Cwm Tyleri 10/7/85, Ochrwyth 27/6/86, Trelleck Bog 25/5/87, Abercarn 26/5/87, Pontllanfraith 10/7/87, Cwmfelinfach 29/6/88.

Sq. **SO** 20,30,40,50,41. **ST** 28,48,19,29,49,59.

2279 *Acronicta aceris* Linn.   The Sycamore
Two Monmouthshire records only, from the west of the county, viz. Risca 11/7/85 (1) to act.l. (MEA) (1st v-c. rec.). Ynysddu 12/7/85 (1) to m.v.l. (MEA).

Sq. **ST** 19,29.

2280 *Acronicta leporina* Linn.   The Miller
This moth, widespread and not uncommon in Monmouthshire, inhabits woodland and damp bushy places, flying from late May to mid-August. All Gwent specimens are of the greyish form ab. *grisea* Cochrane.

Wye Valley, a larva on alder 1906 (Bird 1907). Pontllanfraith, one ab. *bradyporina* Treits. (= f. *grisea*) 1914 (Rait-Smith 1915).

Usk, garden m.v.l. trap, 9/6/66 and noted most years until 1983 eg. 1976 (16) 24/5 – 16/8, 1983 (13) 5/7 – 16/7. Ed. 24/5/76 (2), 1/6/81. Ld. 19/7/68, 16/8/76. Usk (Park Wood) 4/6/68 Wyndcliff 13/7/69 (2). Went-wood 1971 (JMD), 26/7/74. Magor Marsh 1976 25/6 (1), 2/7 (2). Hael Wds 22/6/76. Slade Wds 8/7/83 (svl.). Llansoy 26/7/86. Hendre Wds 1/7/86. (MEA) : Risca 7/7/83, 1984, 1986; Cwm Merddog 7/7/85; Ynysddu 12/7/85; Ochrwyth 27/6/86; Cwmfelinfach 15/7/86; Abercarn 26/5/87; Nant Gwyddon 7/6/88; Pontllanfraith 20/7/87.

Sq. **SO** 10,30,40,50,41. **ST** 28,48,19,29,49,59.

2281 *Acronicta alni* Linn.   Alder Moth
Flying from the middle of May to mid-July in damp woodlands and river valleys, mainly in the eastern half of the county, this moth, though local, is often plentiful where it occurs.

Llandogo dist., "one on oak-trunk" May 9 1892 and three to light on

May 26 1892 (Nesbitt 1892).

Usk, garden m.v.l. trap, 12/7/66 and recorded annually until 1983 eg. 1968 (6) 29/5 – 13/6, 1982 (9) 12/5 – 9/6. Ed. 19/5/80, 12/5/82. Ld. 19/7/81, 12/7/83. Wyndcliff 12/6/67. Usk (Park Wood), usually plentiful to m.v.l. eg. 4/6/68 (8); 9/6/69 (9) (CGMdW, GANH). Prescoed 8/6/69 (1). Wentwood 1970, to m.v.l., 2/6 (2), 6/6 (6) (JMD). Hendre Wds 5/6/71, 19/5/80 (5). Cwm-mawr 2/7/71 (3). Hael Wds 3/7/71 (svl.), 30/5/78 (8). Cwm Coed-y-cerrig 5/6/78. Llansoy 5/6/85 (1), 15/6/86 (1), 27/6/86 (3). (MEA) : Cwmfelinfach 24/5/87, 8/5/88; Abercarn 16/6/88; Dixton Bank 18/6/88.

Sq. **SO** 20,30,40,50,41,51,22. **ST** 19,29,39,49,59.

2283 *Acronicta tridens* D. & S.　　　　Dark Dagger
Skenfrith 1972, two larvae feeding on willow during last week of August, reared. (MJL in litt.) (1st v-c. rec). Magor Marsh 13/7/77 (1) to m.v.l. and Llansoy 12/7/86 (1), both confirmed by genitalia examnination.

Sq. **SO** 40,42. **ST** 48.

2284 *Acronicta psi* Linn.　　　　Grey Dagger
Widespread and common in Monmouthshire, flying from mid-May to August in deciduous woodland, marshes, heaths etc.

Llandogo, one, Sept. 1892 (Nesbitt 1892b). Wye Valley 1906 (Bird 1907). Abertillery dist. 1911, 1912 (Rait-Smith 1912, 1913).

Usk, garden m.v.l. trap, noted every year from 1966 to 1983 and often plentiful eg. 1983 (30) 28/6 – 7/8. Ed. 14/5/76, 13/5/80. Ld. 4/8/70, 8/8/80. Wyndcliff 6/6/67, 25/7/83. Pont-y-saeson 13/6/67 (REMP). Prescoed 31/5/68 (imago); larva on wych elm 18/7/73. Rhyd-y-maen 6/6/70. Hael Wds 3/7/71, 29/5/82. Piercefield Park 18/7/73. Magor Marsh 24/6/75. Cwm Tyleri 27/6/76. Trelleck Bog 16/7/79. Slade Wds 21/6/81. Llansoy 16/6/86. Hendre Wds 11/7/86. Clydach 11/5/90. (MEA) : Risca 7/7/81, Ynysddu 12/7/85, Ochrwyth 12/7/86, Cwmfelinfach 15/7/86, Dixton Bank 18/6/88, Nant Gwyddon 25/6/88.

Sq. **SO** 20,30,40,50,21,41,51. **ST** 28,38,48,19,29,39,49,59.

2286 *Acronicta menyanthidis* Esp.　　　　Light Knot Grass
　　　ssp. *menyanthidis* Esp.
Local and scarce in Monmouthshire

Abertillery dist. (Rait-Smith 1912, 1913, 1915).

Usk, garden m.v.l.trap, 1968 (1) 10/6, 1970 (1) (freshly emerged) 30/5. Cwm Tyleri, to m.v.l. 26/5/73 (2), 27/6/76 (2), 9/7/77 (1), 17/6/79 (1), 18/5/80 (1), 28/6/86 (1). (MEA) : Abercarn 26/5/87, Clydach 4/5/90.

Sq. **SO** 20,30,21. **ST** 29.

2289   *Acronicta rumicis* Linn.   Knot Grass
Widespread and common, frequenting woods and bushy places in general. Double-brooded, it flies from May to July and again in August and the first half of September, the melanic form ab. *salicis* Curtis often occurs.

Llandogo, one, Sept 1892 (Nesbitt 1892b). Wye Valley 1906 (Bird 1907). Abertillery dist. 1912, 1914 (Rait-Smith 1913, 1915); larvae on sallow, 1911, bred inc. ab. *salicis* (Rait-Smith 1912).

Usk, garden m.v.l. trap, noted every year from 1966 to 1983 and often frequent eg. 1982 (18) 24/5 – 19/9 inc. ab. *salicis*. Pont-y-saeson 13/6/67. Wyndcliff 13/7/69. Usk (Park Wood) 3/9/69. Hendre Wds 26/5/70. Magor Marsh 2/6/70, 24/8/74. Wentwood 2/6/70. Cwm-mawr 2/7/71. Trelleck Bog 16/6/73. Cwm Tyleri 17/6/73. Hael Wds 30/5/78. Llansoy 9/7/85. (MEA) : Risca 7/9/79, Trefil 3/7/87, Pontllanfraith 16/8/87, Lower Machen (Park Wood) 13/5/88.

Sq. **SO** 20,30,40,50,11,41. **ST** 28,48,19,29,49,59.

2291   *Craniophora ligustri* D. & S.   The Coronet
Flies during June and July in deciduous woods containing ash trees but local and scarce in Monmouthshire and known only from the eastern half of the county. The melanic variant ab. *coronula* Haw. occurs.

Wyndcliff, occurs regularly eg. 7/6/66 (1), 13/7/69 (10), 3/8/81 (1) and ab. *coronula* Haw. (1), 21/7/87 (3). Usk Castle, one to m.v.l. 15/7/69 (RBH). Redding's Inclosure 7/6/71 (1) (GANH).

Sq. **SO** 30,51. **ST** 59.

2293   *Cryphia domestica* Hufn.   Marbled Beauty
Widespread and common, flying in late June, July and August and is found in damp woods and in the vicinity of walls, old buildings and quarries.

Wye Valley 1904, 1905 (Bird 1907). Abertillery dist. 1911 (Rait-Smith 1912).

Usk, garden m.v.l. trap, recorded every year from 1966 to 1983 eg. 1973 (9) 30/6 – 25/8, 1981 (11) 11/7 – 14/8. Ed. 25/6/76. Ld. 25/8/73. Usk 1968, seen plentifully by day, together with *Cryphia muralis*, on the stone walls of Usk Prison during the second half of July with many in cop. Magor Marsh 23/8/68. Cwm-mawr 2/7/71. Wentwood. Cwm Tyleri 8/7/76 (AR in litt). Wyndcliff 6/7/81. Slade Wds 28/7/81. Hael Wds 2/8/81. Llansoy 11/8/84, 25/6/85. (MEA) : Risca 7/9/79 and annually until 1988 eg. 1984 (19) 24/7 – 17/8; Trelleck 19/7/83; Ynysddu 12/7/85; Ochrwyth 27/6/86; Pontllanfraith 10/7/87; Uskmouth 16/8/88; Clydach 18/7/89.

Sq. **SO** 20,30,40,50,21. **ST** 28,38,48,19,29,49,59.

2295 *Cryphia muralis* Forst.   Marbled Green
       ssp. *muralis* Forst.

Apart from the old Wye Valley records, only three colonies of this attractive little moth associated with lichen-covered walls have been found in Gwent viz. at Usk, Risca and Newport (Gaer Hill). The moth flies during July and August and by day may be seen resting on the walls of buildings.

Wye Valley, "in small numbers" 1904 and 1905 (Bird 1907).

Usk, Plas Newydd, garden m.v.l. trap, recorded every year from 1966 to 1983 eg. 1969 (14) 15/7 – 18/8 and occasionally found by day on walls of the house 16/7/81 etc. Ed. 1/7/71, 7/7/73. Ld. 18/8/69, 23/8/74.

In 1968 and subsequently, I saw it by day in fair numbers, with many in cop., in the second half of July, on the stone walls of Usk prison which was built in Victorian times. 1966, to m.v.l. Usk Castle (RBH). 1983 (June), Gaer Hill (Newport) (CJR). (MEA) : Risca, to act.l. 1984 (9) 5/7 – 15/8; 1986 9/8, 12/8; 1987.

Sq. **SO** 30. **ST** 28,29.

AMPHIPYRINAE

2297 *Amphipyra pyramidea* Linn.   Copper Underwing
Common and widespread, frequenting hedgerows and deciduous woods in August and September.

Llandogo. 1892 (Nesbitt 1892b). Wye Valley 1906 (Bird 1907). Bettws Newydd (Crowther 1933). Cwmyoy, Oct. 1965 (SHR 1966).

Usk, garden m.v.l. trap, 21/8/66 and recorded annually until 1983 eg. 1970 (31) 2/8 – 26/9. Ed. 2/8/70, 4/8/82. Ld. 26/9/68, 30/9/81. Wentwood 24/8/68. Bica Common 27/8/68. Wyndcliff 2/9/69 (plfl), 15/9/81. Usk (Park Wood) 3/9/69. Hendre Wds 23/8/77. Hael Wds 22/8/82. Tal-y-coed 9/9/83. Llansoy 18/8/86. Magor Marsh 12/8/86. (MEA) : Risca 6/9/81, Slade Wds 20/9/85, Cwmfelinfach 26/9/86, Ochrwyth 17/9/88.

Sq. **SO** 30,40,50,31,41,32. **ST** 28,48,19,29,49,59.

2298 *Amphipyra berbera* Rungs   Svensson's Copper Underwing
       ssp. *svenssoni* Fletch.

This species, first separated from *A. pyramidea* in 1968, flies during August and September and, though infrequent, appears to be fairly widely distributed in Monmouthshire.

Raglan 2/7/69 (SC) (1st v-c. rec.). This specimen was exhibited at the 1969 Annual Exhibition of B.E. & N.H.S. 1969.).

Usk, garden m.v.l. trap, 12/8/70 and also noted occasionally in 1972, 78, 79, 80 and 82. Three specimens of Copper Underwing taken in 1967 on 18/8 (1) and 20/8 (2) were subsequently found to be of this species. Slade Wds 15/8/80. Hael Wds 31/7/82 (2). Hendre Wds. Llansoy 30/8/87. (MEA) :

Risca 22/8/83, Ochrwyth 28/9/86, Wyndcliff 10/9/88.

Sq. **SO** 30,40,50,41. **ST** 28,48,29,59.

2299 *Amphipyra tragopoginis* Cl.   Mouse Moth
Widespread and frequent in gardens, woods, moorland etc., flying from early July to mid-September.

Llandogo 1892 (Nesbitt 1892b). Wye Valley 1906 (Bird 1907). Abertillery dist. 1911 (Rait-Smith 1912).

Usk, m.v.l. trap, noted most years from 1966 to 1983 eg. 1971 (26) 2/7 – 22/9. Ed. 2/7/71, 11/7/82. Ld. 15/9/69, 22/9/71. Wyndcliff 2/9/69. Magor Marsh 2/8/70. Cwm Tyleri 13/7/75. Hael Wds 22/8/82. Tal-y-coed 9/9/83. Llansoy 11/8/84. Wentwood. Hendre Wds. (MEA) : Risca 7/9/79, Ochrwyth 9/8/86, Cwmfelinfach 12/8/86, Pontllanfraith 29/8/87, Trefil 7/8/88.

Sq. **SO** 20,30,40,50,11,31,41. **ST** 28,48,19,29,49,59.

2300 *Mormo maura* Linn.   Old Lady
Of infrequent and sporadic occurrence in Monmouthshire, this moth is sometimes seen roosting by day indoors or in outhouses. It flies from late May until the end of August and comes only reluctantly to light.

Wye Valley, Llandogo dist., "is exceedingly common here and comes to sugar in numbers" (Nesbitt 1892).

Usk, several indoors inc. 27/5/58 (1), 18/8/65 (1), 5/7/73 (1). Usk, garden m.v.l. trap, recorded sporadically eg. 5/6/67; 27/8/68; 1970 (6) 17/7 – 3/8; 1982 11/7, 29/7. Tintern 1/8/70, Newport Docks 17/8/76, Tal-y-coed 9/9/83, Undy (Collister Pill) 18/8/87 (all single moths to m.v.l.). Little Mill 29/7/77, one settled on roadway at dusk. Prescoed. Llansoy 3/8/91 (indoors). (MEA) : Mescoed Mawr 12/8/80 (3) to m.v.l.

Sq. **SO** 30,40,50,31. **ST** 38,48,29,39.

2302 *Rusina ferruginea* Esp.   Brown Rustic
Widespread in Monmouthshire but rarely numerous, flying during June and July in woodland, on heaths and in the hilly moorland areas of the north and west of the county.

Llandogo dist. 1892, "to light" 26/5 (Nesbitt 1982). Abertillery dist. 1911, 1912 (Rait-Smith 1912, 1913).

Usk, garden m.v.l. trap, occurred sporadically from 1967 to 1982 eg. 2/6/68, 1971 (4) 7/6 – 9/7. Ed. 28/5/70. Pont-y-saeson 13/6/67. Prescoed 8/6/69. Usk (Park Wood) 9/6/69. Wentwood 21/6/69 (JMD). Redding's Inclosure 15/6/71. Wyndcliff 22/6/71. Cwm-mawr 2/7/71. Hael Wds 2/7/71. Bica Common 13/6/72. Cwm Tyleri 30/7/72, 14/6/74 (plfl.), 26/6/76 (plfl.). Trelleck Bog 16/6/73. Hendre Wds 24/6/79. (MEA) : Cwm Merddog 7/7/85, Ynysddu 12/7/85, Risca 14/7/85, Ochrwyth 12/7/86, Pontllanfraith 20/6/87. Abercarn 27/6/87, Trefil 3/7/87, Cwmfelinfach 6/7/87,

Nant Gwyddon 25/6/88.

Sq. **SO** 10,20,30,50,11,41,51. **ST** 28,19,29,39,49,59.

2303 *Thalpophila matura* Hufn. Straw Underwing
Scarce and local in Monmouthshire flying in grassy places and open woodland during the second half of July and through August.

Usk, garden m.v.l. trap, 1966 16/7 (1), 1967 (2), 1969 21/7 (1) and 1/8 (1), 1970 17/7 (1), 1974 28/7 (1), 1982 24/7 (1). Wentwood, to m.v.l., 1969 (frequent) inc. 25/7 (1), 29/7 (6), 31/7 (4), 2/8 (1) (JMD). Brockwells near Caerwent 11/8/74 (CT). Magor Marsh 24/8/81 (1), 1/8/88 (1). Redding's Inclosure 2/8/90. Hael Wds 3/8/90. Llansoy 3/8/90. (MEA) : Risca 9/9/85.

Sq. **SO** 30,40,50,51. **ST** 48,29,49.

2305 *Euplexia lucipara* Linn. Small Angle Shades
Common and widespread throughout Monmouthshire, flying during June and July in deciduous woods, gardens and bushy places.

Abertillery dist. 1911, 1912 (Rait-Smith 1912, 1913). Bettws Newydd 1934. "one at sugar" 20/7 (RBH).

Usk, garden m.v.l. trap, recorded annually 1967 – 82 eg. 1969 (15) 23/6 – 31/7, 1981 (17) 11/6 – 4/8. Ed. 3/6/78, 20/5/82. Ld. 2/8/80, 4/8/81. Prescoed 8/6/69. Wentwood 22/6/69. Usk (Park Wood) 7/7/69. Magor Marsh 11/7/69. Wyndcliff 13/7/69, 6/7/81. Hendre Wds 5/7/70, 11/7/86. Redding's Inclosure 7/6/71. Hael Wds 12/7/71, 4/7/81. Pont-y-saeson 18/7/72. Cwm Tyleri 30/7/72. Trelleck Bog 16/6/73. St Pierre's Great Woods 13/7/73. Slade Wds 21/6/81. Llansoy 25/6/86. Ysguborwen (Llantrisant) 17/7/86. Kilpale 30/6/87. (MEA) : Risca 7/7/81, Black Vein 3/7/81, Cwm Merddog 7/7/85, Ynysddu 12/7/85, Abercarn 16/7/85, Ochrwyth 12/7/86, Cwmfelinfach 12/8/86, Pontllanfraith 22/6/87, Trefil 3/7/87, Nant Gwyddon 10/6/88, Dixton Bank 18/6/88, Clydach 18/7/89.

Sq. **SO** 10,20,30,40,50,21,41,51. **ST** 28,48,19,29,39,49,59.

2306 *Phlogophora meticulosa* Linn. Angle Shades
Common and widespread in Gwent , flying from May to October.

Wye Valley 1906 (Bird 1907). Abertillery dist. 1911, 1912 (Rait-Smith 1912, 1913). Cwmyoy 1965 (SHR).

Usk, garden m.v.l. trap, recorded annually 1966 – 1983 and sometimes abundant eg. 1976 4/5 to 27/10 inc. 24/8 (25), 19/9 (13), 1/10 (10). Ed. 12/5/75, 4/5/76. Ld. 28/10/70, 27/10/76. In some years late moths were numerous eg. 1971, plentiful from 5/9 to 27/11 with late singletons on 12/12 and 13/12; 1977 14/10 (49). Pont-y-saeson 13/6/67. Coed-y-paen 7/10/68 on ivy flowers, Llanbadoc (d°). Trostrey 25/10/69 plentiful at ivy blossom. Magor Marsh 2/6/70. Gwehelog (Camp Wood) 28/8/70. Undy (Collister Pill) 30/8/70. Wyndcliff 28/9/70. Cwm-mawr 31/5/71. Redding's

Inclosure 15/6/71, 19/10/71. Slade Wds 24/10/71. Hael Wds 22/6/76. Cwm Tyleri 26/6/76. Newport Docks 17/8/76. Llansoy 23/10/85 (6). Trelleck Bog 30/8/86. Kilpale 30/6/87. (MEA) : Risca 8/9/79, Blorenge 7/7/84, Wentwood 9/7/84, Abercarn 16/7/85, Cwmfelinfach 26/9/86, Pontllanfraith 10/7/87, Nant Gwyddon 10/6/88, Uskmouth 16/8/88, Clydach 11/5/90.

Sq. **SO** 20,30,40,50,21,51,32. **ST** 38,48,19,29,39,49,59.

2311   *Ipimorpha retusa* Linn.   Double Kidney
Scarce in Monmouthshire. Usk, garden m.v.l. trap, 1967, 22/8 (1) (GANH) (1st v-c. rec.), 26/8 (1); 1968, 21/8 (1); 1969, 31/7 (1); 1970, 3/8 (1), 4/8 (2); 1971, 30/7 (1); 1972, 23/8 (1); 1981, 3/8 (1), 22/8 (1). Redding's Inclosure 2/8/90 (1). 1991, Cwm Clydach 30/8 (RS).

Sq. **SO** 30,21,51.

2312   *Ipimorpha subtusa* D. & S.   The Olive
Local and scarce in Monmouthshire, flying in woodland and marshy places during July and August.

Wye Valley, larvae on poplar 1906 (Bird 1907). Tintern, one at light 1909 (Bird 1910). Bettws Newydd 1934, one at light 8/8 (RBH). 1969, Raglan, one to m.v.l. 2/7 (SC).

Usk, garden m.v.l. trap, 2/8/70 (1), 2/8/75 (1), 13/8/79 (1), 21/8/81 (1). Usk Castle one to m.v.l. 2/8/81 (RBH). Usk, one on house wall 20/9/82 (DGT pers. comm.). Usk (Cwm Cayo) one to m.v.l. 29/7/82. Magor Marsh 13/7/83 (1).

Sq. **SO** 30,40,50. **ST** 48.

2314   *Enargia ypsillon* D. & S.   Dingy Shears
Local and at low density in Monmouthshire, flying in marshes and river valleys during July and the first half of August.

Usk, garden m.v.l. trap, a few most years from 1970 to 83 eg. 1970, (5) 6/7 – 16/7; 1982, (10) 30/6 – 11/7. Ed. 30/6/82. Ld. 15/8/77. Magor Marsh 6/7/71; 1976 2/7(2), 6/7(1); 1977 (5) 5/7 – 1/8. Magor Marsh 1975 "common and fresh at sugar" (MJL in litt.). (MEA) : Risca 17/8/85, Sugar Loaf (St Mary's Vale) 8/7/87.

Sq. **SO** 30,21. **ST** 48,29,59.

2316   *Cosmia affinis* Linn.   Lesser-spotted Pinion
Local and very scarce in Monmouthshire, flying during July and August.
Tintern 1912, "bred from larva off wych elm" (Bird 1913). Usk 1934, one taken at sugar (RBH).

Rhyd-y-maen 18/7/70 (1) to m.v.l. (GANH). Magor Marsh 3/8/74 (1) to m.v.l. (MJL). Hael Wds 16/8/76 (1) (GANH).

Sq. **SO** 30,40,50. **ST** 48,59.

2318   *Cosmia trapezina* Linn.   The Dun-bar
Common and widespread, frequenting woods, hedges, etc. from early July to mid-September.

Wye Valley 1906 (Bird 1907). Abertillery dist. 1911 and 1912 (Rait-Smith 1912, 1913).

Usk, garden m.v.l. trap, noted most years from 1966 to 1983 and usually plentiful. Ed. 7/7/67, 8/7/82. Ld. 13/9/68, 9/9/71. Wyndcliff 19/8/67. Bica Common 23/8/68. Magor Marsh 24/8/68. Usk (Park Wood) 3/9/69. Rhyd-y-maen 18/7/70. Tintern 1/8/70. Redding's Inclosure 27/8/71. Hendre Wds 23/8/77. Hael Wds 2/8/81. Slade Wds 10/8/81. Llansoy 28/7/84. Trelleck Bog 30/8/86. (MEA) : Mescoed Mawr 3/9/77, Wentwood 17/8/83, Risca 7/8/84, Cwmfelinfach 12/8/86, Ochrwyth 28/9/86, Pontllanfraith 14/8/87, Clydach 18/7/89.

Sq. **SO** 30,40,50,21,41,51. **ST** 28,48,19,29,49,59.

2319   *Cosmia pyralina* D. & S.   Lunar-spotted Pinion
Scarce and local in Monmouthshire.

Usk Castle, one to m.v.l. 14/7/69 (RBH) (1st v-c. rec.). Usk, garden m.v.l. trap, occurred sporadically and sparingly from 1969 to 1983: 1969, (3) 15/7 – 26/7; 1970, (1); 1971, (1); 1972 (1); 1976, 7/7 (3), 8/7 (1); 1977, (1); 1981, (1); 1982, (3); 1983, 23/7 (3). Ed. 7/7/76. Ld. 5/8/72. Magor Marsh 6/7/71 (1). Cwm Tyleri 8/7/76 (1) (BG in litt.). Llansoy 25/7/86.

Sq. **SO** 20,30,40. **ST** 48.

2321   *Apamea monoglypha* Hufn.   Dark Arches
Common and widespread in Monmouthshire, flying in gardens, woodland and grassy places from mid-June to late August. Melanic forms occasionally occur.

Llandogo, September 1892 (Nesbitt). Abertillery dist. 1911 and 1912 (Rait-Smith 1912, 1913).

Usk, garden m.v.l. trap, recorded every year from 1966 to 1983 and often abundant with maximum numbers in mid-July. Ed. 11/6/71, 14/6/76. Ld. 7/9/70, 30/8/71. Melanic f. 14/7/78. Prescoed 5/7/68. Wentwood 8/7/68. Trelleck Common 9/7/68. Pont-y-saeson 25/8/68. Hendre Wds 5/7/70. Hael Wds 5/7/70. Magor Marsh 10/7/70. Rhyd-y-maen 18/7/70. Wyndcliff 22/6/71. Cwm-mawr 2/7/71. Redding's Inclosure 20/7/71. St Pierre's Great Wds 13/7/73. Piercefield Park 18/7/73. Cwm Tyleri 20/8/73. Blaenavon Mountain 25/8/73. Trelleck Bog 16/7/79. Slade Wds 28/7/81. Usk (Park Wood) 29/7/82. Llansoy 26/7/84, 31/7/86 (melanic f.). Pontllanfraith 2/7/87. Kilpale 30/6/87. (MEA) : Llantarnam 24/6/76, Mescoed Mawr 1/7/76, Sugar

Loaf 7/7/77, Risca 9/7/79, Blorenge 7/7/84, Newport Docks 11/7/84, Cwm Merddog 7/7/85, Ynysddu 12/7/85, Abercarn 16/7/85, Ochrwyth 27/6/86, Cwmfelinfach 15/7/86, Trefil 15/8/87, Undy (Collister Pill) 18/8/87, Nant Gwyddon 25/6/88, Uskmouth 16/8/88, Clydach 18/7/89.

Sq. **SO** 10,20,30,40,50,11,21,41,51. **ST** 28,38,48,19,29,39,49,59.

2322 *Apamea lithoxylaea* D. & S. Light Arches
Widespread and not uncommon in Monmouthshire, flying from late June to early August in open woodland and grassy places.
Abertillery dist. 1912 (Rait-Smith 1913).
Usk, garden m.v.l. trap, a few recorded most years from 1967 to 1983 eg. 1970 (7) 26/6 – 27/7. Ed. 26/6/70, 25/6/76. Ld. 3/8/77, 7/8/81. Devauden 5/7/61. Magor Marsh 10/7/70, 6/7/71. Cwm Tyleri 5/8/74. Redding's Inclosure 17/6/74. Hael Wds 28/6/74. Wentwood 26/7/74. Slade Wds 28/7/81. Llansoy 4/7/85. Kilpale 30/6/87. (MEA) : Mescoed Mawr 3/7/76, Ynysyfro 4/7/76, Risca 14/7/79, Wyndcliff 20/7/83, Ynysddu 12/7/85, Ochrwyth 9/8/86, Cwmfelinfach 12/8/86, Pontllanfraith 10/7/87.

Sq. **SO** 20,30,40,50,51. **ST** 28,48,19,29,49,59.

2325 *Apamea oblonga* Haw. Crescent Striped
Magor Marsh, one to m.v.l. 11/7/78 (GANH) (1st v-c. rec.) (Horton 1980).

Sq. **ST** 48.

2326 *Apamea crenata* Hufn. Clouded-bordered Brindle
Common and widespread in Monmouthshire, flying from late May to early August in open woodland and marshes and in the hills and moorland of the northern and western parts of the county. Melanic forms occur, eg. ab. *combusta* Haw.
East Monmouthshire, inc. ab. *combusta* (Bird 1905). Wye Valley, 1906 inc. ab. *combusta* (Bird 1907). Tintern 1912 (Bird 1913). Abertillery dist. 1911 and 1912 inc. ab. *alepecurus* Esp. (Rait-Smith 1912, 1913). Usk, 1955, one 22/6 (RBH).
Usk, garden m.v.l. trap, recorded annually from 1966 to 1983 inc. ab. *combusta* eg. 1970 (4) 24/5 – 8/8. Ed. 24/5/70, 18/5/82. Ld. 8/8/70, 25/7/82. Wyndcliff 23/6/67. Wentwood 23/6/69. Magor Marsh 2/6/70. Rhyd-y-maen 6/6/70. Cwm-mawr 2/7/71. Cwm Tyleri 28/7/72, 26/6/76 (ab. *combusta*). Redding's Inclosure 17/6/74. Hael Wds 28/6/74. Slade Wds 28/7/81 (ab. *combusta*). Llansoy 17/7/86. Kilpale 30/6/87. Pontllanfraith 2/7/87 (1) and ab. *combusta* (1). (MEA) : Black Vein 12/6/81, Risca 14/7/82, Cwm Merddog 8/8/86, Ochrwyth 27/6/86, Cwmfelinfach 12/8/86, Trefil 3/7/87, Nant Gwyddon 10/6/88.

Sq. **SO** 10,20,30,40,50,11,51. **ST** 28,48,19,29,49,59.

2327   *Apamea epomidion* Haw.   Clouded Brindle

In Monmouthshire, this moth is widespread but at low density, flying in woods, gardens etc during June and the first half of July.

Abertillery dist. 1911, 1912 (Rait-Smith 1912, 1913).

Usk, to light 23/6/64 and a few recorded most years from 1966 to 1983 at garden m.v.l. trap. Ed. 7/6/66, 30/5/68. Ld. 8/7/81, 12/7/83. Magor Marsh 2/6/70. Wentwood. Trelleck Bog 16/7/79. Hendre Wds 20/7/79, 11/7/86. Hael Wds 3/7/83. Slade Wds 4/7/84. Llansoy 29/6/86. Dinham 4/6/87. (MEA) : Mescoed Mawr 25/6/76, Black Vein 3/7/81, Wyndcliff 17/7/83, Risca 17/6/84, Cwm Merddog 14/7/86, Cwmfelinfach 10/6/87, Clydach 9/7/89.

Sq. **SO** 10,20,30,40,50,21,41. **ST** 48,19,29,49,59.

2329   *Apamea furva* D. & S.   The Confused
         ssp. *britannica* Cock.

This moth, scarce in Gwent, is found in the hilly moorland areas of the west and north of the county from late June to early August.

Cwm Tyleri 1973, 9/7 (1) to m.v.l. (GANH) (1st. v-c. rec.), 8/8 (1); 1976, 27/6 (1), 8/7 (3) to m.v.l. (AR, GANH); 1986, 28/6 (1). Clydach 9/7/89, 11/5/90. (MEA) : Ochrwyth 1986 27/6 (1), 12/7 (1), Abercarn 27/6/87, Trefil 3/7/87.

Sq. **SO** 20,11,21. **ST** 28,29.

2330   *Apamea remissa* Hb.   Dusky Brocade

Widespread and fairly common in Monmouthshire, flying in woodland, marshes and grassy places from mid-June to early August.

Wye Valley 1906 (Bird 1907). Abertillery dist. 1911 (Rait-Smith 1912).

1961, Devauden 5/7 (GANH). Usk, garden m.v.l. trap, noted sporadically and in small numbers from 1966 to 1983, eg. 1967 (5) 10/6 – 8/7, 1983 (4) 12/7 – 20/7. Ed. 10/6/68, 19/6/71. Ld. 4/8/81. Wentwood 1968 (JMD). Cwm Tyleri 1972 30/7 (1), 7/8(1). Undy (Collister Pill) 13/7/73. Hael Wds 12/7/74. Magor Marsh 21/8/76. Trelleck Bog 16/7/79. Llansoy 27/6/87. Slade Wds. (MEA) : Mescoed Mawr 1/7/76, Risca 6/6/82, Ochrwyth 27/6/86, Cwmfelinfach 12/6/86, Pontllanfraith 2/7/87, Abercarn 27/6/87.

Sq. **SO** 20,30,40,50. **ST** 28,48,19,29,49.

2331   *Apamea unanimis* Hb.   Small Clouded Brindle

Local and scarce in Monmouthshire, flying during June and July.

Usk, garden m.v.l. trap, a few sporadic records: 1967, 9/6 (1), 13/6 (1); 1972, (6) 10/7 – 30/7; 1982, 6/6 (1), 7/6 (2). Pont-y-saeson 1968. Magor Marsh 21/6/76 (1), 19/6/79 (1). Dixton Bank 18/6/88 (2). Llansoy 28/6/88 (1). (MEA) : Nant Gwyddon 25/6/88.

Sq. **SO** 30,40,51. **ST** 48,29,59.

2334   *Apamea sordens* Hufn.   Rustic Shoulder-knot
This moth, flying in grassy places from late May to mid-July, is widely distributed but not very common in Monmouthshire.

Tintern dist. (Bird 1905). Wye Valley 1906 (Bird 1907). Tintern 1912 (Bird 1913). Abertillery dist. 1911 and 1912 (Rait-Smith 1912, 1913).

Usk, garden m.v.l. trap, 8/6/66 and recorded most years until 1982 eg. 1968, (4) 1/6 – 12/6; 1971, (14) 20/5 – 26/6. Ed. 20/5/71, 18/5/80. Ld. 8/7/70, 24/7/81. Pont-y-saeson 13/6/67 (REMP). Hael Wds 21/5/73. Wentwood. Llansoy 7/7/86. Dinham 4/6/87. (MEA) : Allt-yr-yn 23/5/76, Black Vein 12/6/81, Risca 11/5.82, Cwm Merddog 7/7/85, Ochrwyth 27/6/86, Cwm Tyleri 7/7/86, Cwmfelinfach 24/5/87, Pontllanfraith 20/6/87, Slade Wds 11/6/88, Abercarn 16/6/88.

Sq. **SO** 10,20,30,40,50. **ST** 28,38,48,19,29,49,59.

2335   *Apamea scolopacina* Esp.   Slender Brindle
Flying during July and the first half of August this woodland species is widely but sparsely distributed in Monmouthshire.

Tintern dist. "single larva seen 19/6/65" (Mere 1965).

Usk, garden m.v.l. trap, 30/7/66 (1) and noted every year until 1983. eg. 1969 (11) 18/7 – 15/8; 1975 (2); 1982 (16) 9/7 – 23/7. Ed. 9/7/70, 7/7/76. Ld. 19/8/68, 18/8/81. Usk (Park Wood) 23/7/68 (3) to m.v.l. Wentwood 1968, 22/7(3), 24/8(1). Tintern 1/8/70. Hael Wds 6/8/74, 14/8/78, 2/8/81. Hendre Wds 1/8/80. (MEA) : Mescoed Mawr 13/7/82, Wyndcliff 20/7/83, Risca 26/7/84, Cwmfelinfach 13/8/87, Pontllanfraith 16/8/87, Strawberry Wood 14/8/91

Sq. **SO** 30,50,41,32. **ST** 19,29,49,59.

2336   *Apamea ophiogramma* Esp.   Double Lobed
Occurring locally in Monmouthshire this moth is found regularly and sometimes plentifully at Magor Marsh during July and early August. At Usk it appeared only very sporadically and it has been recorded from Wentwood.

Magor Marsh, to m.v.l., 23/8/68 (1); 1969, 22/7 (plfl.); 1970 10/7 (plfl.), 2/8; 1976, July 6, 25, Aug. 3, 6, 8 . Also 1971, 75, 77, 78, 83, and 88. Wentwood, one to m.v.l. 1969 (JMD). Usk, garden m.v.l. trap, 1969, 19/7 (1); 1976, 5/7 (1), 7/7 (2); 1978, 27/7 (1); 1983, 23/7 (1), 7/8 (1).

Sq. **SO** 30. **ST** 48,49.

2337   *Oligia strigilis* Linn.   Marbled Minor
Common and widespread in Monmouthshire, flying throughout June and July in open woods, marshes and grassy places generally, .

Abertillery dist. 1911 – 1914 (Rait-Smith 1912, 13, 15). Usk Castle 1955, one 18/6 (RBH).

Usk, garden m.v.l. trap, recorded annually 1967 to 1982 eg. 1972, (10)

2/7 – 30/7; 1973, (32) 16/6 (12) – 21/7. Ed. 7/6/71, 3/6/82. Ld. 30/7/72, 21/7/73. Wyndcliff 12/6/67 (REMP). Pont-y-saeson 13/6/67 (REMP). Prescoed 5/7/68. Hael Wds 5/7/70. Slade Wds 1/6/71. Cwm-mawr 2/7/71. Magor Marsh 3/7/73. Trelleck Bog 8/7/78. Hendre Wds 20/7/78. Llansoy 27/6/86. Kilpale 30/6/87. (MEA) : Tintern 30/6/78, Mescoed Mawr 5/7/79, Risca 30/6/80, Pant-yr-eos 6/7/85, Cwm Tyleri 10/7/85, Abercarn 16/7/85, Ochrwyth 27/6/86, Cwmfelinfach 15/7/86, Pontllanfraith 2/7/87, Dixton Bank 18/6/88, Nant Gwyddon 25/6/88.

Sq. **SO** 20,30,40,50,21,41,51. **ST** 28,48,19,29,39,49,59.

2338 *Oligia versicolor* Borkh.   Rufous Minor
Locally common in Monmouthshire in open woodland and grassy places during June, July and early August.

Usk, garden m.v.l. trap, 13/6/67 and occurring not infrequently most years until 1982 eg. 1971 (12) 21/6 – 20/7; 1981 (15) 5/6 – 8/8. Ed. 30/5/82. Ld. 8/8/81. Usk (Park Wood) 15/6/67 (REMP), 23/7/68, 7/7/69. Prescoed 5/7/68. Wentwood 1968 8/7, 12/7. Magor Marsh 1/7/69, 22/7/69. Wyndcliff 13/7/69. Cwm-mawr 2/7/71. Hael Wds 3/7/71, 2/8/86. Slade Wds 28/7/81. Llansoy 30/6/86. Kilpale 30/6/87. (MEA) : Cwmfelinfach 6/7/87.

Sq. **SO** 20,30,40,50. **ST** 18,19,39,49,59.

2339 *Oligia latruncula* D. & S.   Tawny Marbled Minor
Widespread and common in open woodlands and grassy places during June and July.

Usk, garden m.v.l. trap, 3/6/66 and recorded annually until 1982 and often numerous. eg 1973 (80) 4/6 – 29/7; 1981 (50) 11/6 – 7/8. Ed. 28/5/82. Ld. 6/8/72, 7/8/81. Wyndcliff 13/7/69. Rhyd-y-macn 6/6/70. Hendre Wds 5/7/70, 1/7/86. Hael Wds 3/7/71, 2/8/86. Redding's Inclosure 20/7/71. Wentwood 29/7/73. Magor Marsh 3/7/73. St Pierre's Great Wds 13/7/73. Piercefield Park 18/7/73. Newport (Coldra) 15/6/75. Trelleck Bog 8/7/78. Llansoy 2/7/85. Slade Wds 22/7/86. Clydach 18/7/89 (MEA) : Mescoed Mawr 5/7/79, Risca 11/7/83, Cwm Merddog 14/7/86, Cwmfelinfach 15/7/86, Nant Gwyddon 25/6/88.

Sq. **SO** 10,20,30,40,50,21,41,51. **ST** 38,48,19,29,49,59.

2340 *Oligia fasciuncula* Haw.   Middle-barred Minor
This common and widespread species, flying during June and July, frequents woodland rides and grassy places generally but is especially plentiful in marshy situations.

East Monmouthshire (Bird 1905). Abertillery dist. 1911 (Rait-Smith 1912).

Usk, garden m.v.l. trap, noted every year from 1966 to 83 eg. 1973 (52) 8/6 – 28/7 inc. (23) on 18/6. Ed. 2/6/80, 1/6/81. Ld. 30/7/72, 28/7/83.

Wyndcliff 12/6/67. Usk (Park Wood) 15/6/67. Magor Marsh 1/7/69, 24/6/75 (abdt.), 19/6/79 (inc. a melanic specimen). Rhyd-y-maen 6/6/70. Cwm-mawr 2/7/71. Cwm Tyleri 1/7/73. Hael Wds 10/7/73. Trelleck Bog 8/7/78. Llansoy 5/6/85. Slade Wds 27/6/85 (melanic f.). Hendre Wds 11/7/86. Dinham 4/6/87. Kilpale 30/6/87. (MEA) : Mescoed Mawr 25/6/76, Blorenge 2/7/76, Wentwood 4/7/77, Risca 8/6/80, Cwm Merddog 14/7/86, Ochrwyth 27/6/86, Cwmfelinfach 15/7/86, Pontllanfraith 2/7/87, Dixton Bank 18/6/88, Nant Gwyddon 25/6/88.

Sq. **SO** 10,20,30,40,50,21,41,51. **ST** 28,48,19,29,49,59.

2341 *Mesoligia furuncula* D. & S. Cloaked Minor
Flying during July and August, this moth is widespread but at low density in Monmouthshire.
    Abertillery dist. 1911 (Rait-Smith 1912).
    Usk, m.v.l. trap, 3/8/70 and sporadically until 1983 but numbers usually small eg. 1973, (9) 30/7 – 20/8; 1983, (6) 6/7 – 5/8; Ed. 2/7/75, 6/7/83; Ld. 20/8/73, 15/8/83. Magor Marsh, 2/8/75 (MJL in litt.); also 1974, 76, 77. Newport (Coldra) 27/6/75. Hendre Wds 20/7/79. Slade Wds 2/8/80. Usk (Park Wood) 29/7/82. Hael Wds 31/7/82. Llansoy 14/7/86. (MEA) : Wentwood 4/7/77, Risca 11/8/81, Abercarn 16/7/85, Ochrwyth 9/8/86, Cwmfelinfach 15/7/86, Pontllanfraith 16/8/87, Uskmouth 16/8/88.

Sq. **SO** 20,30,40,50,41. **ST** 28,38,48,19,29,49.

2342 *Mesoligia literosa* Haw. Rosy Minor
This moth, of local and infrequent occurrence in Monmouthshire, flies during late July and August.
    Abertillery dist. 1911 (Rait-Smith 1912).
    Usk, garden m.v.l. trap, 14/8/67 (2) and several recorded most years until 1983 eg. 1969 (6) 21/7 – 24/7, 1982 (3) 25/7 – 10/8. Ed. 21/7/69. Ld. 22/8/81. Wentwood 10/8/68. Magor Marsh 2/8/69. Trelleck Bog 22/7/78. (MEA) : Risca 6/8/81, Wyndcliff 29/7/84, Pontllanfraith 29/8/87.

Sq. **SO** 20,30,50. **ST** 48,19,29,49,59.

2343 *Mesapamea secalis* Linn. Common Rustic
Widespread and often abundant, flying in open woodland, gardens, waste ground and grassy places in general from late June to mid-September.
    Abertillery dist. 1911 (Rait-Smith 1912). Bettws Newydd (Crowther 1935).
    Usk, m.v.l. trap, every year from 1966 to 1983 and usually abundant with peak numbers in late July and early August eg. 26/7/69, 1/8/76 (41). Ed. 24/6/69, 23/6/81. Ld. 23/9/72, 16/9/81. Wentwood 12/7/68. Magor Marsh 11/7/69. Prescoed 3/8/69. Hendre Wds 5/7/70. Rhyd-y-maen 18/7/70. Redding's Inclosure 27/8/71. Slade Wds 29/8/72. Trelleck Bog 18/9/72. Hael

Woods 10/7/73. Undy (Collister Pill) 13/7/73. Piercefield Park 18/7/73. Cwm Tyleri 6/8/73. Coldra 27/6/75. Wyndcliff 3/8/81. Usk (Cwm Cayo) 29/7/82. Llansoy 26/7/84. (MEA) : Ynysyfro 4/7/76, Risca 12/8/80, Mescoed Mawr 14/7/82, Newport Docks 11/7/84, Cwmfelinfach 15/7/86, Ochrwyth 9/8/86, Pontllanfraith 10/7/87, Uskmouth 16/8/88.

Sq. SO 20,30,40,50,41,51. ST 28,38,48,19,29,39,49,59.

2345   *Photedes minima* Haw.   Small Dotted Buff
Common and widespread, flying from late June to early August in open woodland, marshes and damp grassy places.

Wye Valley 1906 (Bird 1907). Abertillery 1911, 1912 (Rait-Smith 1912, 1913).

Usk, garden m.v.l. trap, 1967 and noted every year until 1983 eg. 1973 (7) 24/6 – 24/7, 1982 (5) 11/7 – 9/8. Ed. 1/7/71, 24/6/73. Ld. 11/8/75, 14/8/76. Pont-y-saeson 13/6/67. Wyndcliff 13/7/69. Rhyd-y-maen 18/7/70. Redding's Inclosure 15/6/71, 24/7/83. Magor Marsh 6/7/71. Hael Wds 10/7/73. Piercefield Park 18/7/73. Cwm Tyleri 12/7/77. Hendre Wds 24/6/79. Slade Wds 28/7/81. Llansoy 26/7/84. (MEA) : Mescoed Mawr 3/7/78, Ochrwyth 12/7/86, Pontllanfraith 2/7/87, Dinham 7/7/87.

Sq. SO 20,30,40,50,41,51. ST 28,48,19,29,49,59.

2349   *Photedes fluxa* Hb.   Mere Wainscot
In Monmouthshire an established colony exists in a wood in the south where the grass *Calamagrostis epigejos* flourishes.

S. Monmouthshire, 1980, several to m.v.l. 2/8 (GANH, REMP) (1st v-c. rec. and the first Welsh record for this species). 1981, 10/8 (1), 11/8 (2); 1989, 20/7/89 (1) (GANH). No other sites are known in the county for this moth.

2350   *Photedes pygmina* Haw.   Small Wainscot
Occurs locally and at low density in Monmouthshire in damp woods and marshy situations, flying during August and September.

Llandogo 1906 (Bird 1907).

Usk, garden m.v.l. trap, 1967 13/9 (1), 21/9 (1). A further ten specimens noted from 1968 to 1976. Fd. 11/8/70. Ld. 25/9/70. Wentwood 24/8/68 (1). Bica Common 24/9/68 (1), 19/9/69 (1). Usk (Park Wood) 7/7/69 (1). Magor Marsh 11/7/69 (1), 29/9/70 (2). Hendre Wds 19/9/78 (2). Tal-y-coed 9/9/83. (MEA) : Pontllanfraith 1/9/87, Wyndcliff 10/9/88.

Sq. SO 30,50,31,41. ST 48,19,49,59.

2352   *Eremobia ochroleuca* D. & S.   Dusky Sallow
This moth, common in eastern and southern England, is gradually extending

its range westwards. It had reached Somerset (Brean Down) by 1972 (pers. obsv.) and in 1976 it was recorded at Tidenham in the west of Gloucestershire, just inside the boundary with Monmouthshire (*Macro lep.* in *Glos.* 1984. 116). By 1984 it had reached Monmouthshire and is now (1989) locally common in the south of the county.

Wyndcliff 1984 one to m.v.l. 26/7 (MEA) (1st v-c. rec.). Magor Marsh 1/8/88 (1) to m.v.l. (GANH, MEA). Magor Pill, 18/7/89 (by day) two on thistle flowers on sea-wall; 19/7/89 a number on thistle flowers at dusk and (10) to m.v.l. (MEA, GANH).

Sq. **ST** 48,59.

2353 *Luperina testacea* D. & S.   Flounced Rustic
Locally common in grassy places during August and September.

Llandogo 1892 (Nesbitt 1892b). Wye Valley 1906 (Bird 1907). Tintern 1912 (Bird 1913).

Usk, garden m.v.l. trap, plentiful most years from 1966 to 1982 eg. 1982 (95) 1/8 – 18/9. Ed. 10/8/68, 30/7/81. Ld. 15/9/71, 18/9/82. Magor Marsh 29/8/70. Wentwood 29/8/70. Undy (Collister Pill) 30/8/70. Cwm Tyleri 22/8/72. Rhyd-y-maen 1/9/73. Usk (Park Wood). Llansoy 1/10/84. (MEA) : Wyndcliff 12/9/77, Risca 7/9/79, Uskmouth 16/8/88.

Sq. **SO** 20,30,40,50. **ST** 38,48,29,49,59.

2358 *Amphipoea fucosa* Freyer   Saltern Ear
   ssp. *paludis* Tutt
Occurs sparingly in the small areas of salt-marsh on the Severn Estuary littoral.

Undy (Collister Pill) 1970, two to m.v.l. 30/8 (GANH) (1st v-c. rec). Magor Pill (sea-wall) 6/8/89 (3) to m.v.l. (GANH, MEA.).

Sq. **ST** 48.

2360 *Amphipoea oculea* Linn.   Ear Moth
Local and scarce in Monmouthshire, flying during the second half of July and in August.

Wye Valley 1906 (Bird 1907).

Usk, garden m.v.l. trap, recorded every year from 1966 to 1983 except 1967 with a total of (101) over the eighteen years. Ed. 15/7/73, 9/7/82. Ld. 3/9/72, 31/8/83. Newport Docks 17/8/76 (1). (MEA) : Wyndcliff 26/7/84, Risca 26/7/84 (1), 27/7/84 (1); Ochrwyth 28/9/86.

Sq. **SO** 30. **ST** 28,38,29,59.

2361 *Hydraecia micacea* Esp.   Rosy Rustic
Common and widespread in open woods, marshes, waste-ground etc. from

early August to the beginning of November.

Llandogo (Nesbitt 1892).

Usk, garden m.v.l. trap, recorded most years from 1966 to 1983 and often plentiful eg. 1968, (50) 10/8 – 23/10; 1970, (70) 6/8 – 4/10; 1982, (150) 21/7 – 3/11. Ed. 21/7/82. Ld. 3/11/82. Magor Marsh 29/8/70, 1976, 1978, 1979 etc. Rhyd-y-maen 1/9/73 (10) to m.v.l. Newport Docks 17/8/76. Tal-y-coed 9/9/83. Llansoy 15/8/85. Usk (Park Wood). Wentwood. (MEA) : Risca 25/9/79, Undy (Collister Pill) 18/8/87, Magor Pill 6/8/89.

Sq. **SO** 30,40,50,31. **ST** 38,48,29,49.

2364 *Gortyna flavago* D. & S.   Frosted Orange
Widespread and common in Monmouthshire flying from late August to mid-October in marshes, waste-ground, rough woodland, neglected fields etc.

Llandogo 1892 (Nesbitt 1892b). Wye Valley 1906, "pupae in thistle and foxglove stems" (Bird 1907). Tintern 1912 (Bird 1913). Abertillery dist. 1911, a pupa in marsh thistle stem (Rait-Smith 1912).

Usk, garden m.v.l. trap, recorded annually from 1966 to 1982 eg. 1982, (54) 29/8 – 19/9. Ed. 21/8/76. Ld. 11/10/71. Magor Marsh 23/9/69. Redding's Inclosure 19/10/71. Rhyd-y-maen 3/9/73. Hendre Wds 8/10/79. Tal-y-coed 9/9/83. Llansoy 7/10/86. Wentwood. (MEA) : Risca 7/9/79, Cwmfelinfach 26/9/86.

Sq. **SO** 20,30,40,50,31,41,51. **ST** 48,19,29,49.

2368 *Celaena leucostigma* Hb.   The Crescent
      ssp. *leucostigma* Hb.
Scarce and local in Monmouthshire.

Usk, 1974, ♀ to garden m.v.l. trap, 21/8 (GANH) (1st v-c. rec.). 1982, singles on 5/8, 21/8, 2/9; 1983 15/7 (1). Magor Marsh 2/8/75 (1) (MJL), 4/8/75 (1) (GANH). The only records.

Sq. **SO** 30. **ST** 48.

2369 *Nonagria typhae* Thunb.   Bulrush Wainscot
Frequents marshes and river valleys from late July to September but scarce in Monmouthshire.

Usk, garden m.v.l. trap, 21/8/68 (1), and single specimens in 1969, 1970 and 1983; 1975 (4) 2/8 – 6/9. Wentwood 13/8/69 (1) (JMD). Magor Marsh 2/8/75 ♀ (1) (MJL). Usk (Park Wood) 29/7/82 (1). (MEA) : Risca 19/9/87 (1) to act.l.

Sq. **SO** 30. **ST** 48,29,49.

2370 *Archanara geminipuncta* Haw.   Twin-spotted Wainscot
A reed-bed inhabitant occurring locally in the Gwent Levels.

Magor Marsh 23/8/68, several netted (GANH) (1st v-c. rec.); 2/8/75 one to m.v.l and four more netted round reed-bed (MJL in litt.); 1976, to m.v.l. 3/8 (2), 8/8 (2) (GANH). (MEA) : Magor Pill 19/7/89 (1).

Sq. **ST** 48.

2371 *Archanara dissoluta* Treit.   Brown-veined Wainscot
Newport Docks (reed-bed) 1976, one to m.v.l. 17/8/76 (GANH) (1st v-c. rec.). Magor Marsh 24/7/90 one netted at dusk (MEA).

Sq. **ST** 38,48.

2375 *Rhizedra lutosa* Hb.   Large Wainscot
Occurs infrequently in Monmouthshire near reed-beds and in river valleys, flying from late July to November.

Usk, garden m.v.l. trap (sited 100 yds. from River Usk), 27/9/67 and sporadically until 1983 eg. 1970, (5) 3/8 – 28/9; 1982, (1) 3/11. Ed. 3/8/70. Ld. 7/11/83. Magor Marsh, to m.v.l., 23/9/69, 16/10/79 (7) (GANH); 19/7/75 (2) (MJL). Llansoy 16/10/85 (1) (a vagrant).

Sq. **SO** 30,40. **ST** 48.

2379 *Coenobia rufa* Haw.   Small Rufous
Of local occurrence, flying in marshes and on boggy heaths from mid-July to September.

Abertillery dist. 1912 (Rait-Smith 1913).

Magor Marsh, to m.v.l., 23/8/68 (svl.), 22/7/69 (plfl.), 2/8/70 (plfl.), 1975, 1976 (GANH). Magor Marsh 19/7/75 (2) (MJL). (MEA) : Mescoed Mawr 11/7/77, Cwmfelinfach 26/9/86, Pontllanfraith 16/8/87 (abdt.), 15/7/88.

Sq. **SO** 20. **ST** 48,19,29.

2380 *Charanyca trigrammica* Hufn.   Treble Lines
Widespread and common, inhabiting open woods, field margins, rough ground, waysides etc. from mid-May to late July.

Llandogo dist. "to light" 26/5/92 (Nesbitt 1892). East Monmouthshire, and ab. *bilinea* (Bird 1905). Wye Valley 1906 (Bird 1907). Bettws Newydd 1935, "at light" 29/5 and "in abundance" 31/5 (RBH).

Usk, garden m.v.l. trap, most years 1966 – 82, numbers very variable but sometimes plentiful eg. 1971 (130) 19/5 – 5/7 with peak numbers on 7/6 (13), 11/6 (15), 13/6 (19). Ed. 15/5/82. Ld. 5/7/71. Wyndcliff 12/6/67. Pont-y-saeson 13/6/67. Usk (Park Wood) 15/6/67. Wentwood 3/6/70. Cwm-mawr 2/7/71. Hael Wds 3/7/71. Cwm Tyleri 28/7/72. Newport (Coldra) 15/6/75. Magor Marsh 21/6/76. Hendre Wds 8/6/79. Slade Wds 30/6/84. Llansoy 4/6/85. Dinham 4/6/87 (plfl.). (MEA) : Risca 31/5/80, Ochrwyth 27/6/86, Cwmfelinfach 24/5/87, Abercarn 26/5/87, Pontllanfraith 20/6/87, Dixton

Bank 18/6/88.

Sq. SO 20,30,40,50,41,51. ST 28,38,48,19,29,49,59.

2381 *Hoplodrina alsines* Brahm   The Uncertain

Locally common in Monmouthshire, flying from mid-June to mid-August, in gardens, waste places and rough open woodland.

Usk, garden m.v.l. trap, 30/7/66 and recorded most years until 1982 and some years plentiful eg. 1970, 28/6 – 11/8; 1981, 24/6 – 15/8. Ed. 19/6/69, 9/6/82. Ld. 21/8/71, 15/8/81. Hael Woods 5/7/70. Rhyd-y-maen 18/7/70. Magor Marsh 19/7/71. Trelleck Bog 17/7/79. Llansoy 26/7/84. Wentwood. Wyndcliff 4/7/87 (RFMcC in litt.).

Sq. SO 30,40,50. ST 48,49,59.

2382 *Hoplodrina blanda* D. & S.   The Rustic

Common and widespread, flying from mid-June to mid-August, in gardens, open woods, waste ground etc.

Usk, garden m.v.l. trap, 13/6/67 and every year until 1983 eg. 1970, 24/6 – 3/8; 1981, 22/6 – 13/8. Ed. 13/6/67, 17/6/76. Ld. 3/8/70, 13/8/81. Prescoed 5/7/68. Wentwood 8/7/68. Wyndcliff 13/7/69. Hendre Wds 5/7/70. Hael Wds 5/7/70. Rhyd-y-maen 18/7/70. Magor Marsh 2/8/70. Newport (Coldra) 27/6/75. Trelleck Bog 16/7/79. Llansoy 1/7/85. Kilpale 30/6/87. (MEA) : Ynysyfro 4/7/76, Newport Docks 11/7/84, Risca 26/6/85, Abercarn 16/7/85, Ynysddu 12/7/85, Cwm Tyleri 7/7/86, Cwm Merddog 14/7/86, Cwmfelinfach 15/7/86, Pontllanfraith 10/7/87, Clydach 26/7/90.

Sq. SO 10,20,30,40,50,21,41. ST 38,48,19,29,39,49,59.

2384 *Hoplodrina ambigua* D. & S.   Vine's Rustic

The only definite Monmouthshire records are given below but it had undoubtedly been overlooked for some time and probably occurs here more often than these records would suggest.

Usk, garden m.v.l. trap, 1979 one 3/10. (GANH) (1st v-c. rec.) (detn. confirmed by DSF). I had seen this species for several years previously but doubting its identity had failed to record it. (MEA) : Risca to act.l., 1984 Aug. 13,15,17,27,31; 1985 Sept. 8,10,11,18; 1988 Aug. 15; 1990 Aug. 23,28,30; 1991 July 26, Aug. 21, Sept. 2,10,14.

Sq. SO 30. ST 29.

2385 *Spodoptera exigua* Hb.   Small Mottled Willow

An immigrant species of sporadic occurrence.

Usk, garden m.v.l. trap, 1969 3/10 (1); 1982 6/9 (1), 10/9 (1). Llansoy 5/7/86 (1).

Sq. SO 30,40.

2387 *Caradrina morpheus* Hufn.   Mottled Rustic
Common and widespread, flying from mid-June to early August in open woodland, rough fields and pastures, waste ground etc.

Usk, garden m.v.l. trap, 17/7/66 and recorded annually until 1982 and usually plentiful eg. 1969, 20/6 – 4/8; 1971, 11/6 – 8/8 with a single specimen on 15/9, probably a second generation moth. Ed. 11/6/71. Ld. 9/8/70. Prescoed 5/7/68. Usk (Park Wood) 7/7/69. Magor Marsh 11/7/69. Wentwood 11/7/69. Hendre Wds 5/7/70. Rhyd-y-maen 18/7/70. Hael Wds 12/7/71. Redding's Inclosure 20/7/71. Newport (Coldra) 15/6/75. Cwm Tyleri 12/7/77. Llansoy 28/7/84. Trelleck Bog 24/7/86. Kilpale 30/6/87. Wyndcliff 4/7/87 (RFMcC.). (MEA) : Risca 14/7/80, Cwm Merddog 14/7/86.

Sq. **SO** 10,20,30,40,50,41,51. **ST** 38,48,29,39,49,59.

2389 *Caradrina clavipalpis* Scop.   Pale Mottled Willow
Of sporadic and infrequent occurrence in Monmouthshire.

Abertillery dist. 1911 and 1912 (Rait-Smith 1912, 1913).

Usk, garden m.v.l. trap, 10/9/66 and a few most years until 1983 eg. 1969 (7) 25/6 – 9/9, 1982 (9) 1/6 – 17/9. Ed. 8/6/75, 1/6/82. Ld. 2/10/76, 14/10/77. Magor Marsh 21/8/76 (2). Usk Castle 22/9/84(5) (RBH). Llansoy 1985 9/7(1), 11/10(1); 1986 3/10, 8/10; 1987 27/6, 17/9; 1989 19/6 (6), 22/7 (1). Wyndcliff 1987 11/7, 13/7 (RFMcC in litt.). (MEA) : Risca 3/7/83, 6/9; 1984, 86, 87, 88. Pontllanfraith 16/8/87; Trelleck Bog 17/8/88.

Sq. **SO** 20,30,40,50. **ST** 48,19,29,59.

2394 *Stilbia anomala* Haw.   The Anomalous
Found locally in damp woods, on boggy heaths, and in the hills and moorland of the western half of the county. Flies in late August and September and the males come readily to light.

Near Abertillery 1912 and 1913.(Rait-Smith 1913, 1914).

Bica Common 23/8/68, several to m.v.l. Wentwood (Foresters' Oaks) 27/8/68 (5), 19/8/69 (1) (GANH). Usk, garden m.v.l. trap, 2/9/71 (1), 23/8/82 (1). Cwm Tyleri 1973 8/8 (1), 20/8 (2); 23/8/74 (2); 1986, 15/9 a female on heather at dusk and males plentiful to m.v.l. Trelleck Bog 30/8/86 (5). Blorenge Mountain 16/9/86 (1). Llansoy 24/8/89 ♀ (1) to u.v.l.. (MEA) : Cwmfelinfach 5/9/86, Risca 7/9/86, Trefil 7/8/88, Nant Gwyddon 15/8/88.

Sq. **SO** 20,30,40,50,11,21. **ST** 19,29,49.

2397 *Panemeria tenebrata* Scop.   Small Yellow Underwing
Local but not uncommon in Monmouthshire, mainly in the south and east of the county, flying by day in May and early June and frequenting flowery meadows and banks, and sunny woodland rides and waysides.

Llandogo (Nesbitt 1893). Wye Valley (Barraud 1906). Wye Valley

1906 (Bird 1907). Abertillery dist. 1911 and 1912 (Rait-Smith 1912, 1913). Tintern, "several noticed by day 1n 1946" (Blathwayt 1967). Tintern 31/5/68 (JMC-H 1969).

Cicelyford 9/6/69, one visiting buttercup flowers in morning sunshine (CGMdW, GANH). Wentwood (Foresters' Oaks) 3/6/70 (1) (JMD). Magor Marsh 13/5/75, pair in cop. on hogweed leaf. Whitson 19/5/75, one flying in hot afternoon sunshine. Hendre Wds 2/6/79 (1). Llantrisant 3/6/79 (1). Pont-y-saeson 1/6/81, (7) flying in flowery woodland ride. Fedw Fach (Trelleck) 1988: 21/5 (1), 27/5 (4). (MEA) : Blackwood (Woodfieldside) 28/5/88, Pontllanfraith 11/6/88.

Sq. **SO** 20,40,50,41. **ST** 38,48,19,39,49,59.

HELIOTHINAE

2399 *Pyrrhia umbra* Hufn.   Bordered Sallow

Magor Pill (sea-wall), 1/6/90, one to m.v.l. (1st v-c. rec.) (MEA); 14/6/90, two to m.v.l. (MEA, GANH). Uskmouth, 17/7/90 (CJR in litt.).

Sq. **ST** 38,48.

ACONTIINAE

2410 *Lithacodia pygarga* Hufn.   Marbled White Spot

Locally common in moist deciduous woods in the eastern half of the county, especially the Wye Valley. Flying during June and July, it comes readily to light and is sometimes abundant.

Llandogo, one, 1907 (Bird 1907). Tintern, "a few at light in 1947" (Blathwayt 1967).

Usk, garden m.v.l. trap, 1983 16/7 (2), 20/7 (1) – the only records in the eighteen years from 1966 to 1983. Pont-y-saeson, 13/6/67 (numerous), 13/7/72. Wyndcliff 13/6/67, 25/7/83. Hendre Wds 5/7/70 (plfl.). Hael Wds 5/7/70 (abdt.), 28/6/74, 22/6/76 (plfl.), 26/6/85. Redding's Inclosure, 7/6/71 (abdt.), 20/7/71. Wentwood 29/6/73. Trelleck Bog 16/7/79. Dixton Bank 18/6/88

Sq. **SO** 30,50,41,51. **ST** 49,59.

2412 *Eustrotia uncula* Cl.   Silver Hook

Local and scarce in Monmouthshire and recorded only from boggy heaths in the far west of the county.

Ochrwyth, 1986, one to m.v.l. 27/6 (MEA) (1st v-c. rec.). Pontllanfraith 20/6/87, (3) by day (MEA, GANH), 2/7/87 (1), 3/7/87 (1).

Sq. **ST** 28,19.

CHLOEPHORINAE

2418 *Earias clorana* Linn.   Cream-bordered Green Pea
Usk, 1983, one to garden m.v.l. trap 20/6 (GANH). (1st v-c. rec. and new to Wales).

Sq. **SO** 30.

2421 *Bena prasinana* Linn.   Scarce Silver-lines
Flying in late June and through July, this moth occurs sporadically and at low density in oakwoods mainly in the eastern half of the county.

Usk, garden m.v.l. trap, 19/7/68 (1) and sporadically until 1983 eg. 1969 (5) 20/7 – 23/7; 1983 28/6 (1), 2/7 (1). Ed 26/6/76. Ld. 25/7/82. Wentwood 8/7/68 (1) (GANH), 7/6/69 (1) (JMD). Hael Wds 5/7/70, 2/8/86. Magor Marsh 2/7/76 (1). Hendre Wds 11/7/86. Llansoy 13/7/86, 28/6/87. MEA: Wyndcliff 6/7/84; Risca (to act.l.) 13/7/83, 5/7/84, 16/7/86; Cwmfelinfach 12/8/86; Pontllanfraith 10/7/87.

Sq. **SO** 30,40,50,41. **ST** 48,19,29,49,59.

2422 *Pseudoips fagana* Fabr.   Green Silver-lines
      ssp. *britannica* Warr.
Not uncommon in Monmouthshire, this moth flies in deciduous woodland from late May to mid-July and comes to light.

Llandogo dist. "to light" 26/5/92 (Nesbitt 1892). Wye Valley (Barraud 1906).

Usk, garden m.v.l. trap, 9/6/66 ♀ (1) and sporadically until 1983 eg. 1974, (3) 20/6 – 4/7; 1975, (5) 12/6 – 23/6; 1981 (1). Ed. 9/6/82. Ld. 12/7/83. Usk (Park Wood) 4/6/68 (2) to m.v.l. Wyndcliff 7/6/66, 1976 (plfl.), 25/6/86. Redding's Inclosure 7/6/71. Hael Wds 3/7/71, 20/5/87 (20).

Pont-y-saeson 13/7/72. Wentwood 29/6/73. Cwm Coed-y-cerrig 5/6/78. Hendre Wds 16/6/79. Dinham 4/6/87. (MEA) : Risca 13/7/83, Ynysddu 12/7/85, Cwmfelinfach 12/8/86, Cwm Merddog 20/6/86, Ochrwyth 27/6/86 Abercarn 26/5/87, Pontllanfraith 22/6/87, Nant Gwyddon 25/6/88.

Sq. **SO** 10,30,50,41,51,22. **ST** 28,19,29,49,59.

SARROTHRIPINAE

2423 *Nycteola revayana* Scop.   Oak Nycteoline
Fairly frequent in oakwoods and parkland in the eastern half of the county, flying from late August to mid-October and again, after hibernation, from February to June.

Tintern dist. (Bird 1905, 1907). Tintern 1912 (Bird 1913).

Usk, garden m.v.l. trap, 1968 28/6, 22/7, 10/10 and sporadically until 1983 (16/3). Usk 19/3/73, one flying in sunshine. Usk (Park Wood) 26/4/68 (3) to m.v.l. Bica Common 1968 23/8, 27/8. Wentwood 1970 (JMD).

Redding's Inclosure 27/2/72, 13/5/79, 11/10/79 (melanic form). Hendre Wds 4/3/80 one beaten from foliage of yew-tree, 30/4/83 one to m.v.l. Hael Wds 20/5/82, 1/5/86. Llansoy 11/5/88 (1), 14/7/90 (2). Wyndcliff 22/5/91.

(MEA) : Risca 10/8/84, Magor Marsh 2/10/86, Hilston Park 4/10/86.

Sq. **SO** 30,40,50,41,51. **ST** 48,29,49,59.

PANTHEINAE

2425 *Colocasia coryli* Linn.   Nut-tree Tussock

Widespread and often abundant in deciduous woodland, flying from mid-April to mid-June and as a second brood from mid-July through August.

Llandogo dist. "to light May 26th 1892" (Nesbitt 1892). Wye Valley 1906, larvae on beech (Bird 1907). Bettws Newydd (Crowther 1933). Bettws Newydd "four to light" 31/5/35 (RBH). Tintern 1946-48, "fairly common to light in all three years" (Blathwayt 1967).

Usk, garden m.v.l. trap, recorded every year from 1966 to 1983 and usually plentiful eg. 1970 frequent from 3/5 to 4/6 and from 27/7 to 7/8. Ed. 16/4/72. Ld. 18/8/81. Wyndcliff 6/6/67. Pont-y-saeson 13/7/67, 27/4/80. Hael Wds 25/5/70, 21/5/73 (abdt.). Wentwood 2/6/70. Cwm-mawr 31/5/71. Slade Wds 1/6/71. Hendre Wds 5/6/71. Redding's Inclosure 7/6/71. Prescoed 16/4/72. St Pierre's Great Wds 15/5/73. Trelleck Bog 22/5/73. Cwm Tyleri 26/5/73, 20/8/73. Cwm Coed-y-cerrig 5/6/73. Mynyddislwyn 10/6/74. Slade Wds 10/3/81. Llansoy 29/5/85. (MEA) : Mescoed Mawr 19/5/77, Risca 16/6/83, Cwmfelinfach 14/4/87, Ochrwyth 24/4/87, Lasgarn 20/5/87, Abercarn 26/5/87, Pontllanfraith 1/9/87, Nant Gwyddon 7/6/88, Clydach 11/5/90.

Sq. **SO** 20,30,40,50,21,41,51,22. **ST** 28,48,19,29,39,49,59.

PLUSIINAE

2434 *Diachrysia chrysitis* Linn.   Burnished Brass

Widespread and common, frequenting gardens, waste ground, marshy places etc., this moth is double-brooded, flying in June and July and again in August and September.

Wye Valley 1906 (Bird 1907). Abertillery dist. 1911 (Rait-Smith 1912).

Usk, garden m.v.l. trap, recorded every year from 1966 to 1983 and sometimes plentiful eg. 1971 from 2/6 to 29/7 and from 27/8 to 16/9. Ed. 24/5/80. Ld. 24/9/70. Usk (Park Wood) 15/6/67. Wyndcliff 16/6/67. Prescoed 5/7/68. Wentwood 22/7/68. Hendre Wds 5/7/70. Rhyd-y-maen 18/7/70. Magor Marsh 29/8/70. Cwm Tyleri 22/8/72. Newport Docks 17/8/76. Hael Wds 4/7/81. Slade Wds 28/7/81. Pont-y-saeson 13/7/81. Llansoy 5/7/86. Trelleck Bog 18/7/86. Kilpale 30/6/87. Sugar Loaf (St Mary's Vale) 8/7/87. (MEA) : Risca 18/7/79, Mescoed Mawr 13/7/82, Ochrwyth 12/7/86, Cwmfelinfach 12/8/86, Dixton Bank 18/6/88, Pontllanfraith 15/7/88, Nant

Gwyddon 15/8/88.

Sq. **SO** 20,30,40,50,21,41,51. **ST** 28,38,48,19,29,39,49,59.

2435 *Diachrysia chryson* Esp.   Scarce Burnished Brass
Tintern dist. (Piffard 1859). No recent records.

Sq. **SO** 50.

2437 *Polychrysia moneta* Fabr.   Golden Plusia
Occurs locally and sparingly in Monmouthshire, frequenting gardens from mid-June to mid-August.

Monmouth 1902, "a mile from Monmouth close to River Wye", in. colln. H. Green (Monmouth), (Thornewill 1904). Usk 1933, one in Castle gardens 7/7 (RBH). Bettws Newydd (Crowther 1935).

Usk, garden m.v.l. trap, recorded most years from 1967 to 1983. eg. 1967: (4), 24/6 – 23/8; 1968: (5), 13/6 – 19/7; 1983: (1), 23/7. Ed. 13/6/68. Ld. 23/8/67. Wentwood 1969 several to m.v.l. inc. 25/7 and 31/7 (JMD). Caerwent 1/7/74 (CT). Llansoy 1986 (3) 12/7 – 14/7. Risca 17/8/87 (MEA).

Sq. **SO** 30,40,51. **ST** 29,49.

2439 *Plusia festucae* Linn.   Gold Spot
Occurs locally in Monmouthshire, flying from mid-June to the middle of September in damp woodland, marshes and on boggy heaths and moors.

Wye Valley 1892 (Nesbitt 1892a). Abertillery dist. "usually abundant but scarce in 1912" (Rait-Smith 1912, 1913, 1915). Bettws Newydd 1933 (Crowther). Usk 1933, "(3) taken over lavender and (3) at red valerian in Usk Castle gardens on 22/7" (RBH).

Usk (Plas Newydd), garden m.v.l. trap, recorded in small numbers most years from 1966 to 1983 eg. 1966, (1), 10/9 ; 1970, (6) 26/6 – 28/8. Ed. 6/6/71. Ld. 10/9/66. Magor Marsh 23/8/68 (8) to m.v.l. and subsequently noted every year until 1976 (GANH), 2/9/87 (4) (MEA). Wentwood 23/8/69 (JMD). Caldicot (The Neddern) 26/8/70 (CT). Cwm Tyleri 30/7/72. Slade Woods 2/8/82. Llansoy 24/7/89. Magor Pill (sea-wall) 6/8/89 (9) to m.v.l. (MEA) : Risca 21/8/83, The Blorenge 7/7/84, Pontllanfraith 2/7/87, Trefil 3/7/87, 27/7/91 (abdt.), Uskmouth 16/8/88.

Sq. **SO** 20,30,40,11,21. **ST** 38,48,19,29,49.

2441 *Autographa gamma* Linn.   Silver Y
An immigrant moth which produces one or two generations after arrival in this country. Its numbers vary widely from year to year and it is sometimes scarce as in 1972, but in other years it is abundant and virtually ubiquitous as in 1982 and in the autumn of 1987. Usually noted from mid-May to late November and may often be seen visiting flowers in the daytime. The small

form ab. *gammina* Stdgr. occurs not uncommonly.

Llandogo (Nesbitt 1893). Wye Valley 1906 (Bird 1907). Tintern, Dec. 7th 1912 (Bird 1913). Abertillery dist. 1911 and 1914 (Rait-Smith 1911, 1915). Cwmyoy Oct. 1965 (SHR). Tintern 31/5/68 (JMC-H 1969).

Usk, garden m.v.l. trap, recorded every year from 1966 to 1983 eg. 1970 9/5 – 2/11 inc. (74) on 24/9; 1982, (abdt.), 15/5 – 10/11 (7) with peak numbers on 14/7 (93) and 19/9; Ed. 9/5/70, 15/5/82; Ld. 10/11/82, 25/11/75; the small ab. *gammina* appeared not infrequently eg. 1966, 11/8 and 3/9, 1969 26/7 and 16/8. Pont-y-saeson 13/6/67. Usk (Park Wood) 15/6/67. Prescoed 5/7/68. Wentwood 8/7/68. Trelleck Common 9/7/68. St Pierre's Great Wds 23/7/68. Bica Common 24/9/68. Rhyd-y-maen 18/7/70. Hendre Wds 2/8/70. Magor Marsh 29/8/70. Redding's Inclosure 27/8/71. Hael Wds 22/10/71. Slade Wds 24/10/71. Piercefield Park 1/7/73. Trelleck Bog 3/7/73. Llanvapley 18/8/73. Cwm Tyleri 20/8/73. Mynyddislwyn 25/8/73. Blaenavon Mountain 25/8/73. Peterstone Wentlooge 26/8/73. Newport Docks 17/8/76. Cwm Coed-y-cerrig 5/6/78. Tal-y-coed 9/9/83. Llansoy 5/6/85. Pontllanfraith 2/7/87. (MEA) : Ynysyfro 4/7/76, Llantarnam 10/6/76, Mescoed Mawr 5/9/77, Wyndcliff 12/9/77, Risca 8/10/79, Ochrwyth 27/6/86, Cwmfelinfach 26/9/86, Dinham 4/6/87, Uskmouth 16/8/88, Trefil 9/9/88.

Sq. **SO** 20,30,40,50,11,21,31,41,51,22,32. **ST** 28,38,48,19,29,39,49,59.

2442 *Autographa pulchrina* Haw.   Beautiful Golden Y

Widespread and fairly common, flying in woods and gardens during June and July.

Wye Valley 1906 (Bird 1907). Abertillery dist. 1911 and 1914 (Rait-Smith 1912, 1915). Usk, 1955, "five specimens taken at light and on red valerian at Usk Castle June 18 – 27" (RBH).

Usk, garden m.v.l.trap, recorded most years from 1966 to 1983 eg. 1970 (4) 26/6 – 17/7, 1981 (28) 11/6 – 7/7. Ed. 29/5/82. Ld. 20/7/71. Pont-y-saeson 13/6/67. Usk (Park Wood) 15/6/67. Wentwood 3/7/68. Prescoed 5/7/68. Bica Common 28/6/69. Magor Marsh 1/7/69. Redding's Inclosure 7/6/71. Hael Wds 3/7/71. Trelleck Bog 16/6/73. Newport (Coldra) 15/6/75. Slade Wds 21/6/81. Hendre Wds 18/7/81. Llansoy 22/6/86. (MEA) : Wyndcliff 7/7/83, Mescoed Mawr 8/7/84, Pant-yr-eos 6/7/85, Cwm Tyleri 7/7/86, Cwmfelinfach 12/8/86, Pontllanfraith 22/6/87, Abercarn 27/6/87, Dinham 7/7/87, Risca 10/7/87, Ochrwyth 17/6/88, Dixton Bank 18/6/88, Nant Gwyddon 25/6/88.

Sq. **SO** 20,30,40,50,41,51. **ST** 28,38,48,19,29,39,49,59.

2443 *Autographa jota* Linn.   Plain Golden Y

A common moth flying from late May to mid-August in gardens, open woods and rough places.

Llandogo dist., "to light 26/5/92" (Nesbitt 1892). Wye Valley 1906

(Bird 1907). Usk Castle 8/7/33, "two on red valerian" (RBH).

Usk dist. 1964, frequent. Usk, garden m.v.l. trap, recorded every year from 1966 to 1983 eg. 1973, (115) 27/5 – 31/7 with a single late moth on 22/8; 1981, (26) 21/6 – 18/8. Ed. 27/5/73. Ld. 22/8/73, 18/8/81. Pont-y-saeson 13/6/67. Usk (Park Wood) 4/6/68. Prescoed 5/7/68. Wentwood 8/7/68. Magor Marsh 1/7/69. Rhyd-y-maen 6/6/70. Hendre Wds 5/7/70. Cwm-mawr 2/7/71. Hael Wds 3/7/71. Redding's Inclosure 20/7/71. Newport (Coldra) 15/6/75. Cwm Tyleri 12/7/76. Wyndcliff 3/8/81. Tal-y-coed 9/9/83 (very late date). Llansoy 15/7/86. Trelleck Bog 18/7/86. Slade Wds 13/6/87. (MEA) : Mescoed Mawr 1/7/76, Ynysyfro 4/7/76, Risca 7/7/81, Cwm Merddog 14/7/86, Cwmfelinfach 12/8/86, Pontllanfraith 10/7/87, Cwmcarn 11/7/87, Dixton Bank 18/6/88.

Sq. **SO** 10,20,30,40,50,31,41,51. **ST** 38,48,19,29,39,49,59.

2444 *Autographa bractea* D. & S.   Gold Spangle
In Monmouthshire, this northern species is near the southern limits of its range in Britain. First recorded here in 1969, it has since been noted sporadically and sparingly, flying from early July to mid-August, in open woods, gardens, and in the moorlands of the north.

Usk Castle 1969, to m.v.l., 13/7 (2), 14/7 (1) (RBH) (1st v-c. rec.).

Usk (Plas Newydd), garden m.v.l. trap, 1969 15/7 (1), 1/8 (1); 1972 (3); 1973 (3); 1975 (1); 1976 (5); 1982 (1). Ed. 1/7/76. Ld. 14/8/73. Wentwood 1969 (4) 30/7 – 12/8 (JMD). Hael Wds 10/7/73, Cwm Tyleri 13/7/75 (GANH).

Sq. **SO** 20,30,50. **ST** 49.

2447 *Syngrapha interrogationis* Linn.   Scarce Silver Y
This northern moorland species occurs in the hills and moors in the north of the county and is almost certainly resident.

Cwm Tyleri 1976, 29/6 (2) to m.v.l. (GANH) (1st v-c. rec.); 8/7 (2) (AR, BG). Cwm Merddog 14/7/86 (MEA). Sugar Loaf (St Mary's Vale) 8/7/87 (1) (MEA, GANH). Trefil 7/8/88, 27/7/91; Cwm Tyleri 28/7/91 (2), 12/8/91 (MEA).

Sq. **SO** 10,20,11,21.

2449 *Abrostola trigemina* Werneb.   Dark Spectacle
This moth noted at Usk from early June to the beginning of September is local and scarce in Monmouthshire. During the eighteen years from 1966 to 1983 a total of (72) was recorded at my Usk garden m.v.l. trap at Plas Newydd but its numbers declined considerably over this period and I have not found it in any other locality in this county nor do I know of any other recent records.

Wye Valley 1906 (Bird 1907). Abertillery dist. 1911 (Rait-Smith 1912).

Usk, garden m.v.l. trap, recorded annually from 1966 to 1983 giving a total of (72) in this eight-year period eg. 1967 (10) 5/7 – 12/8; 1969 (21) 3/6 – 4/8 inc. (5) on 15/7; 1974 (7) 6/6 – 7/8. In the following nine years from 1975 to 1983 only (9) were seen: 1976 (3), 1978 (1), 1979 (3), 1980 (1) and 1981 (1). Ed. 31/5/68. Ld. 9/9/71.

Sq. **SO** 20,30,50.

2450 *Abrostola triplasia* Linn.   The Spectacle

Common throughout Gwent, flying from late April to early September in gardens, waste-ground, open woodland etc.

Wye Valley 1906 (Bird 1907). Tintern 1913 (Bird 1913). Abertillery dist. 1911 (Rait-Smith 1912).

Usk, garden m.v.l. trap, recorded annually from 1966 to 1983 and often plentiful eg. 1971 9/5 – 2/9, 1982 30/4 – 23/8. Ed. 28/4/72. Ld. 2/9/71. Wyndcliff 12/6/67. Usk (Park Wood) 15/6/67. Craig-y-Master 21/5/68. Prescoed 31/5/68. Wentwood 24/8/68. Rhyd-y-maen 23/5/70. Magor Marsh 2/6/70. Cwm-mawr 2/7/71. Redding's Inclosure 20/7/71. Hael Wds 21/7/71. Mynyddislwyn 10/6/74. Cwm Tyleri 29/6/76. Newport Docks 17/8/76. Llansoy 1/7/85. Hendre Wds 11/7/86. Dinham 4/6/87. Slade Wds 13/6/87. Kilpale 30/6/87. Clydach 11/5/90. (MEA) . Risca 14/7/80, Ynysddu 12/7/85, Ochrwyth 27/6/86, Cwmfelinfach 12/8/86, Trelleck Bog 25/5/87, Abercarn 27/6/87, Pontllanfraith 10/7/87, Dixton Bank 18/6/88, Nant Gwyddon 25/6/88.

Sq. **SO** 20,30,40,50,21,41,51. **ST** 28,38,48,19,29,39,49,59.

CATOCALINAE

2452 *Catocala nupta* Linn.   Red Underwing

Widespread and not uncommon in Gwent flying in wooded localities and marshy areas from early August to the end of October.

Llandogo 1892 (Nesbitt 1892b). Near Monmouth "on telegraph poles" 1906 (Bird 1907). Usk 1934, 6/8 (1), and another "at sugar" 23/8 (RBH). Caldicot (The Nedern) 7/9/62, Caerwent (Brockwells) 12/10/62, Caerwent Quarries 23/8/69 (CT in litt.).

Usk, noted most years from 1966 to 1983 either flying round buildings, resting on house walls by day or coming to garden m.v.l. trap at night eg. 1969 (10) 19/8 – 25/10, 1970 (21) 6/8 – 30/9, 1973 (3) 4/9 – 14/9. Ed. 6/8/70. Ld. 30/10/72. Wyndcliff 2/9/69. Magor Marsh 29/8/70 (2) to m.v.l. Undy (Collister Pill) 30/8/70. Rhyd-y-maen 26/8/71 to m.v.l. and one on trunk of dead elm 6/9/76. Llansoy 2/8/76, 1/10/86. Tal-y-coed 9/9/83. (MEA) : Risca 10/8/84, Cwmfelinfach 26/9/86.

Sq. **SO** 30,40,31,51. **ST** 48,19,29,39,49,59.

2462   *Callistege mi* Cl.   Mother Shipton
Flies by day during May and June in flowery meadows and woodland rides and on sunny embankments and hillsides but is local and scarce in Monmouthshire.

Llandogo (Nesbitt 1893). Wye Valley (Barraud 1906). Wye Valley 1906, imagines and larvae (Bird 1907). Abertillery dist. 1911 and 1912 (Rait-Smith 1912, 1913).

Wentwood (Foresters' Oaks) June 1968, 28/6/69 (1) (JMD). Wentwood 25/6/74 (CT). Tintern (Old Furnace) 9/6/69 (1) (CGMdW). Slade Wds 12/6/76 (MEA). Panta Arch (Pont-y-saeson) 27/5/78 (svl.), 5/6/78 (GANH). Gaer Hill, Newport 16/6/87 (1); Trelleck (Fedw Fach) 27/5/88 (1); Pontllanfraith 1988 1/6, 17/6, 30/6 (GANH, MEA).

Sq. **SO** 20,50. **ST** 28,48,19,49,59.

2463   *Euclidia glyphica* Linn.   Burnet Companion
Flying by day in May and June, this species frequents flowery situations such as sunny meadows and embankments. Local in Monmouthshire but occasionally plentiful where it occurs.

Llandogo (Nesbitt 1893). Wye Valley (Barraud 1906). Wye Valley 1906 (Bird 1907). Abertillery dist. 1912 (Rait-Smith 1913). Tintern 31/5/68 (JMC-H 1969).

Pont-y-saeson June 1958, 27/6/78. Wentwood (Foresters' Oaks) 10/6/69. Cicelyford 9/6/69. Tintern (Old Furnace) 9/6/69 (plfl.). Slade Wds 30/5/71 (2). Hendre Wds 9/6/79. Fleur-de-lis 17/6/88, "swarmed" in huge numbers on roadside verges and on old railway embankment (MEA, GANH). (MEA) : Gaer Hill, Newport 13/6/87; Pontllanfraith 20/6/87, 3/7/87; Dixton Bank 18/6/88.

Sq. **SO** 20,50,41,51. **ST** 28,48,19,49,59.

OPHIDERINAE

2466   *Lygephila pastinum* Treit.   The Blackneck
This moth, flying in late June and July, is very scarce and of sporadic occurrence in Monmouthshire.

Usk, garden m.v.l. trap, 10/7/67 (1) (GANH) (1st v-c. rec.). In the eighteen years from 1966 to 1983 a total of (13) was recorded: 1967, (5) 10/7 – 15/7; 1968, (2); 1969, (2); 1972, (2); and single moths in 1971 and 1975. Ed. 30/6/68. Ld. 26/7/71. Wentwood (Foresters' Oaks) 4/7/70 (JMD). Magor Marsh 1976 2/7 (1), 6/7 (1); 1989 2/7 (1). Llansoy 6/7/87 (1). (MEA) : Mescoed Mawr 11/7/79; Risca 10/7/79, 1984, 1985; Dinham 7/7/87.

Sq. **SO** 30,40. **ST** 48,29,49.

2469   *Scoliopteryx libatrix* Linn.   The Herald
Widespread and fairly common in Monmouthshire, inhabiting gardens, woodland etc. and flying in July and August and through to November. In the winter months these moths congregate in numbers in caves, outhouses, cellars, etc. until they reappear in April and May after hibernation.

Wye Valley 1906 (Bird 1907). Tintern 1946 – 1948, "several in all three years by day and, at light by night" (Blathwayt 1967). Caerwent 19/9/63; Piercefield House cellars 2/2/74 (6), 12/1/75 (15), 17/8/75 (12); cave at Graig-y-Garcoed (Usk) 24/3/74 (3) (CT in litt.).

Usk, 1961, and noted annually at garden m.v.l. trap from 1966 to 1983 eg. 1969 8/4(1), 7/6(2), 8/6(1), then (9) from 16/7 to 30/7; 1973 16/5 (1), 25/5 (1), 5/6 (1), then (5) from 24/7 to 3/8, 23/10 (1), 1/11 (2). Hendre Wds 26/5/70. Hael Wds 21/5/73. Magor Marsh 1974 (MJL), 12/8/87 (GANH). Llansoy 16/6/86. (MEA) : Rhiwderin 21/5/76, Mescoed Mawr 3/7/76, Ynysyfro 4/7/76, Wentwood 8/9/77, Wyndcliff 26/5/77, Black Vein 7/5/81, Cwm Tyleri 29/9/86, Cwmfelinfach 9/10/86, Nant Gwyddon 25/6/88.

Sq. **SO** 20,30,40,50,41. **ST** 28,48,19,29,49,59.

2470   *Phytometra viridaria* Cl.   Small Purple-barred
Local and scarce in Gwent, flying from late May to mid-August in open woods and woodland rides and on heaths and moorland.

Abertillery dist. 1911 and 1912 (Rait-Smith 1912, 1913). Tintern 1968, 31/5 (JMC-H 1969).

Pont-y-saeson 1967, to m.v.l. (REMP). Wentwood (Foresters' Oaks), 1969 to m.v.l. trap 21/7 and 9/8 (JMD). Magor Marsh 22/7/69, Slade Wds (by day) 30/5 (1) (GANH). Blorenge Mountain (WEK). (MEA) : Wyndcliff 20/7/83, 6/7/84; Pontllanfraith 16/8/87 (2) to m.v.l; Brynawel (Sirhowy Valley) 12/7/89.

Sq. **SO** 20,50,21. **ST** 48,19,29,49,59.

2473   *Laspeyria flexula* D. & S.   Beautiful Hook-tip
Local but not uncommon in the eastern half of the county, flying in damp woodland from late June to mid-August.

Usk, garden m.v.l. trap, a few recorded every year from 1968 to 1983 eg. 1968 (4) 13/7 – 22/7; 1972 (4) 19/7 – 23/7. Ed. 5/6/82. Ld. 12/8/70 and a very late moth on 30/9/83. Wyndcliff 10/7/67, 21/7/87 (plfl.). Magor Marsh 22/7/69. Piercefield Park 18/7/73. Trelleck Bog 16/7/79. Slade Wds 28/7/81. Redding's Inclosure 24/7/83. Llansoy 26/7/84, 1987 (11) 29/6 – 13/7. Hendre Wds 11/7/86. Hael Wds 2/8/86. (MEA) : Risca 12/7/86.

Sq. **SO** 30,40,50,41,51. **ST** 48,29,59.

2474   *Rivula sericealis* Scop.   Straw Dot
Occurs locally and at low density in Monmouthshire, flying from late June to

early August in damp woodland and marshy situations.

Llandogo, one, July 1906 (Bird 1907).

Usk, garden m.v.l. trap, 1969: (7), 22/7 – 30/8; 1981: (1), 26/7; 1982: (2), 13/7 and 22/7; these being the only records in the eighteen years from 1966 to 1983. Wentwood (Foresters' Oaks) 35/7/69 (JMD). Redding's Inclosure 20/7/71, 24/7/83. Magor Marsh 11/8/71 (svl.), 21/6/76. St Pierre's Great Wds 13/7/73. Hendre Wds 1/8/80, 11/7/86. Slade Wds 2/8/80, 6/8/84. Wyndcliff 3/8/81. Hael Wds 31/7/82. Fforest Coal Pit 7/7/78 (plfl.).

Sq. **SO** 30,50,41,51,22. **ST** 48,49,59.

2475   *Parascotia fuliginaria* Linn.   Waved Black

Usk (Plas Newydd), 1970, one to garden m.v.l. trap on 3/8 (1st v-c. rec.), and another on 30/7/73 (GANH). These are the only Monmouthshire records of this rare species. Elsewhere in Wales, it has only been reported from one other site, in the south-west of the Pricipality.

Sq. **SO** 30.

HYPENINAE

2476   *Hypena crassalis* Fabr.   Beautiful Snout

This moth, in flight during June and July, is locally common in the eastern half of Monmouthshire, especially the Wye Valley, in woodland where whinberry (bilberry) grows. It has also recently been recorded from the hills of western Gwent. The male is easily disturbed from bilberry bushes during the day, while at night the females but, in my experience, never the males, come to light.

Tintern dist. (Piffard 1859). Wye Valley 1892 (Nesbitt 1892a). Wye Valley 1906, larvae common on *Vaccinium myrtillus* July 4 to last week of August (Bird 1907). Tintern dist. (Mere 1965).

Whitelye July 1947, males abundant and easily disturbed by day among whinberry bushes (GANH). Trelleck Common 14/6/67 (5) males by day (REMP, GANH). Wentwood 20/6/69 (♀♀ 2) to m.v.l. (GANH) and singletons on 25/7, 26/7 and 9/8 (JMD). Wyndcliff 13/7/69 (♀). Hael Wds 5/7/70 etc. Usk, garden m.v.l. trap, 12/7/83 ♀ (1) (the only Usk record in eighteen years). Trelleck Bog 19/7/86. (MEA) : Cwm Merddog 14/7/86, Cwmfelinfach 15/7/86, Abercarn 27/6/87, Nant Gwyddon 25/6/88.

Sq. **SO** 10,30,50. **ST** 19,29,49,59.

2477   *Hypena proboscidalis* Linn.   The Snout

Common and widespread, inhabiting gardens, woodland, waste-ground etc., the Snout flies from mid-June to mid-August and again, as a second brood, in September and October.

Wye Valley 1906 (Bird 1907). Abertillery dist. 1911 (Rait-Smith 1912).

Usk, garden m.v.l. trap, noted every year from 1966 to 1983 eg. 1982 plentiful from 30/6 to 4/8 and from 3/9 to 18/9. Ed. 15/6/76. Ld. 29/9/70. Wyndcliff 22/7/68. Hael Wds 5/7/70. Rhyd-y-maen 18/7/70. Tintern 1/8/70. Magor Marsh 2/8/70. Redding's Inclosure 20/7/71. Pont-y-saeson 18/7/72. Cwm Tyleri 28/7/72. Trelleck Bog 16/6/73. Wentwood 29/6/73. Hendre Wds 20/7/79. Llansoy 26/7/84. Ysguborwen (Llantrisant) 17/7/86. Slade Wds 22/7/86. Kilpale 30/6/87. (MEA) : Llantarnam 24/6/76, Mescoed Mawr 25/6/76, Ynysyfro 28/6/76, Black Vein 3/7/81, Risca 19/6/82, Newport Docks 11/7/84, Ynysddu 12/7/85, Cwmfelinfach 15/7/86, Sugar Loaf (St Mary's Vale) 8/7/87, Pontllanfraith 10/7/87, Nant Gwyddon 25/6/88.

Sq. **SO** 20,30,40,50,21,41,51. **ST** 28,38,48,19,29,39,49,59.

2480 *Hypena rostralis* Linn.   Buttoned Snout
Scarce in Monmouthshire, the few records of this species being of individuals noted in the Spring after hibernation.

Tintern, May 24 1912 (Bird 1913).

Usk, 1967, one to garden m.v.l. trap 29/5; 1968, one indoors 5/3 and one to m.v.l. trap, 10/5 (GANH).

Sq. **SO** 30. **ST** 50.

2482 *Schrankia taenialis* Hb.   White-line Snout
This moth described as "local" but with a "widespread distribution" in southern Britain has, to date, only been recorded with certainty from one woodland site in the south of the county.

Slade Wds 1982, one to m.v.l., 2/8 (1st v-c. rec.) (GANH); 1983, 21/7 (3); 1987, 17/7 (1).

Sq. **ST** 48,59.

2484 *Schrankia costaestrigulis* Steph.   Pinion-streaked Snout
Scarce and local in Monmouthshire, this species inhabiting damp woods, fens, bogs and river valleys flies from July to October and comes to light.

Wye Valley 1906 (Bird 1907).

Usk, garden m.v.l. trap, 24/9/70 (1), 23/7/71 (3) and single moths in 1974, 79 and 81. Magor Marsh 10/7/70 (1), 29/9/70 (1), 2/10/86 (1). Trelleck Bog 19/7/86. (MEA) : Wentwood 4/8/84, Wyndcliff 1984, Risca 13/8/86.

Sq. **SO** 30,50. **ST** 48,29,49,59.

2485 *Hypenodes humidalis* Doubl.   Marsh Oblique-barred
       = *turfosalis* Wocke
This small moth occurring locally in the British Isles on boggy heaths has so far only been noted at one site in the east of the county.

Trelleck Bog 1986, one to m.v.l. 19/7 (1st v-c. rec.) (GANH), 24/7 (2);

1988 17/8 (1).

Sq. **SO** 50.

2489 *Herminia tarsipennalis* Treit.　　The Fan-foot
Widespread and common in Monmouthshire, flying in gardens, hedgerows and woodland from mid-June to late August.
　　Wye Valley 1906 (Bird 1907.).
　　Usk, garden m.v.l. trap, recorded annually from 1966 to 1983 and often plentiful eg. 1972, 6/7 – 22/8 and a late, possibly second generation, moth on 23/9; 1973, 17/6 – 29/8. Ed. 15/6/76. Ld. 29/8/73, 15/8/77. Prescoed 5/7/68. Wentwood 8/7/68. Trelleck Common 9/7/68. Magor Marsh 1/7/69. Wyndcliff 3/7/69. Hendre Wds 5/7/70. Hael Wds 5/7/70. Redding's Inclosure 20/7/71. St Pierre's Great Wds 13/7/73. Piercefield Park 18/7/73. Usk (Park Wood) 29/7/82. Llansoy 7/7/85. Trelleck Bog 19/7/86. (MEA) : Slade Wds 11/6/76, Ynysyfro 22/6/76, Llantarnam 24/6/76, Mescoed Mawr 3/7/76, Black Vein 3/7/81, Risca 6/7/83, Ynysddu 12/7/85, Ochrwyth 12/7/86, Cwmfelinfach 15/7/86, Pontllanfraith 22/6/87, Abercarn 16/6/88.

Sq. **SO** 30,40,50,41,51. **ST** 28,48,19,29,39,49,59.

2492　*Herminia grisealis* D. & S.　　Small Fan-foot
　　= *nemoralis* Fabr.
Common throughout Gwent inhabiting woods, gardens, hedgerows etc. and flying from early June to mid-August or, in some years, even later.
　　Wye Valley (Barraud 1906). Wye Valley 1906 (Bird 1907).
　　Usk, garden m.v.l. trap, noted every year from 1966 to 1983 and usually plentiful eg. 1972, 14/7 – 20/8 and a late moth on 13/9; 1973 10/6 – 16/8. Ed. 29/5/68. Ld. 13/9/72, 16/8/73. Usk (Park Wood) 15/6/67. Wyndcliff 23/6/67. Wentwood 28/6/69. Hendre Wds 5/7/70. Hael Wds 5/7/70. Redding's Inclosure 20/7/71. Pont-y-saeson 18/7/72. Piercefield Park 18/7/73. Slade Wds 23/7/79. Llansoy 1/7/85. Ysguborwen (Llantrisant) 17/7/86. Kilpale 30/6/87. (MEA) : Llantarnam 10/6/76, Allt-yr-yn 20/6/76, Ynysyfro 22/6/76, Mescoed Mawr 3/7/76, Tintern 30/6/78, Risca 2/7/80, Black Vein 3/7/81, Pant-yr-eos 16/6/83, Trelleck 19/7/83, Ynysddu 12/7/85, Abercarn 16/7/85, Ochrwyth 27/6/86, Cwmfelinfach 15/7/86, Pontllanfraith 10/7/87, Dixton Bank 18/6/88, Nant Gwyddon 25/6/88.

Sq. **SO** 30,40,50,41,51. **ST** 28,48,19,29,39,49,59.

Plate 1: The Gwent Levels Drainage Reens.

Plate 2: The Severn Estuary. Salt marsh at Magor Pill.

Plate 3: Magor Marsh Reserve.

Plate 4: Reed-bed at Magor Reserve.

Plate 5: Reen at Magor.

Plate 6: The contracting Trellech Bog.

Plate 7: Hael Woods, haunt of the White Admiral and Great Oak Beauty *(Boarmia roboraria)*.

Plate 8: Ancient Woodland in the Lower Wye Gorge.

Plate 9: Ancient Woodland on the Carboniferous limestone hills and cliffs of the lower Wye Valley.

Plate 10: Looking north over the Olway flood-plain and Vale of Usk to the distant Black Mountains.

Plate 11: North-west Monmouthshire - the haunt of the Silurian.

Plate 12: The first British specimen of *Eriopygodes imbecilla* (The Silurian) was captured in 1972 in the quarry on the right of the photograph.

Plate 13: Gully (1,550 ft) where the Silurian flew in hot afternoon sunshine. The stone walls in centre of photograph (1,400 ft) harbour the Pyralid moth *Eudonia murana*.

Plate 14: The head of the "Silurian Gully" at 1,700 ft, haunt of the Wood tiger *(Parasemia plantaginis)* and site of an encounter with a vagrant Marbled White butterfly.

Plate 15: Woodlands in the Clydach Gorge NNR notable for their mature self-regenerating beeches.

Plate 16: Clydach. Disused quarry on Carboniferous Limestone. Haunt of the moths Thyme Pug, Light Feathered Rustic and *Pyrausta cingulata*.

Plate 17: North-west Monmouthsire - disused quarries and moorland near Trefil. *Anarta myrtilli* flies here.

Plate 18: Moorland in the extreme north-west of the county looking towards the distant Brecon Beacons.

Plate 19: Six-spot Burnet and Narrow-bordered Five-spot Burnet Moths on flowers of tufted vetch. Dixton Bank, Monmouth 1988.

Plate 20: Comma butterfly St Pierre's Great Woods 1991.

Plate 21: White Admiral on bramble flower. Hendre Woods.

Plate 22: Small Pearl-bordered Fritillary on marsh thistle Fforest Coal Pit 1988.

Plate 23

Plate 24

Plate 25

Plate 26

Plate 27

Section 2: The smaller moths (microlepidoptera)

MICROPTERIGIDAE

1 *Micropterix tunbergella* Fabr.
= *thunbergella* auct.
Usk, old railway line, 11/5/80, several flying in evening sunshine. Slade Wds 2/6/80 abdt. by day, 3/6 d°.

Sq. **SO** 30. **ST** 48.

3 *Micropterix aureatella* Scop.
Pont-y-saeson (Panta Arch), 5/6/78 flying in numbers in sunshine. (GANH). Hael Wds 29/5/82, to m.v.l. (JDB, GANH).

Sq. **SO** 50. **ST** 59.

4 *Micropterix aruncella* Scop.
Tintern, 31/5/68 (JMC-H 1969). Usk, old railway line, 20/5/82, one on hawthorn flowers. (GANH).

Sq. **SO** 30. **ST** 50.

5 *Micropterix calthella* Linn.
Usk, 1982, abundant on buttercups 20/5 – 31/5. Penorth Mill near Trelleck, 24/5/82. Hendre Wds 29/5/82, "swarming" on buttercups.

Sq. **SO** 30,40,41.

ERIOCRANIIDAE

6 *Eriocrania subpurpurella* Haw.
Redding's Inclosure, 11/4/72, to m.v.l. (plfl.) Wyndcliff 26/4/76. Hendre Wds 27/4/79, 1980 abundant 24/4 – 29/4. Pont-y-saeson 27/4/80. Hael Wds 26/4/82. Llansoy 23/4/87.

Sq. **SO** 50,41,51. **ST** 59

8 *Eriocrania unimaculella* Zett.
MBGBI 1: Map 8 indicates v-c. 35 (Monmouthshire).

11 *Eriocrania haworthi* Bradl.
MBGBI 1: Map 11 indicates v-c. 35 (Monmouthshire).

NEPTICULIDAE

40 *Bohemannia pulverosella* Stt.
Llandogo, 8/9/73 (AME).

Sq. **SO** 50.

23 *Ectoedemia argyropeza* Zell.
Hendre Wds, 9/11/76, mines on aspen with larvae feeding, (coll. GANH, det. AME). (1st v-c. rec.) Hendre Wds 2/6/80 to m.v.l., and imago taken by day on 29/5/82 (JDB). Dixton Bank 4/11/84 (AME).

Sq. **SO** 41,51.

28 *Ectoedemia angulifasciella* Stt.
Trelleck Bog, 8/9/73, mines on *Rosa* spp. (AME).

Sq. **SO** 50.

29 *Ectoedemia atricollis* Stt.
Wye Valley, 7/9/73, (AME).

Sq. **SO** 50.

30 *Ectoedemia arcuatella* H.-S.
Wyndcliff, 19/9/90, tenanted mines on *Fragaria*. "New to Wales." (AME in litt.).

Sq. **ST** 59.

34 *Ectoedemia occultella* Linn.
 = *argentipedella* Zell.
Hendre Wds, 2/6/80, imago to m.v.l. (JDB, GANH). Dixton Bank, 4/11/84, mines on birch (AME).

Sq. **SO** 41.51.

35 *Ectoedemia minimella* Zett.
 = *mediofasciella* auct.

Trelleck Bog, 8/9/73, mines on birch (AME).

Sq. **SO** 50.

37 *Ectoedemia albifasciella* Hein.
Tintern, 7/9/73, larvae on oak (AME). Hendre Wds, 9/11/76 mines on oak (GANH). Near Monmouth 4/11/84 (AME).

Sq. **SO** 50,41.

38 *Ectoedemia subbimaculella* Haw.
Hendre Wds, 9/11/76, larval mines on oak, (coll. GANH, bred and det. AME) (1st v-c. rec.). Dixton Bank, 4/11/84 (AME). Wyndcliff 23/6/86, Llansoy 25/6/86 (1), Hael Wds 26/6/86 imagines abdt. to m.v.l. (JDB, GANH).

Sq. **SO** 40,50,41,51. **ST** 59.

39 *Ectoedmia heringi* Toll
     = *quercifoliae* Toll
Hendre Wds, 9/11/76 on oak (two vacated mines in one leaf and two larvae feeding in another) (coll. GANH, det. AME) (1st v-c. rec.). Dixton Bank 4/11/84 (AME).

Sq. **SO** 41,51.

42 *Fomoria septembrella* Stt.
Pont-y-saeson, 11/10/81, several mines in leaves of *Hypericum perforatum*, bred (GANH). (1st v-c. rec.). Dixton Bank 4/11/84 (AME). Wyndcliff 19/9/90 (AME).

Sq. **SO** 51. **ST** 59.

50 *Stigmella aurella* Fabr.
Redding's Inclosure, 31/5/68, larval mines on *Rubus*, and Deri Fach 2/6/68 (JMC-H 1969, as *Nepticula aurella*). Tintern 7/9/73, Llandogo 8/9/73,
     Trelleck Bog 8/9/73, Dixton Bank 4/11/84 (AME). Pont-y-saeson, larval mines on *Rubus* 8/2/86 (GANH). Wyndcliff 19/9/90 (AME).

Sq. **SO** 50,21,51. **ST** 59.

51 *Stigmella gei* Wocke
Llandogo, 8/9/73, on *Geum rivale* (AME).

Sq. **SO** 50.

52 *Stigmella dulcella* Hein.
Llandogo, 8/9/73, (AME).

Sq. **SO** 50.

53 *Stigmella splendidissimella* H.-S.
Wyndcliff, 19/9/90, vacated mines on *Fragaria* and *Rubus* spp (1st v-c. rec.) (AME in litt.).

Sq. **ST** 59.

61   *Stigmella serella* Stt.
Trelleck Bog, 8/9/73, "on *Potentilla erecta*, moths bred" (AME).

Sq. **SO** 50.

63   *Stigmella marginicolella* Stt.
Wye Valley, 7/9/73 and Dixton Bank, 4/11/84 on elm (AME).
    Wyndcliff, on *Ulmus glabra*, 19/9/90 (AME).

Sq. **SO** 50,51. **ST** 59.

64   *Stigmella continuella* Stt.
Trelleck Bog, 8/9/73, on birch (AME).

Sq. **SO** 50.

65   *Stigmella speciosa* Frey
Dixton Bank, 4/11/84, on sycamore (AME) (1st v-c. and Welsh rec.).
    Wyndcliff, 19/9/90 (AME).

Sq. **SO** 51. **ST** 59.

66   *Stigmella sorbi* Stt.
Trelleck Bog, 8/9/73, on rowan, (AME).

Sq. **SO** 50.

67   *Stigmella plagicolella* Stt.
Llandogo, 8/9/73, (AME).

Sq. **SO** 50.

68   *Stigmella salicis* Stt.
Llandogo, 8/9/73 (AME). Hendre Wds, 9/11/76, three larvae feeding in one sallow leaf (GANH).

Sq. **SO** 50.41.

72   *Stigmella myrtillella* Stt.
Trelleck Bog, 8/9/73 on *Vaccinium myrtillus*, "moths bred" (AME).

Sq. **SO** 50.

75   *Stigmella floslactella* Haw.
Llandogo 8/9/73, and Dixton Bank 4/11/84, on hazel (AME). Wyndcliff, 19/9/90 (AME).

Sq. **SO** 50,51. **ST** 59.

77   *Stigmella tityrella* Stt.
Llandogo, 8/9/73 on beech (AME). Wyndcliff, 19/9/90 (AME).

Sq. **SO** 50. **ST** 59.

79   *Stigmella perpygmaeella* Doubl.
       = *pygmaeella* Haw.
Llandogo, 8/9/73 on hawthorn (AME).

Sq. **SO** 50.

80   *Stigmella ulmivora* Fol.
Tintern, 7/9/73 on elm (AME).

Sq. **SO** 50.

81   *Stigmella hemargyrella* Koll.
Llandogo, 8/9/73 on beech (AME).

Sq. **SO** 50.

83   *Stigmella atricapitella* Haw.
Hendre Wds, 9/11/76, larvae feeding in oak leaves (coll. GANH; bred and det. AME) (1st v c. rec.). Also Hendre Wds, mine collected 13/8/81, bred, imago emerged 18/4/82 (GANH). Dixton Bank 4/11/84 (AME).

Sq. **SO** 41,51.

84   *Stigmella ruficapitella* Haw.
Dixton Bank, 4/11/84, mines in oak leaves (AME). Wyndcliff, 19/9/90 (AME).

Sq. **SO** 50. **ST** 59.

90   *Stigmella tiliae* Frey
St Pierre's Great Wds, 31/10/76, mines in leaves of *Tilia europaea* (GANH) (1st v-c. rec.). Detn confirmed by AME who wrote in litt. "The second British record this year for this "nep" on common lime". It appears that hitherto in Britain this locally occurring species was only known on the small-leaved lime (*Tilia cordata*).

   Wyndcliff, common 19/9/90 (AME). AME in litt. Oct. 1990 wrote "*S. tiliae* has now adapted to common lime and occurs on that food-plant in many localities where small-leaved lime is about."

Sq. **ST** 59.

92   *Stigmella anomalella* Goeze
Trelleck Bog 8/9/73, on *Rosa* spp. and Dixton Bank 4/11/84 (AME).
    Wyndcliff, 19/9/90 (AME).

Sq. **SO** 50,51. **ST** 59.

95   *Stigmella viscerella* Stt.
Near Monmouth 7/10/76, on elm (AME). Dixton Bank 4/11/84 (AME).

Sq. **SO** 41,51.

97   *Stigmella malella* Stt.
Llandogo 8/9/73, on apple (AME).

Sq. **SO** 50.

99   *Stigmella hybnerella* Hb.
Llandogo, on hawthorn, 8/9/73 and Dixton Bank 4/11/84 (AME).

Sq. **SO** 50,51.

99a   *Stigmella mespilicola* Frey
Wyndcliff, 19/9/90, vacated mines on Sorbus torminalis, "new to Wales" (JRL, per AME in litt.).

Sq. **ST** 59.

100   *Stigmella oxyacanthella* Stt.
    = *aeneella* auct.
Near Monmouth 7/10/76 and Dixton Bank 4/11/84 (AME).

Sq. **SO** 41,50.

103   *Stigmella nylandriella* Tengst.
    = *aucupariae* Frey
MBGBI 1: Map 98 indicates v-c. 35 (Monmouthshire). AME in litt. writes, "found almost wherever there is rowan, and occurs in v-c. 35."

104   *Stigmella magdalenae* Klim.
    = *nylandriella* sensu auct.

Trelleck Bog, on rowan, 8/9/73 (AME). Mines in colln. AME. (pers. comm.).

Sq. **SO** 50.

108   *Stigmella crataegella* Klim.
Llandogo, 8/9/73, on hawthorn (AME).

Sq. **SO** 50.

110 *Stigmella betulicola* Stt.
Trelleck Bog, 8/9/73, on birch (AME).

Sq. **SO** 50.

111 *Stigmella microtheriella* Stt.
Llandogo, 8/9/73 and Dixton Bank 4/11/84 (AME). Wyndcliff, 19/9/90 (AME).

Sq. **SO** 50,51. **ST** 59

112 *Stigmella luteella* Stt.
Trelleck Bog, on birch, 8/9/73 and Dixton Bank 4/11/84 (AME). St Pierre's Great Wds 31/10/76 and Hendre Wds 9/11/76 (coll. GANH, det. AME).

Sq. **SO** 50,41,51. **ST** 59.

113 *Stigmella sakhalinella* Puplesis
   = *distinguenda* Hein.
Dixton Bank, 4/11/84, on birch (AME).

Sq. **SO** 51.

114 *Stigmella glutinosae* Stt.
Near Monmouth, on alder, 7/10/76 and Dixton Bank 4/11/84 (AME).

Sq. **SO** 41,51.

116 *Stigmella lapponica* Wocke
Trelleck Bog, on birch, 8/9/73 (AME).

Sq. **SO** 50.

117 *Stigmella confusella* Wood
Trelleck Bog, on birch, 8/9/73 (AME).

Sq. **SO** 50.

TISCHERIIDAE

123 *Tischeria ekebladella* Bjerk.
Tintern 7/9/73 (AME). Usk (Plas Newydd garden) 7/11/76 larvae feeding in oak leaves (coll. GANH, det. AME). Wentwood and Pont-y-saeson, 11/10/80 (GANH). Dixton Bank, 4/11/84 (AME). Hael Wds 26/6/80, imago to m.v.l. (JDB). Wyndcliff, 17/9/90 (AME).

Sq. **SO** 30,50,51. **ST** 49,59.

124 *Tischeria dodonaea* Stt.
Hendre Wds 9/11/76, larva in oak leaf, feeding. (coll. GANH, det. AME). (1st v-c. rec.).

Sq. **SO** 41.

125 *Tischeria marginea* Haw.
Dixton Bank, 4/11/84, mines (AME) (1st v-c. rec.). Hael Woods, 26/6/86, one imago to m.v.l. (JDB).

Sq. **SO** 50,51.

127 *Tischeria angusticollella* Dup.
Trelleck Bog, 8/9/73 (AME).

Sq. **SO** 50.

INCURVARIIDAE
INCURVARIINAE

128 *Phylloporia bistrigella* Haw.
MBGBI 1: Map 123 indicates v-c. 35 (Monmouthshire).

129 *Incurvaria pectinea* Haw.
MBGBI 1: Map 124 indicates v-c. 35 (Monmouthshire).

130 *Incurvaria masculella* D. & S.
Mountain Air (Pontypool), 25/5/80 abdt. by day with many in cop. sitting on boulders among the bilberry bushes. Llansoy 27/4/87. Hendre Wds 9/5/87, one to m.v.l.

Sq. **SO** 40,41. **ST** 29.

PRODOXINAE

131 *Lampronia oehlmanniella* Hb.
Trelleck Beacon 18/5/87, one on bilberry (1st v-c. rec) (GANH).

Sq. **SO** 50.

135 *Lampronia luzella* Hb.
Panta Arch (Angidy Valley) 6/6/72, one by day (1st v-c. rec) (GANH). Slade Woods 3/6/80. Hendre Wds 29/5/82, one (JDB). Wyndcliff, one to m.v.l. 23/6/86 (JDB).

Sq. **SO** 41. **ST** 48,59.

140 *Nematopogon swammerdamella* Linn.
Wye Valley (as *Nemophora swammerdamella* Linn.) (Barraud 1906). Aberbeeg (Rait-Smith 1913). Hael Wds 29/4/78 (1). St Pierre's Great Wds 13/5/80, several at dusk (det. JDB). Cwmfelinfach 9/5/87 (1).

Sq. SO 20,50. ST 19,59.

141 *Nematopogon schwarziellus* Zell.
 = *panzerella* auctt.
Wye Valley (as *Nemophora schwarzella*) (Barraud 1906). Deri-fach 2/6/68 and Coed-y-Bwnydd 5/6/68 (JMC-H 1969). Hael Wds 29/5/82 several to m.v.l. (JDB, GANH). Wyndcliff 25/6/86, one to m.v.l. (JDB). St Pierre's Great Wds, 19/5/87, common. Llansoy, 10/6, 13/6/90.

Sq. SO 30,40,50,21. ST 59.

143 *Nematopogon metaxella* Hb.
Hendre Wds 2/6/79, one, by day (GANH) (1st v-c. rec.) (det. JDB). Hael Wds 29/5/82 (3) to m.v.l. (JDB, GANH). Trelleck Bog 28/5/87.

Sq. SO 50,41.

ADELINAE

147 *Nemophora metallica* Poda
 = *scabiosella* Scop.
Runston 25/7/78, numbers flying by day around the flowers of a clump of field scabious (*Knautia arvensis*) (GANH) (1st v-c. and Welsh rec.).

Sq. ST 49.

148 *Nemophora degeerella* Linn.
Coed-y-Bwnydd 5/6/68 (JMC-H 1969). Pont-y-saeson 15/6/71 (abdt.), 21/6/77. Hendre Wds 2/6/80. Hael Wds 30/6/76 (MWH in litt.).

Sq. SO 30,50,41. ST 59.

150 *Adela reaumurella* Linn.
Abertillery dist. (as *Adela viridella*) (Rait-Smith 1912). Tintern and Redding's Inclosure 31/5/68, Deri fach 2/6/68 (JMC-H 1969).
 Usk, garden m.v.l. trap, 18/5/73 (♀ 1). Usk, old railway line, by day 28/4/77 (plfl.). Hael Wds 21/5/73, 29/5/82 (to m.v.l.). Pont y-saeson, 26/5/73, swarming round bushes in sunshine; 30/4/77 (abdt.). Slade Wds 29/5/79. Llandogo 26/5/79 (plfl.). Trostrey Common 11/5/80 (2). St Pierre's Great Wds 19/5/87. Llangeview 19/5/91. Llantrisant 21/5/91.

Sq. SO 20,30,50,21,51. ST 48,39,59.

151 *Adela croesella* Scop.
Magor Marsh 13/7/83, one to m.v.l.. Wyndcliff, 23/6/86 and Hael Wds, 26/6/86 (JDB, GANH).

Sq. **SO** 50. **ST** 48,59.

152 *Adela rufimitrella* Scop.
Magor Marsh, 27/5//82, one netted by day (JDB) (1st v-c. rec.).

Sq. **ST** 48.

HELIOZELIDAE

154 *Heliozela sericella* Haw.
"Monmouthshire, widespread" (AME in litt.).

156 *Heliozela resplendella* Stt.
Near Monmouth 7/10/76 (AME).

Sq. **SO** 41.

157 *Heliozela hammoniella* Sorh.
    = *betulae* Stt.
Trelleck Bog 8/9/73 (AME).

Sq. **SO** 50.

159 *Antispila petryi* Mart.
"Monmouthshire" (AME in litt. 1976). Slade Wds 26/5/80, one imago by day (coll. GANH, det. JDB). Wyndcliff, 19/9/90 (AME).

Sq. **ST** 48,59.

PSYCHIDAE
TALEPORIINAE

180 *Diplodoma herminata* Geoff.
MBGBI **2**: Map 17 indicates v-c. 35 (Monmouthshire). ["? Tutt J.W. 1890, the most likely source", AME in litt.].

181 *Taleporia tubulosa* Retz.
MBGBI **2**: Map 22 indicates v-c. 35 (Monmouthshire). ["? Tutt J.W. 1890, the most likely source", AME in litt.].

PSYCHINAE

186 *Psyche casta* Pallas
Angidy Valley near Panta Arch, 1978, numerous larval cases on lichen-covered fence posts in damp woodland, 5/6 and 27/6 (reared); found here plentifully most years since, eg. 1992 (23/5). A single case found on undergrowth at Slade Wds.

Sq. **ST** 48,59.

TINEIDAE
SCARDIINAE

196 *Morophaga choragella* D. & S.
  = *boleti* Fabr.
Hendre Wds, 9/6/79, several freshly-emerged imagines on rotting tree-stumps (GANH) (1st v-c. rec.).

Sq. **SO** 41.

TEICHOBIINAE

199 *Psychoides verhuella* Bru.
Larvae on hart's-tongue fern (*Phyllitis scolopendrium* L.) recorded at Glascoed and Llangeview in early 1970s.

Sq. **SO** 30. **ST** 39.

NEMAPOGONINAE

216 *Nemapogon cloacella* Haw.   Cork Moth
Wyndcliff 25/6/86 and Hael Wds 26/6/86, to m.v.l. (JDB, GANH). Llansoy 5/7/87, to u.v.l. St Pierre's Great Wds 5/7/88 (JDB).

Sq. **SO** 40,50. **ST** 59.

217 *Nemapogon wolffiella* Karsh. & Niel.
  = *albipunctella* Haw.
MBGBI 2: Map 50 indicates v-c. 35 (Monmouthshire).

220 *Nemapogon clematella* Fabr.
  = *arcella* auct.
Usk, garden m.v.l. trap, 5/7/79, one.

Sq. **SO** 30.

221 *Nemapogon picarella* Cl.
MBGBI **2**:181 states "an uncommon species . . . Recorded from

Monmouthshire (Gwent) . . ."

224 *Triaxomera parasitella* Hb.
Hendre Wds, 9/6/79, numerous freshly-emerged imagines on rotting tree-stumps (GANH) (1st. v-c. rec.). Llansoy, 23/5/88.

Sq. **SO** 40,41.

225 *Triaxomera fulvimitrella* Sodof.
Hael Wds 29/5/82, one to m.v.l. (JDB, GANH) (1st v-c. rec.).

Sq. **SO** 50.

TINEINAE

227 *Monopis laevigella* D. & S.   Skin Moth
   = *rusticella* Hb.
Ysguborwen (Llantrisant), April 1981, many bred from the debris collected from birds' nest-boxes in January.

Sq. **ST** 49.

228 *Monopis weaverella* Scott
MBGBI **2**: Map 62 indicates v-c. 35 (Monmouthshire).

238 *Niditinea piercella* Bent.
MBGBI **2**: Map 71 indicates v-c. 35 (Monmouthshire).

246 *Tinea semifulvella* Haw.
Wyndcliff 24/6/71, to m.v.l. (JMC-H, GANH), 26/6/86 (JDB). Usk, garden m.v.l. trap, 15/6/76.

Sq. **SO** 30. **ST** 59.

247 *Tinea trinotella* Thunb.
Tintern 31/5/68 (JMC-H 1969). Usk, garden m.v.l. trap, recorded frequently eg. 21/8/77, 31/7/80, 3/6/82. Wyndcliff 26/6/86, several to m.v.l. Llansoy 4/5/90.

Sq. **SO** 30,40,50. **ST** 59.

LYONETIIDAE
CEMIOSTOMINAE

255 *Leucoptera wailesella* Stt.
Llangeview 1/9/79 several on dyer's greenweed (*Genista tinctoria* Linn.) (GANH) (1st v-c. rec.). At same site, abundant on 4/9/79.

Sq. **ST** 40.

256 *Leucoptera spartifoliella* Hb.
Hael Wds, 30/6/76, larvae on broom (MWH in litt.) (1st v-c. rec.). Hendre Wds, 8/7/79 imagines abundant on broom (GANH).

Sq. **SO** 50,41.

LYONETIINAE

263 *Lyonetia clerkella* Linn.
Llandogo, bred from *Malus*, 3/10/64 (AME).

Sq. **SO** 50.

BEDELLIINAE

264 *Bedellia somnulentella* Zell.
Monmouthshire. Recorded by MWH.

BUCCULATRIGINAE

266 *Bucculatrix nigricomella* Zell.
Monmouthshire. Recorded by ECP-C.

271 *Bucculatrix albedinella* Zell.
Near Monmouth, 7/10/76 (AME).

Sq. **SO** 41.

273 *Bucculatrix thoracella* Thunb.
St Pierre's Great Wds, 31/10/76, mines in leaves of *Tilia europaea* (coll. GANH, det. AME) (1st v-c .rec.). Wyndcliff, common 19/9/90 (AME).

Sq. **ST** 59.

274 *Bucculatrix ulmella* Zell.
Hendre Wds, 9/11/76, larval mine in oak (coll. GANH, det. AME) (1st v c. rec.).

Sq. **SO** 41.

275 *Bucculatrix crataegi* Zell.
Dixton Bank 4/11/84 (AME in litt.) (1st v-c. rec.). Wyndcliff, 25/6/86, one to m.v.l. (JDB).

Sq. **SO** 51. **ST** 59

GRACILLARIIDAE
GRACILLARIINAE

282 *Caloptilia elongella* Linn.
Near Monmouth, 7/10/76, (AME in litt.). Usk, garden m.v.l. trap, 13/4/79 (1). Llansoy, 12/10/90, one.

Sq. **SO** 30,40,41.

286 *Caloptilia alchimiella* Scop.
Llandogo, 28/5/65, (AME in litt.). Llansoy, 4/5/90 (2).

Sq. **SO** 40,50.

287 *Caloptilia robustella* Jäckh
Monmouthshire, 1983 (ECP-C). Wyndcliff, 23/6/86, several to m.v.l. (JDB, GANH). Hael Wds, 26/6/86, two to m.v.l. (JDB, GANH).

Sq. **SO** 50. **ST** 59.

288 *Caloptilia stigmatella* Fabr.
Usk, Plas Newydd garden m.v.l. trap, 8/7/81 (1) (GANH).

Sq. **SO** 30.

290 *Caloptilia semifascia* Haw.
MBGBI 2: Map 114 indicates v-c. 35 (Monmouthshire). See *Proc. Trans. Br. Ent. nat. Hist. Soc*. 3:21 (ECP-C). Wyndcliff, old larval feedings, 19/9/90 (AME in litt.).

Sq. **ST** 59.

293 *Caloptilia syringella* Fabr.
Tintern, 31/5/68, by day (JMC-H 1969) (1st v-c. rec.). Wyndcliff 24/6/71 to m.v.l. (JMC-H, GANH). Slade Wds 21/7/79, larvae on ash leaves (JDB, GANH). Usk (old railway line), 1980, plfl. on 29/5 and 1/6 (GANH). Hendre Wds, 29/5/82, by day (JDB). Llandogo (AME). Wyndcliff, 19/9/90 (AME).

Sq. **SO** 30,50,41. **ST** 48,59.

294 *Aspilapteryx tringipennella* Zell.
"Monmouthshire" (HWNM per AME), no details.

297 *Calybites auroguttella* Steph.
Slade Wds, 21/7/79, pupae on leaves of *Hypericum perforatum* (JDB, GANH). (1st v-c. rec.). Dixton Bank, 4/11/84 (AME in litt.).

Sq. **SO** 51. **ST** 48.

301 *Parornix betulae* Stt.
Trelleck Bog 8/9/73 (AME).

Sq. **SO** 50.

302 *Parornix fagivora* Frey
Wyndcliff, 19/9/90, larval feeding in beech, "2nd Welsh record" (AME in litt.).

Sq. **ST** 59.

303 *Parornix anglicella* Stt.
Near Monmouth 7/10/76 (AME). Dixton Bank 4/11/84, Wyndcliff 19/9/90 (AME in litt.).

Sq. **SO** 41,51. **ST** 59.

304 *Parornix devoniella* Stt.
Llandogo, 8/9/73 (AME). Dixton Bank 4/11/84, Wyndcliff 19/9/90 (AME in litt).

Sq. **SO** 50,51. **ST** 59

305 *Parornix scoticella* Stt.
Wyndcliff, 19/9/90, larval feeding on rowan, (AME in litt.).

Sq. **ST** 59.

308 *Parornix finitimella* Zell.
Wyndcliff, 19/9/90, mines and feeding on blackthorn (AME). "Monmouthshire" (undated) G. Bryan (per AME).

Sq. **ST** 59.

309 *Parornix torquillella* Zell.
Pont-y-saeson 23/6/86, ♀ (1), by day (JDB) (1st v-c. rec.).

Sq. **ST** 59.

313 *Acrocercops brongniardella* Fabr.
Tintern dist., 19/6/65, "larvae on oak" (Mere. *Proc. S. Lond. Ent. Soc.* 1965: 88).

Sq. **ST** 59.

LITHOCOLLETINAE

315 *Phyllonorycter harrisella* Linn.
Hendre Wds, mined oak leaves collected 13/10 and 27/10/81 (GANH), reared, several imagines emerged April. (1st v-c. rec.).

Sq. **SO** 41.

317 *Phyllonorycter heegeriella* Zell.
Hendre Wds, mined oak leaves coll. 11/11/80 (GANH), (6) imagines emerged 10/4 – 23/4/81. (1st v-c. rec.).

Sq. **SO** 41.

318 *Phyllonorycter tenerella* Joann.
Slade Wds, 24/6/86, to m.v.l. (1st v-c. rec.) and Hael Wds, 26/6/86 (2) to m.v.l. (JDB, GANH).

Sq. **SO** 50. **ST** 48.

320 *Phyllonorycter quercifoliella* Zell.
Hendre Wds, mined oak leaves coll. 11/11/80, one imago emerged 4/4/81; mines coll. 27/10/81 bred, several imagines emerged in March and early April (GANH).

Sq. **SO** 41.

321 *Phyllonorycter messaniella* Zell.
Pont-y-saeson, mine in oak leaf 11/10/81, bred, (GANH) (1st v-c. rec.). Dixton Bank 4/11/84, mines (AME). Wyndcliff 19/9/90 (AME).

Sq. **SO** 51. **ST** 59.

322 *Phyllonorycter muelleriella* Zell.
Hael Wds, 29/5/82, one to m.v.l. (JDB) (1st v-c. rec.).

Sq. **SO** 50.

323 *Phyllonorycter oxyacanthae* Frey
Rogiet, mined hawthorn leaves collected 24/10/81, imagines emerged March (GANH) (1st v-c. rec.). Dixton Bank, 4/11/84 (AME).

Sq. **SO** 51. **ST** 48.

326 *Phyllonorycter blancardella* Fabr.
St Pierre's Great Wds 13/5/80, one netted at dusk (GANH) (1st v-c. rec.). Dixton Bank, 4/11/84 (AME).

Sq. **SO** 51. **ST** 59.

328 *Phyllonorycter junoniella* Zell.
MBGBI **2**: Map 151 gives v-c. 35 (Monmouthshire); recorded by I.A. Watkinson (per AME).

329 *Phyllonorycter spinicolella* Zell.
MBGBI **2**: Map 152 gives v-c. 35 (Monmouthshire). "Widespread" (AME).

330 *Phyllonorycter cerasicolella* H.-S.
Dixton Bank, 4/11/84, (AME in litt.). (1st v-c. rec.).

Sq. **SO** 51.

332 *Phyllonorycter corylifoliella* Hb.
Dixton Bank, 4/11/84, (AME in litt.).

Sq. **SO** 51.

337 *Phyllonorycter spinolella* Dup.
Hael Wds 30/6/76 (MWH in litt.) (1st v-c. rec.). Hael Wds, 26/6/86, two to m.v.l. (JDB, GANH).

Sq. **SO** 50.

338 *Phyllonorycter cavella* Zell.
Hael Wds, 26/6/86, to m.v.l. (JDB) (1st v-c. rec.).

Sq. **SO** 50.

340 *Phyllonorycter scopariella* Zell.
Hael Wds, 30/6/76, "out of broom" (MWH in litt.) (1st v-c. rec.). Hendre Wds, 8/7/79, beaten in abundance from broom bushes (GANH).

Sq. **SO** 50,41.

341 *Phyllonorycter maestingella* Müll.
MBGBI **2**: Map 165 gives v.c. 35 (Monmouthshire), "ubiquitous" (AME). Wyndcliff, 19/9/90 (AME).

Sq. **ST** 59.

342 *Phyllonorycter coryli* Nic.
Dixton Bank, 4/11/84, (AME in litt.). Wyndcliff, 19/9/90 (AME).

Sq. **SO** 51. **ST** 59.

345 *Phyllonorycter rajella* Linn.
 = *alnifoliella* Hb.

Hendre Wds 1981, mines on alder 13/10 and 27/10, bred (GANH).

Sq. **SO** 41.

346 *Phyllonorycter distentella* Zell.
MBGBI **2**:341 states "very local and uncommon in old-established Oak woodlands of Gloucestershire, Monmouthshire and Herefordshire and in Kent".

348 *Phyllonorycter quinqueguttella* Stt.
St Pierre's Great Wds 5/7/88 (JDB).

Sq. **ST** 59.

351 *Phyllonorycter lautella* Zell.
Hendre Wds, 11/11/80, mines in oak leaves, bred, (20) imagines emerged 31/3 – 24/4/81; St Pierre's Great Wds, 2/10/81, mines, bred, emerged 1/4; Llangeview, mines 7/11/81, emerged 2/4 (GANH). Dixton Bank, 4/11/84 (AME).

Sq. **SO** 41,51. **ST** 39,59.

352 *Phyllonorycter schreberella* Fabr.
Dixton Bank, on elm, 4/11/84 (AME in litt.) (1st v-c. rec.).

Sq. **SO** 51.

353 *Phyllonorycter ulmifoliella* Hb.
Hendre Wds, 11/11/80, mined birch leaves, imago emerged 13/4/81 (GANH). Dixton Bank 4/11/84 (AME).

Sq. **SO** 41,51

356 *Phyllonorycter tristrigella* Haw.
MBGBI **2**: Map 180 indicates v.c. 35 (Monmouthshire).

357 *Phyllonorycter stettinensis* Nic.
Dixton Bank, 4/11/84, (AME in litt.).

Sq. **SO** 51.

359 *Phyllonorycter nicellii* Stt.
Dixton Bank, 4/11/84, (AME) (1st v-c. rec.). Wyndcliff, 19/9/90 (AME).

Sq. **SO** 51. **ST** 59.

360 *Phyllonorycter kleemannella* Fabr.
Hendre Wds 13/10/81, mines on alder, bred, emerged 18/4/82 (GANH) (1st v-c. rec.). Dixton Bank 4/11/84 (AME in litt.).

Sq. **SO** 41,51.

361 *Phyllonorycter trifasciella* Haw.
Usk Castle, 25/6/86, (JDB).

Sq. **SO** 30.

362 *Phyllonorycter acerifoliella* Zell.
= *sylvella* Haw.
Hael Wds, 29/5/82, one to m.v.l. (JDB) (1st v-c. rec.). Wyndcliff, 19/9/90 (AME).

Sq. **SO** 50. **ST** 59.

364 *Phyllonorycter geniculella* Rag.
Dixton Bank, 4/11/84 (AME in litt.) (1st v-c. rec.).

Sq. **SO** 51.

PHYLLOCNISTINAE

368 Phyllocnistis unipunctella Steph.
Near Monmouth, 7/10/76 (AME in litt.).

Sq. **SO** 41.

CHOREUTIDAE

385 *Anthophila fabriciana* Linn.
Widespread and abundant among nettles.
Tintern 31/5/68 and Deri-fach 2/8/68 (JMC-H 1969).
Magor Marsh 26/9/78. Hendre Wds 23/6/79. Ochrwyth 20/6/87. Cwmfelinfach 20/6/87. Pontllanfraith 23/6/87. Llantrisant 27/6/87, Llansoy 6/7/87.

Sq. **SO** 30,40,50,21,41. **ST** 28,48,19,39.

388 *Prochoreutis myllerana* Fabr.
Magor Marsh, 27/5/82 (JDB).

Sq. **ST** 48.

GLYPHIPTERIGIDAE

391 *Glyphipteryx simpliciella* Steph.   Cocksfoot Moth
Tintern and Redding's Inclosure, 31/5/68 and Coed-y-Bwynydd, 5/6/69 (JMC-H 1969 as *G. cramerella*.

Usk (Plas Newydd) 31/5/80, pair in cop. on house wall. Slade Woods 3/6/80, one on flower of wood spurge. Pont-y-saeson, 1/6/81. Magor Marsh, 27/5/82, one netted by day (JDB). Usk (old railway line) 25/5/82, three on ox-eye daisy 25/5/82. Hendre Wds, 29/5/82 (JDB).

Sq. **SO** 30,50,41,51. **ST** 48,59.

394 *Glyphipterix forsterella* Fabr.
MBGBI **2**: Map 218 indicates v-c. 35 (Monmouthshire). "Recorded by ECP-C" (AME in litt.).

396 *Glyphipterix fuscoviridella* Haw.
Wye Valley (Barraud 1906). Tintern and Redding's Inclosure 31/5/68 (JMC-H 1969).

Mountain-Air (Pontypool), 25/5/80. Pont-y-saeson, 1/6/81 (abdt.). Hendre Wds, 29/5/82 (by day). Hael Wds, 29/5/82 to m.v.l.

Sq. **SO** 50,41,51. **ST** 29,59.

397 *Glyphipterix thrasonella* Scop.
Inhabits rushy bogs and is usually abundant where it is found.

Gwernesney 23/6/71, (abdt.), (GANH). Fforest Coal Pit 1988, 15/6 and 24/6 (abdt.). Pontllanfraith 6/7/88 (abdt.). Goetre (Wern Fawr) 10/6/89.

Sq. **SO** 30,40,22. **ST** 19.

YPONOMEUTIDAE
ARGYRESTHIINAE

401 *Argyresthia laevigatella* H.-S.
Wyndcliff 25/6/86, Hael Woods 28/6/86 (JDB, GANH)

Sq. **SO** 50. **ST** 59.

410 *Argyresthia brockeella* Hb.
Slade Wds, 21/7/79 by day, (abdt.). Trelleck Bog 19/7/86, to m.v.l. Wyndcliff 3/7/91, (1) to m.v.l.

Sq. **SO** 50. **ST** 48,59.

411 *Argyresthia goedartella* Linn.
Magor Marsh 13/7/77. Usk 11/8/77. Slade Wds 21/7/79, by day (abdt.).

Sq. **SO** 30. **ST** 48.

412 *Argyresthia pygmaeella* Hb.
Hendre Wds 20/7/79, to m.v.l. (JDB, GANH).

Sq. **SO** 41.

413 *Argyresthia sorbiella* Treit.
Monmouthshire. Recorded by DJLA (per AME).

414 *Argyresthia curvella* Linn.
Usk, garden m.v.l. trap, 12/7/75, 28/7/79. Llandogo, 30/6/76, imagines in *Malus* on river bank (MWH in litt.). Usk, 14/7/82, in bushes along old railway line (abdt.).

Sq. **SO** 30,50.

415 *Argyresthia retinella* Zell.
Slade Wds 21/7/79, netted by day (JDB).

Sq. **ST** 48.

418 *Argyresthia conjugella* Zell.
Usk, garden m.v.l. trap, 11/9/82. Pont-y-saeson 23/6/86, by day (JDB).

Sq. **SO** 30. **ST** 59.

420 *Argyresthia pruniella* Cl.
Slade Wds 18/7/81 (by day). Usk 14/7/82, large numbers beaten from bushes along disused railway line.

Sq. **SO** 30. **ST** 48.

422 *Argyresthia albistria* Haw.
Usk, garden m.v.l. trap, 4/8/82.

Sq. **SO** 30.

YPONOMEUTINAE

424 **Yponomeuta evonymella** Linn.   Bird-cherry Ermine
Usk, garden m.v.l. trap, 25/7/69 and plentiful most years eg. 1979 27/7 – 1/9. Hendre Wds 4/9/78. Slade Wds 28/7/81. Wyndcliff 3/8/81. Llansoy 7/8/88. Magor Marsh 24/7/90.

Sq. **SO** 30,40,41. **ST** 48,59.

425 **Yponomeuta padella** Linn.   Orchard Ermine
Usk, garden m.v.l. trap, 19/7/69 and frequent most years eg. 1977 2/8 – 21/8;

1982 20/7 – 5/8. Hendre Wds 1/8/80. Llansoy, 25/8/90.

Sq. **SO** 30,40,41.

427 *Yponomeuta cagnagella* Hb.   Spindle Ermine
Usk, garden m.v.l. trap, 21/8/66 and recorded most years until 1982 eg. 1975, 11/6 – 1/8; 1978, 29/7 – 27/8. Ed. 11/6/75. Ld. 8/9/77. Slade Wds 21/8/79. Hael Wds 31/7/82. Llansoy 7/7/89. 1990, Llangeview, Llansoy, etc., larval webs numerous on spindle (*Euonymus europaeus*) growing in hedgerows.

Sq. **SO** 30,40,50. **ST** 48,39.

430 *Yponomeuta plumbella* D. & S.
Usk, 5/6/68, numerous larval webs on spindle, reared (JMC-H 1969). Usk, garden m.v.l. trap, 11/8/82, 7/8/83. Llansoy 20/8/86 (to u.v.l.).

Sq. **SO** 30,40.

435 *Zelleria hepariella* Stt.
Usk, garden m.v.l. trap, 28/7/79 (1), (det. JDB).

Sq. **SO** 30.

436 *Pseudoswammerdamia combinella* Hb.
Usk, garden m.v.l. trap, 23/5/82 (1), (det. JDB).

Sq. **SO** 30.

437 *Swammerdamia caesiella* Hb.
Usk, garden m.v.l. trap, 11/8/77. Llansoy 27/6/86 (1).

Sq. **SO** 30,40.

438 *Swammerdamia pyrella* Vill.
Usk, garden m.v.l. trap, 25/7/79 (1), 30/7/82 (1).

Sq. **SO** 30.

441 *Paraswammerdamia lutarea* Haw.
Usk, garden m.v.l. trap, 2/8/77. Llansoy 14/7/86. Trelleck Bog 24/7/86.

Sq. **SO** 30,40,50.

447 *Roeslerstammia erxlebella* Fabr.
Usk, garden m.v.l. trap, 31/8/79 (1) (det. JDB). Wyndcliff, 19/9/90, larval feedings (AME).

Sq. **SO** 30. **ST** 59.

449 *Prays fraxinella* Bjerk.   Ash Bud Moth
Wyndcliff 13/7/69. Slade Wds 30/6/81. Hael Wds 31/7/82. Llansoy 25/6/86. Clydach 9/7/89.

Sq. **SO** 40,50,21. **ST** 48,59.

450 *Scythropia crataegella* Linn.   Hawthorn Moth
Llansoy, to u.v.l. trap, 2/7/91 (1), 6/7/91 plentiful (GANH) (1st. v-c. rec.). 1993, Llansoy 20/6, 28/6. AME (in litt. 1993) writes "known from adjoining counties v-c 34 (West Gloucestershire) and v-c. 41 (Glamorgan) but not from Herefordshire (v-c. 36)".

Sq. **SO** 40.

PLUTELLINAE

451 *Ypsolopha mucronella* Scop.
Usk, garden m.v.l. trap, 2/4/72, 24/3/76, 19/4/79 (2), etc.

Sq. **SO** 30.

453 *Ypsolopha dentella* Fabr.   Honeysuckle Moth
Pont-y-saeson 16/8/66. Usk, 2/8/77, 29/7/79, 31/8/79 etc. Hael Wds 5/8/82.

Sq. **SO** 30,50. **ST** 59.

455 *Ypsolopha scabrella* Linn.
Usk, garden m.v.l. trap, 19/8/69 (2), 6/8/82 – 18/9/82, etc.
    Llansoy, 29/7/92, one to u.v.l.

Sq. **SO** 30,40.

457 *Ypsolopha lucella* Fabr.
Pont-y-saeson (Panta Arch)16/8/66, three netted at dusk (GANH) (1st v-c. rec.).

Sq. **ST** 59.

459 *Ypsolopha sylvella* Fabr.
Llansoy, 9/9/86, one to u.v.l. (GANH). (1st v-c. rec.).

Sq. **SO** 40.

460 *Ypsolopha parenthesella* Linn.
Usk, garden m.v.l. trap, common, 21/8/72 (2) etc.; Ed. 9/8/80; Ld. 2/10/79. Hael Wds 18/8/75. Hendre Wds 20/7/79, 8/10/79. Pont-y-saeson 18/9/79 (2). Wyndcliff 11/8/82.

Sq. **SO** 30,50,41. **ST** 59.

461 *Ypsolopha ustella* Cl.
Usk, garden m.v.l. trap, 23/2/76, 8/7/78, 11/8/82. Pont-y-saeson 28/2/76. Hael Wds 11/8/82, 22/8/82. Wyndcliff, two imagines 19/9/90 (JRL per AME).

Sq. **SO** 30,50. **ST** 59.

462 *Ypsolopha sequella* Cl.
Usk, garden m.v.l. trap, occurs frequently eg. 1968, Aug 19 (2); 1971, Aug 17,20,21,24; 1978, 29/8; Ed. 6/8/80; Ld. 18/10/79. Hendre Wds 8/10/79. Hael Wds 15/9/80. Clydach 9/7/89.

Sq. **SO** 30,50,21,41.

463 *Ypsolopha vitella* Linn.
Monmouthshire. Recorded by DJLA (per AME).

464 *Plutella xylostella* Linn.   Diamond-back Moth
Usk, garden m.v.l. trap, 2/8/77, 28/8/80, 19/6/83 etc. Hael Wds Aug. 1982. Wyndcliff 25/6/86. Llansoy, 1986, frequent 26/6 – 15/7.

Sq. **SO** 30,40,50. **ST** 59.

465 *Plutella porrectella* Linn.
Usk, m.v.l. trap, 13/8/78 (1), 2/6/79 (1) (det. JDB).

Sq. **SO** 30.

469 *Eidophasia messingiella* F.v.R.
A local and scarce species recorded from several sites in Monmouthshire.
    Crumlyn (= Crumlin), 1860 (Scott, *Ent. Wkly Intell.* **8**:131).
    Hendre Wds, 20/7/79, to m.v.l. (JDB, GANH). Llansoy 24/6/87, 6/8/87.
Magor Pill (sea-wall), 14/6/90, one to m.v.l.

Sq. **SO** 20,40,41. **ST** 48.

ACROLEPIINAE

476 *Acrolepia autumnitella* Curtis
   = *pygmeana* Haw.

Magor Marsh 18/4/76 (1) (coll. GANH, det. JDB).

Sq. **ST** 48.

EPERMENIIDAE

478 *Phaulernis fulviguttella* Zell.
Pont-y-saeson (Panta Arch), 22/8/82, several by day on flower heads of

Angelica sylvestris (GANH, det. JDB).

Sq. **ST** 59.

483 *Epermenia chaerophyllella* Goeze
Wyndcliff, 19/9/90, larval feeding on *Heracleum* (AME in litt.).

Sq. **ST** 59.

SCHRECKENSTEINIIDAE

485 *Schreckensteinia festaliella* Hb.
Llansoy 12/7/86 to m.v.l. (det. JDB).

Sq. **SO** 40.

COLEOPHORIDAE

487 *Metriotes lutarea* Haw.
Pont-y-saeson 22/5/72, one netted by day (coll. GANH, det. JDB). Monmouthshire, (ECP-C per AME).

Sq. **ST** 59.

490 *Coleophora lutipennella* Zell.
Usk (Plas Newydd), garden m.v.l. trap, 13/8/77, 13/7/79. Hendre Wds, 20/7/79, to m.v.l. (JDB). Hael Wds 6/7/88 (JDB).

Sq. **SO** 30,50,41.

491 *Coleophora gryphipennella* Hb.
Slade Wds, 21/7/79, one netted by day (JDB). Slade Wds, 8/11/80 larval cases on *Rosa* spp. Hendre Wds, 11/11/80, cases abundant on *Rosa* spp. St Pierre's Great Wds, 5/7/88, imago by day.

Sq. **SO** 41. **ST** 48,59.

492 *Coleophora flavipennella* Dup.
Hendre Wds, 20/7/79 to m.v.l. (JDB, GANH). Usk, garden m.v.l.trap, 21/7/79 (one).

Sq. **SO** 30,41.

493 *Coleophora serratella* Linn.
    = *fuscodinella* Zell.
Usk dist. (Gwernesney Bog), 23/6/71, larva on alder, reared (GANH). Dixton Bank, 4/11/84 (AME in litt.).

Sq. **SO** 40,51.

495 *Coleophora spinella* Schr.
   = *cerasivorella* Pack.
Slade Wds 24/6/86, larval case on hawthorn, reared, imago emerged 4/7/86 (JDB).

Sq. **ST** 48.

497 *Coleophora badiipennella* Dup.
Monmouthshire, (ECP-C, per AME).

499 *Coleophora limosipennella* Dup.
Dixton Bank, 4/11/84, (AME in litt.).

Sq. **SO** 51.

503 *Coleophora fuscocuprella* H.-S.
Monmouthshire, (RWJU, per AME).

504 *Coleophora viminetella* Zell.
Monmouthshire, (ECP-C, per AME).

524 *Coleophora lithargyrinella* Zell.
Monmouthshire, (RWJU, per AME).

526 *Coleophora laricella* Hb.
Usk dist. (Llanbadoc), 14/4/81, larval cases abundant on larch. Wyndcliff, 23/6/86, imagines to m.v.l. (JDB, GANH). Hael Wds, 26/6/86, abundant at m.v.l.

Sq. **SO** 30,50. **ST** 59.

533 *Coleophora anatipennella* Hb.
Usk, garden m.v.l. trap, 22/6/73. Trelleck Bog, 24/7/86, to m.v.l. (coll. GANH, det. JDB). Tintern 5/7/88 (JDB).

Sq. **SO** 30,50.

534 *Coleophora currucipennella* Zell.
Hendre Wds, 20/7/79, to m.v.l. (JDB, GANH). Usk, garden m.v.l. trap, 26/7/79.

Sq. **SO** 30,41.

535 *Coleophora ardeaepennella* Scott
Hael Wds, 3/7/83 to m.v.l. (det JDB).

Sq. **SO** 50.

537 *Coleophora palliatella* Zinck.
Hael Wds, 11/7/81, larval case on oak (MWH in litt.).

Sq. **SO** 50.

544 *Coleophora albicosta* Haw.
Hael Wds, 29/5/82, one to m.v.l. (JDB).

Sq. **SO** 50.

553 *Coleophora striatipennella* Nyl.
Monmouthshire, (ECP-C, per AME).

555 *Coleophora follicularis* Vall.
= *troglodytella* Dup.

Monmouthshire, (ECP-C, per AME).

556 *Coleophora trochilella* Dup.
Llansoy, 25/6/86, to u.v.l. (det. JDB).

Sq. **SO** 40.

560 *Coleophora paripennella* Zell.
Monmouthshire, (ECP-C, per AME).

580 *Coleophora sylvaticella* Wood
Wye Valley, May 1965, several imagines swept from *Luzula sylvatica* (AME). Tintern dist., Mere 1965. Wye Valley, imagines reared in April 1967 from cases which appeared in Aug. 1965 on seeds of L. sylvatica collected on 20/6/65 (Uffen, *Proc. S. Lond. Ent. Nat. Hist. Soc.* 1967: 79). Tintern 31/5/68, imagines disturbed from L. *sylvatica* (JMC-H *Ent. Rec.* **81**:41.

Sq. **SO** 50. **ST** 59.

582 *Coleophora glaucicolella* Wood
Monmouthshire, (RWJU, per AME).

584 *Coleophora alticolella* Zell.
Blorenge 1982, cases on *Juncus* spp. (abdt.) bred (GANH). Hael Wds, 26/6/86, to m.v.l. (det. JDB). Pontllanfraith 6/7/88 (JDB).

Sq. **SO** 50,21. **ST** 19.

587 *Coleophora caespititiella* Zell.
Hael Wds, to m.v.l. 29/5/82 and 26/6/86 (abdt.) (JDB). Wyndcliff, 25/6/86, to

m.v.l. Pont-y-saeson, 23/6/86 (by day).

Sq. **SO** 50. **ST** 59.

ELACHISTIDAE

593 *Elachista regificella* Sirc.
Tintern 1970, larvae on *Luzula sylvatica*, reared (ECP-C, per AME).

Sq. **SO** 50.

596 *Elachista poae* Stt.
Monmouthshire, (ECP-C per AME).

597 *Elachista atricomella* Stt.
Usk, garden m.v.l. trap, 30/7/77, 30/7/79, 7/8/82 etc.

Sq. **SO** 30.

600 *Elachista luticomella* Zell.
Monmouthshire, (ECP-C per AME).

601 *Elachista albifrontella* Hb
Monmouthshire, (ECP-C per AME).

602 *Elachista apicipunctella* Stt.
Pont-y-saeson, 8/7/78 one netted by day (GANH, det. JDB).

Sq. **ST** 59.

606 *Elachista humilis* Frey
Monmouthshire, (ECP-C per AME).

607 *Elachista canapennella* Hb.
     = *pulchella* Haw.
     = *obscurella* Stt.
Monmouthshire, (ECP-C per AME).

608 *Elachista rufocinerea* Haw.
St Pierre's Great Wds, 13/5/80, one netted at dusk. (GANH, det JDB).

Sq. **ST** 59.

609 *Elachista cerusella* Hb.
Magor Marsh, 8/8/76, abundant at m.v.l. (GANH, det. JDB).

Sq. **ST** 48.

610 *Elachista argentella* Cl.
Tintern, 31/5/68, (JMC-H. Ent.Rec. **81**:41).

Sq. **ST** 50.

622 *Elachista revinctella* Zell.
= *adscitella* Stt.
Hendre Wds, 20/7/79, to m.v.l. (JDB, GANH).

Sq. **SO** 41.

624 *Biselachista trapeziella* Stt.
Tintern, "adults reared from *Luzula sylvatica* 14/5 & 22/5/70" (AME in litt.).

Sq. **SO** 50.

OECOPHORIDAE
OECOPHORINAE

642 *Batia unitella* Hb.
Hendre Wds, 20/7/79, to m.v.l. (JDB). Wyndcliff, 11/8/82, to m.v.l. (GANH).

Sq. **SO** 41. **ST** 59.

646 *Telechrysis tripuncta* Haw.
Llansoy, one to u.v.l. 22/5/92 (GANH).

Sq **SO** 40.

647 *Hofmannophila pseudospretella* Stt. Brown House-moth
Usk, frequent indoors eg. 18/7/78, 12/6/81 etc. Llansoy, 1985 – 90, common.

Sq. **SO** 30,40.

648 *Endrosis sarcitrella* Linn.   White-shouldered House-moth
Usk, to m.v.l. 8/5/67, 16/6/79 etc. Ysguborwen (Llantrisant), May 1981, bred from debris collected in January from Blue Tit's nest-box. Wyndcliff, numerous at m.v.l. 15/9/81. Llansoy, frequent 1984 – 1990.

Sq. **SO** 30,40. **ST** 49,59.

649 *Esperia sulphurella* Fabr.
Tintern 31/5/68 and Coed-y-Bwnydd 5/6/68 (JMC-H, *Ent. Rec.* **81**:41). Usk, Plas Newydd garden, 9/5/76, flying in abundance in morning sunshine; 21/6/78 d°. Hendre Wds 6/6/78, 29/5/82 etc. Llansoy, 3/5/90.

Sq. **SO** 30,40,41. **ST** 50.

651 *Oecophora bractella* Linn.
Rare and local in Monmouthshire.

S.N.A. Jacobs, 1948, wrote "Very rare in Britain, reported from Nottinghamshire, Monmouth, Durham, Newcastle and Gateshead. Abroad in Central and South Europe and in Asia Minor."

Crumlyn (= Crumlin), 1860, as *Harpella bractella* (Scott, *Ent. Wkly. Intell.* 8:131). "Crumlyn, Monmouthshire, in the little lane across the bridge" (F.O. Morris, 1891). The bridge alluded to was the high-level railway viaduct demolished in the 1950s or 60s. Meyrick, 1895 gives "*H. bractella*, Monmouth, Durham, very local".

Hael Wds, 3/7/71 several to m.v.l. (GANH). Trelleck Bog, 16/7/79 (1) (GANH).

Trelleck 6/3/83, larvae under bark of *Larix, Pinus* and *Tsuga*, bred. (JRL, DHS).

Sq. **SO** 50. **ST** 29.

652 *Alabonia geoffrella* Linn.
Fairly common in the eastern half of Monmouthshire.

Wye Valley (Barraud 1906, as *Harpella geofrella*). Tintern, 31/5/68 (JMC-H, *Ent. Rec.* 81:41).

Usk, 11/6/68. Slade Wds, 30/5/71 numerous by day. Pont-y-saeson 5/6/78. Hendre Wds 6/6/78. Goetre (Wern Fawr) 10/6/89. St Pierre's Great Wds 3/6/88.

Sq. **SO** 30,41,50. **ST** 48,59.

658 *Carcina quercana* Fabr.
Usk, garden m.v.l. trap, 19/7/69 (several) and very common most years. Slade Wds 29/7/86. Llansoy, 7/8/88 and usually common. Wyndcliff 7/8/89. Hael Wds 3/8/90

Sq. **SO** 30,40,50. **ST** 48,59.

660 *Pseudatemelia josephinae* Toll
Hael Wds, 10/7/73 (1) to m.v.l. (GANH) (1st v-c. rec.). Hael Wds, 11/7/81, three at dawn (MWH in litt.).

Sq. **SO** 50.

661 *Pseudatemelia flavifrontella* D. & S.
Llansoy, 25/6/86, to u.v.l. (det. JDB).

Sq. **SO** 40.

662 *Pseudatemelia subochreella* Doubl.
Hael Wds, 26/6/86, to m.v.l. (JDB, GANH).

Sq. **SO** 50.

CHIMABACHINAE

663 *Diurnea fagella* D. & S.
Usk (Plas Newydd), garden m.v.l. trap, abundant most years eg. 1967 4/3 – 25/3; 1974 16/3, 23/3; 1980 29/3. Hendre Wds 15/4/79 (SWNH). Hael Wds 19/3/83. Llansoy, 12/4/87.

Sq. **SO** 30,40,50,41.

664 *Diurnea phryganella* Hb.
A specimen labelled "Monmouthshire" in the National Collection at BMNH (teste MWH in litt.).

DEPRESSARIINAE

666 *Semioscopis avellanella* Hb.
Redding's Inclosure (Forest of Dean, Mon.), 20/3/72 (GANH) (1st v-c. rec.). Wyndcliff, 22/3/73, plentiful to m.v.l. Wyndcliff, larval feeding on *Tilia cordata* 19/9/90 (JRL per AME).

Sq. **SO** 51. **ST** 59.

667 *Semioscopis steinkellneriana* D. & S.
Usk (Plas Newydd), garden m.v.l. trap, 23/3/72 (1), (GANH) (1st v-c. rec.).

Sq. **SO** 30.

670 *Depressaria daucella* D. & S.
= *nervosa* auctt.
Newchurch West 11/3/76 (GANH). Whitebrook, 30/6/76, larvae on hemlock water dropwort (*Oenanthe crocata* L.) (MWH in litt.).

Sq. **SO** 50. **ST** 49.

672 *Depressaria pastinacella* Dup.   Parsnip Moth
= *heracliana* auctt.
Usk, garden m.v.l. trap, May 1979, one.

Sq. **SO** 30.

674 *Depressaria badiella* Hb.
= ***brunneella*** Rag.
Clydach, 18/7/89, one to m.v.l. (GANH).

Sq. **SO** 21.

676 *Depressaria pulcherrimella* Stt.
"Monmouthshire". Record attributed to DWHF (per MWH).

688 *Agonopterix heracliana* Linn.
  = *applana* Fabr.
Usk, garden m.v.l. trap, frequently recorded eg. 1976 25/2; 1978 1/8, 14/8 (2), 14/11; 1982 7/8, 11/8.

Sq. **SO** 30.

689 *Agonopterix ciliella* Stt.
Usk, garden m.v.l. trap, 10/9/77 (1), 14/4/79 (1).

Sq. **SO** 30.

692 *Agonopterix subpropinquella* Stt.
Magor Marsh, one to m.v.l., 25/7/77.

Sq. **ST** 48.

695 *Agonopterix alstroemeriana* Cl.
Usk, garden m.v.l. trap, 2/4/76, 21/8/77, 14/10/78 etc. Llandogo, 30/6/76, larvae on hemlock (*Conium maculatum*) on bank of R. Wye, (MWH in litt.).

Sq. **SO** 30,50.

696 *Agonopterix propinquella* Treit.
Usk, garden m.v.l. trap, 18/4/80 (1).

Sq. **SO** 30.

697 *Agonopterix arenella* D. & S.
Usk, garden m.v.l. trap, recorded frequently eg. 31/3/76, 30/8/77, 10/10/78, 6/8/82. Llansoy 19/8/91.

Sq. **SO** 30,40.

698 *Agonopterix kaekeritziana* Linn.
  = *liturella* D. & S.
Monmouthshire (JRL per AME).

701 *Agonopterix ocellana* Fabr.
Usk, garden m.v.l. trap, 1/3/77. Hendre Wds 21/4/80.

Sq. **SO** 30,41.

703 *Agonopterix atomella* D. & S.
   = *pulverella* Hb.
Usk (Plas Newydd), to m.v.l., 13/8/77 (det. JDB). Magor Marsh 20/8/77.

Sq. **SO** 30. **ST** 48.

704 *Agonopterix scopariella* Hein.
Hael Wds, 30/6/76 (MWH in litt.).

Sq. **SO** 50.

706 *Agonopterix nervosa* Haw.
   = *costosa* Haw.
Usk, garden m.v.l. trap, 13/8/77 (1).

Sq. **SO** 30.

709 *Agonopterix literosa* Haw.
Slade Wds 15/8/80, to m.v.l. (GANH, det. JDB).

Sq. **ST** 48.

713 *Agonopterix angelicella* Hb.
Usk (Plas Newydd), garden m.v.l. trap, 4/8/83 (GANH, det. JDB).

Sq. **SO** 30.

ETHMIIDAE

718 *Ethmia dodecea* Haw.
Wyndcliff, 13/7/69, one to m.v.l. (GANH). (New county rec.).

Sq. **ST** 59.

GELECHIIDAE
GELECHIINAE

728 *Paltodora cytisella* Curt.
Trelleck Bog, one to m.v.l., 24/7/86 (GANH, det. JDB).

Sq. **SO** 50.

731 *Eulamprotes atrella* D. & S.
Hendre Wds, 20/7/79 (JDB).

Sq. **SO** 41.

735 *Monochroa tenebrella* Hb.
Usk (Plas Newydd) 21/7/79 (JDB).

Sq. **SO** 30.

736 *Monochroa lucidella* Steph.
Monmouthshire (DJLA per PAS).

742 *Monochroa lutulentella* Zell.
Monmouthshire (DJLA per PAS).

756 *Parachronistis albiceps* Zell.
Hendre Wds 20/7/79 (JDB).

Sq. **SO** 41.

758 *Recurvaria leucatella* Cl.
Llansoy, to u.v.l., 10/7/87 (1) (det. JDB).

Sq. **SO** 40.

760 *Exoteleia dodecella* Linn.
Hendre Wds 20/7/79 (JDB). Hael Wds 3/7/83, to m.v.l.

Sq. **SO** 50,41.

762 *Rhynchopacha mouffetella* Linn.
Usk (Plas Newydd), garden m.v.l. trap, 1977, 30/7, 11/8; 1978, 21/7; 1979, 12/8 (2); 1982, 7/7. Hendre Wds 20/7/79 (JDB).

Sq. **SO** 30,41.

763 *Xenolechia aethiops* Humph. & Westw.
Wyndcliff, 26/4/76, to m.v.l. (GANH, det. JDB).

Sq. **ST** 59.

765 *Teleiodes vulgella* Hb.
Hendre Wds 20/7/79 (JDB).

Sq. **SO** 41.

770 *Teleiodes proximella* Hb.
Wentwood 26/5/80, to m.v.l. (GANH, det. JDB).

Sq. **ST** 49.

772 *Teleiodes fugitivella* Zell.
Redding's Inclosure 21/4/81, to m.v.l. (GANH, det. JDB).

Sq. **SO** 51.

774 *Teleiodes luculella* Hb.
Hael Wds 26/6/86, two to m.v.l. Llansoy 1986, to u.v.l., 29/6 (2), 5/7 (1). Hendre Wds 11/7/86, 14/5/88.

Sq. **SO** 40,50,41.

776 *Teleiopsis diffinis* Haw.
Monmouthshire (ECP-C per PAS.)

780 *Bryotropha similis* Stt.
Llansoy, 23/6/87, one to u.v.l. (det. JDB).

Sq. **SO** 40.

789 *Bryotropha domestica* Haw.
Usk, garden m.v.l. trap, 11/8/77 (det. JDB).

Sq. **SO** 30.

792 *Mirificarma mulinella* Zell.
Usk, garden m.v.l. trap, 11/8/82. Hael Wds 11/8/82.

Sq. **SO** 30,50.

797 *Neofaculta ericetella* Geyer
Wye Valley, as *Gelechia ericetella* (Barraud 1906).
    Trelleck Bog 22/5/73. Cwm Tyleri 26/5/73 (GANH), 12/7/77 (ECP-C).

Sq. **SO** 20,50.

802a *Gelechia sororculella* Hb.
Magor Marsh 21/7/79 (JDB). Usk, garden m.v.l. trap, 20/7/83.

Sq. **SO** 30. **ST** 48.

819 *Scrobipalpa costella* Humph. & Westw.
Monmouthshire (ECP-C per PAS) per AME).

822 *Scrobipalpa acuminatella* Sirc.
Magor Marsh 27/5/82, netted by day (JDB). Hendre Wds 29/5/82, d° (JDB).

Sq. **SO** 41. **ST** 48.

834 *Caryocolum tricolorella* Haw.
Tintern 1970, larvae on *Stellaria holostea*, reared. Imagines emerged June 6 – 11 1970 (AME in litt.). Hael Wds 29/4/78, larvae plentiful in spun shoots of greater stitchwort (*Stellaria holostea* L.) (JRL).

Sq. **SO** 50.

ANACAMPSINAE

854 *Anacampsis blattariella* Hb.
Hendre Wds 29/5/82, numerous larvae in rolled leaves of birch.

Sq. **SO** 41.

855 *Acompsia cinerella* Hb.
Monmouthshire (ECP-C per PAS).

CHELARIINAE

858 *Hypatima rhomboidella* Linn.
Wyndcliff 17/9/79, to m.v.l. Pont-y-saeson 18/9/79, to m.v.l.

Sq. **ST** 59.

859 *Psoricoptera gibbosella* Zell.
Usk (Plas Newydd), one to garden m.v.l. trap, Sept 1978 (det. JDB). Hael Wds 31/7/82.

Sq. **SO** 30,50.

DICHOMERINAE

862 *Dichomeris marginella* Fabr.
Llansoy 26/6/90, one to u.v.l.

Sq. **SO** 40.

866 *Brachmia blandella* Fabr.
    = *gerronella* Zell.
Usk, garden m.v.l. trap, 7/8/80.

Sq. **SO** 30.

868 *Brachmia rufescens* Haw.
Usk, garden m.v.l. trap, 28/7/80.

Sq. **SO** 30.

SYMMOCINAE

870 *Oegoconia quadripuncta* Haw.
Usk (Plas Newydd), to garden m.v.l. trap, 12/8/79 (1), 8/8/81 (1).

Sq. **SO** 30.

BLASTOBASIDAE

873 *Blastobasis lignea* Wals.
Wyndcliff, 17/9/79, plentiful to m.v.l. (GANH) (det. JDB).

Sq. **ST** 59.

MOMPHIDAE
BATRACHEDRINAE

879 *Batrachedra pinicolella* Zell.
Slade Wds, 21/7/79, (JDB).

Sq. **ST** 48.

MOMPHINAE

880 *Mompha langiella* Hb.
 = *epilobiella* Roem.
Hael Wds 30/6/76, larvae on *Epilobium montanum* (MWH in litt.).

Sq. **SO** 50.

882 *Mompha locupletella* D. & S.
 – *schrankella* Hb.
Magor Marsh 2/8/75, one swept from herbage by day (GANH).

Sq. **ST** 48.

883 *Mompha raschkiella* Zell.
Wyndcliff 23/6/86 one to m.v.l. (JDB, GANH). Wyndcliff 19/9/90, mines common on rosebay willowherb (AME in litt.).

Sq. **ST** 59.

884 *Mompha miscella* D. & S.
Monmouthshire, (DJLA per. AME).

886 *Mompha ochraceella* Curt.
Usk, garden m.v.l. trap, 18/7/78 (1) (det. JDB).

Sq. **SO** 30.

888 *Mompha propinquella* Stt.
Usk (Plas Newydd), garden m.v.l. trap, 30/7/81 (2), 7/8/83 (1).

Sq. **SO** 30.

893 *Mompha epilobiella* D. & S.
= *fulvescens* Haw.
Usk, garden m.v.l. trap, 1/4/78, one (GANH).

Sq. **SO** 30.

COSMOPTERIGIDAE
ANTIQUERINAE

899 *Pancalia leuwenhoekella* Linn.
Slade Wds 26/5/80, one netted by day (GANH) (1st v-c. rec) and possibly new to Wales.

Sq. **ST** 48.

BLASTODACNINAE

905 *Spuleria hellerella* Dup.
Usk, garden m.v.l. trap, 1979 19/7, 20/7, 22/7 (4), 23/7. Llansoy, to u.v.l. 28/6/86, 6/7/87.

Sq. **SO** 30,40.

TORTRICIDAE
COCHYLINAE

924 *Hysterophora maculosana* Haw.
Redding's Inclosure 31/5/68 (JMC-H 1969). (1st v-c. rec.).
    Pont-y-saeson 6/6/72, Slade Wds 26/5/80, St Pierre's Great Wds 19/5/87; all flying by day.

Sq. **SO** 51. **ST** 48,59.

925 *Phtheochroa rugosana* Hb.
Usk, garden m.v.l. trap, 1973 30/6; 1974 25/6, 7/7; 1981 20/6.

Sq. **SO** 30.

930 *Piercea alismana* Rag.
Magor Marsh 21/7/79, one to m.v.l. (JDB). The only v-c. record.

Sq. **ST** 48.

933 *Phalonidia gilvicomana* Zell.
Wyndcliff 29/30 June 1985 (MAF. per JDB) (1st v-c. rec.).

Sq. **ST** 59.

937 *Agapeta hamana* Linn.
Widespread and common.

Usk, garden m.v.l. trap, 10/7/69 and recorded annually to 1983 eg. 1975 21/6 – 4/8. Ed. 3/6/78. Ld. 5/9/71. Usk (Park Wood) 7/7/69. Wentwood 1968, 21/6/69. Magor Marsh 1/7/69, 2/7/76 (abdt.). Redding's Inclosure 20/7/71. Piercefield Park 18/7/73. Wyndcliff 2/7/74. Cwm Tyleri 12/7/77. Trelleck Bog 8/7/78. Hendre Wds 9/7/78. Pont-y-saeson 3/7/79. Slade Wds 2/8/80, 17/7/87. Llansoy 4/6/85, 6/7/87. Kilpale 30/6/87. Abergavenny (St Mary's Vale) 8/7/87. Clydach 18/7/89. Magor Pill 6/8/89.

Sq. **SO** 20,30,40,50,21,41,51. **ST** 48,49,59.

938 *Agapeta zoegana* Linn
Uncommon in Gwent.

Usk, garden m.v.l. trap, recorded infrequently, eg. 7/7/67, 2/8/77, 2/8/82 etc. Rhyd-y-maen 18/7/70. Hael Wds 14/8/78. Slade Wds 28/8/79, 19/6/85, 29/7/86, 17/7/87. Llansoy 24/7/86, 22/7/87.

Sq. **SO** 30,40,50. **ST** 48.

939 *Aethes tesserana* D. & S.
Bradley in *British Tortricoid Moths* (1973 **1**: 54) writes "has probably been overlooked in Wales, since the only known records are from Anglesey, Cardiganshire and Pembrokeshire". It does in fact occur locally on the Carboniferous Limestone in south-east Monmouthshire.

Tintern dist. (Piffard 1859. as *Phalonia tesserana* Schiff.).
Slade Wds 1980 26/5 – 24/6 abundant on verges of forestry roads.

Sq. **ST** 48,59.

944 *Aethes williana* Brahm
Bradley (1973 **1**: 60) gives "Monmouthshire (Chepstow)".

945 *Aethes cnicana* Westw
Frequent and widespread.

Prescoed 21/6/71 (2) to m.v.l. Wentwood 29/6/73. Magor Marsh 3/7/73, plentiful at m.v.l.. Usk, garden m.v.l. trap, recorded occasionally eg. 13/8/77, 27/7/71. Trelleck Bog 16/7/79. Slade Wds 23/7/79. Cwmfelinfach 20/6/87. Clydach 26/7/90.

Sq. **SO** 30,50,21. **ST** 48,19,39,49.

974 *Aethes smeathmanniana* Fabr.
Abertillery dist. 1914 (Rait-Smith 1915, as *Cochylis smeathmanniana*).
Llansoy 1992, garden u.v.l. trap 31/7/92 (1) (GANH).

Sq. **SO** 20,40.

954 *Eupoecilia angustana* Hb.
ssp. *angustana* Hb.

Pontllanfraith 1987, several flying by day 10/8 (GANH). (1st v-c rec).

Sq. **ST** 19.

959 *Cochylidia rupicola* Curt.
Tintern dist. 19/6/65 (Mere 1965).
  Hendre Wds 1978, plfl. on 27/6 and 9/7, flying in sunshine. Slade Wds 8/7/80. Usk, two at dusk on old railway line 13/7/82. St Pierre's Great Wds 5/7/88.

Sq. **SO** 30,41. **ST** 48,59.

960 *Falseuncaria ruficiliana* Haw.
Usk, garden m.v.l. trap, 20/8/77 (1) (det. JDB).

Sq. **SO** 30.

964 *Cochylis dubitana* Hb.
Wyndcliff 25/7/83 (1) to m.v.l. (GANH) (1st v-c. rec.); 10/8/83 (1). Gaer Hill, Newport, 1987 several by day 15/6.

Sq. **ST** 28,59.

965 *Cochylis hybridella* Hb.
Usk, garden m.v.l. trap, 8/8/80 (1) (1st v-c rec.). Slade Woods 22/7/86 (2) to m.v.l.

Sq. **SO** 30. **ST** 48.

966 *Cochylis atricapitana* Steph.
Slade Wds 26/5/80 (2) by day (GANH) (1st v-c. rec.).

Sq **ST** 48.

968 *Cochylis nana* Haw.
Cicelyford 17/6/87, several flying by day. (GANH) (new v-c. rec.).

Sq. **SO** 50.

TORTRICINAE

969 *Pandemis corylana* Fabr.   Chequered Fruit-tree Tortrix
Common in woodland from July to September.
  Usk, garden m.v.l. trap, 20/8/73 (plfl.) and recorded most years. Hael Wds 10/7/73. Hendre Wds 29/8/78. Slade Wds 21/8/79. Pont-y-saeson

18/9/79. Wyndcliff 6/7/81. Trelleck Bog 30/8/86.

Sq. **SO** 30,50,41. **ST** 48,59.

970 *Pandemis cerasana* Hb.   Barred Fruit-tree Tortrix
Common in deciduous woodland, flying from June to August.
   1968, Deri-fach 2/6, larva on oak, reared (JMC-H 1969).
   Usk, garden m.v.l. trap, 1973, plentiful in June and recorded annually to 1983. Wyndcliff 22/6/71, 19/6/87 etc. Wentwood 29/6/73. Hael Wds 12/7/84 (abdt.). Cwm Tyleri 5/8/74. Trelleck Bog 8/7/78 (plfl.). Magor Marsh 9/7/78. Hendre Wds 24/6/79 (abdt.), 4/7/80 (d°). Slade Wds 28/7/80, 15/8/80. Redding's Inclosure 23/6/81. Pont-y-saeson 13/7/81. Llansoy 27/6/86. Kilpale 30/6/87. Dixton Bank 18/6/88. Pontllanfraith 2/7/87.

Sq. **SO** 20,30,40,50,21,41,51. **ST** 48,19,49,59.

971 *Pandemis cinnamomeana* Treit.
Widespread and fairly common in woods in the eastern half of the county. Has been noted from June to early August and again in September and on one occasion, in October (v.i.).
   Tintern dist. (Piffard 1859).
   Usk, garden m.v.l. trap, often recorded eg. 1973 July 8, 9, 12, 15; 1978 27/7 and a very late male on 10/10 (detn. confirmed by JDB); 1982, 10/9, 15/9. Wyndcliff 13/7/69, 21/7/87. Cwm Tyleri 9/7/69. Trelleck Bog 16/7/79, (abdt. to m.v.l.), 19/7/86. Slade Wds, 1980, 24/6 (1) by day, 28/7 (to m.v.l.); 1982, 9/9, 16/9 (to m.v.l.). Hael Wds, 1982 to m.v.l. 5/7, 18/9. Ysguborwen (Lantrisant) 17/7/86. Llansoy 25/7/86. Kilpale 30/6/87.

Sq. **SO** 20,30,40,50. **ST** 48,49,59.

972 *Pandemis heparana* D. & S.   Dark Fruit-tree Tortrix
Common in woods, gardens and bushy places, it flies from late June until the end of August.
   Usk, garden m.v.l. trap, frequently noted eg. 3/8/74; 1982, 7/7 – 8/9. Magor Marsh 9/7/78. Hendre Wds 20/7/79. Hael Wds 5/7/82. Slade Wds 21/7/83. Redding's Inclosure 24/7/83. Llansoy 25/6/86. Wyndcliff 5/7/88.

Sq. **SO** 30,40,50,41,51. **ST** 48,59.

977 *Archips podana* Scop.   Large Fruit-tree Tortrix
Common in woods and gardens, flying from late June to the middle of September. Melanic specimens occur.
   1968, Deri-fach 2/6, larva on oak, reared (JMC-H 1969).
   Usk, garden m.v.l. trap, 7/7/67 and recorded annually to 1983. Ed. 20/6/79. Ld. 19/9/82. Redding's Inclosure 20/7/71. Hael Wds 10/7/73. St Pierre's Great Wds 13/7/73. Magor Marsh 1/7/74, 11/7/78 (melanic

specimen). Trelleck Bog 16/7/79. Hendre Wds 4/7/80, 5/7/86 (plfl.). Slade Wds 28/7/80. Wyndcliff 6/7/81. Pont-y-saeson 13/7/81. Llansoy 2/7/85. Kilpale 30/6/87.

Sq. **SO** 30,40,50,21,41,51. **ST** 48,49,59.

979 *Archips crataegana* Hb.   Brown Oak Tortrix

Occurs locally in mixed deciduous woodland in the eastern half of the county but rarely numerous. Bradley (1973 **1**:106) states that Monmouthshire represents the western limits of its range and elsewhere in the Principality it is known only from Flintshire, North Wales. It flies in July and August.

Hendre Wds 17/7/79, 20/7/79 (frqt.), 11/7/86 etc. Hael Wds 5/7/82, 3/7/83 (fairly plentiful at m.v.l.). Llansoy 12/7/86.

Sq. **SO** 40,50,41.

980 *Archips xylosteana* Linn.   Variegated Golden Tortrix

Fairly common in gardens and mixed woodland during July.

Wyndcliff 13/7/69. Hael Wds, to m.v.l. 10/7/73 (plfl.), 2/8/71 etc.; pupa in rolled leaf of young stool oak 30/6/83 (emerged 2/7/83). Usk, garden m.v.l. trap, 4/7/74; 1979 8/7 – 21/7. Pont-y-saeson 8/7/78. Magor Marsh 9/7/78. Slade Wds 23/7/83. Llansoy 11/7/86. Kilpale 30/6/87.

Sq. **SO** 30,40,50,41. **ST** 48,49,59.

983 *Choristoneura hebenstreitella* Müll.

This woodland species, flying in June and July, occurs locally and usually sparingly in some woods in the east of the county.

Hael Wds 3/7/83 one to m.v.l. (GANH) (1st v-c. rec.); 1985 24/6 (1), 26/6 (1); 1986 24/6 (1), 5/7 (plentiful to m.v.l.). Hendre Wds 11/7/86.

Sq. **SO** 50,41.

986 *Syndemis musculana* Hb.
    ssp. *musculana* Hb.

Common and widespread, flying in open woods from May to early July.

1968, Deri-fach 2/6, "flying in numbers over alders" (JMC-H. *Ent. Rec.* **81**:42). Coed-y-Bwynydd 5/6 (idem).

Pont-y-saeson 11/5/68 (by day). Wyndcliff 22/6/71. Slade Wds 3/6/72. Redding's Inclosure 30/5/77. Usk, garden m.v.l. trap, 24/5/78. Hael Wds 30/5/78. Magor Marsh 1/6/79. Hendre Wds 27/5/79. Wentwood 20/5/80. Wyndcliff 25/6/86. St Pierre's Great Wds 19/5/87, plentiful by day. Trelleck Bog 25/5/87. Llansoy 13/7/87.

Sq. **SO** 30,40,50,21,41,51. **ST** 48,49,59.

987 *Ptycholomoides aeriferanus* H.-S.
Flying in July and August, mainly in woodland containing larches, this species was first recorded in Britain (Kent) in 1951. First noted in Monmouthshire in 1973 (Horton 1975). It has since become widespread and locally common in the eastern half of the county.

Hael Woods 10/7/73 abundant at m.v.l. (GANH) (1st v-c. and Welsh rec.), 31/7/82. Magor Marsh 1/7/75 (1) to m.v.l., 13/7/83. Usk, garden m.v.l. trap, 13/8/77, 12/7/81, 1982, 1983. Redding's Inclosure 30/5/77. Trelleck Bog 16/7/79 (plfl.). Slade Wds 28/7/80 (plfl.), 21/7/83, 20/7/89. Ysguborwen (Llantrisant) 17/7/86.

Sq. **SO** 30,50,51. **ST** 48,49.

988 *Aphelia viburnana* D. & S.   Bilberry Tortrix
This moth occurs in the hills of north-west Gwent, it is readily disturbed from the moorland herbage by day and at night comes to light.

Cwm Tylori 29/7/72, 13/7/75, 12/7/77 (abdt.).

Sq. **SO** 20.

989 *Aphelia paleana* Hb.   Timothy Tortrix
Usk, garden m.v.l. trap, 6/7/74 (1), 1979 (4) 12/8 to 17/9, 10/7/80 (1). Pontllanfraith 6/7/88 (plfl.).

Sq. **SO** 30. **ST** 19.

993 *Clepsis spectrana* Treit.   Cyclamen Tortrix
Usk, garden m.v.l. trap, recorded occasionally eg. 24/7/71, 3/7/80 etc. Magor Marsh frequently noted eg. 25/6/76, 24/8/81 (plfl.). Llansoy 29/6/86.

Sq. **SO** 30,40. **ST** 48.

994 *Clepsis consimilana* Hb.
Usk, garden m.v.l. trap 20/7/79; one at dusk 13/7/82. Hendre Wds (1) 20/7/79. Trostrey Common 18/6/93, one by day.

Sq. **SO** 30,41.

1000   *Ptycholoma lecheana* Linn.
Magor Marsh 19/6/79 one, (melanic f.) Hael Wds 26/6/86 (1).

Sq. **SO** 50. **ST** 48.

1001   *Lozotaeniodes formosanus* Geyer
A species associated with Pinus sylvestris and first recorded in Britain (Surrey) in 1945.

Slade Wds 23/7/79, several to m.v.l. (GANH) (1st v-c. rec.). Also 5/7/80, 1982, 8/7/83 (plfl.), 17/7/87 etc. Usk, garden m.v.l. trap, 17/7/82. Risca, to act. l. 28/6/90 (1) (MEA).

Sq. **SO** 30. **ST** 48,29.

1002 *Lozotaenia forsterana* Fabr.
Widespread and locally fairly common in Monmouthshire, flying in gardens and woods during the second half of June and July.
Usk, garden m.v.l. trap, eg. 19/6/71, 1973 (12) 10/6 – 11/7. Ed. 10/6/73. Ld. 30/7/78. Hael Wds 12/7/71. Pont-y-saeson 18/7/72. Cwm Tyleri 29/6/76. Slade Wds 23/7/79. Llansoy 5/7/85.

Sq. **SO** 20,30,40,50. **ST** 48,59.

1006 *Epagoge grotiana* Fabr.
Occurs locally in woodland during July.
Hael Wds 10/7/73 (GANH); 11/7/81 (MWH in litt.); 2/8/86. Usk, to m.v.l., 25/7/80, 18/7/83. Slade Wds 15/7/80. 1986 Llansoy 6/7, Hendre Wds 11/7.

Sq. **SO** 30,40,50,41. **ST** 48.

1007 *Capua vulgana* Fról.
Fairly common in Gwent, flying in woodland in May, June and early July and also in the moorland areas of the north and west.
1968 (as *C. favillaceana* Hb.) Tintern 31/5, Redding's Inclosure 31/5, Derifach 2/6 (JMC-H 1969).
Wyndcliff 22/6/71 to m.v.l. (JMC-H, GANH). Hael Wds 30/6/76 (MWH pers. comm.); 29/5/82. Pontypool (Mountain Air) 25/5/80. Pont-y-saeson 8/7/80. Trelleck Bog 12/6/87.

Sq. **SO** 50,21,51 **ST** 29,59.

1008 *Philedone gerningana* D. & S.
To date (1992) one record only, from the hilly moorlands in the north of the county, viz. Cwm-Tyleri (1400ft.) 12/7/77 (2) to m.v.l.

Sq. **SO** 20.

1010 *Ditula angustoriana* Haw.   Red-barred Tortrix
Widespread and fairly common in the eastern half of Gwent, flying in gardens and woodland during July.
Usk, garden m.v.l. trap, 29/7/77, 30/7/83 etc. Hendre Wds, plentiful to m.v.l. 20/7/79 and 1/8/80. Slade Wds 28/7/81, 22/7/86. Hael Wds 5/7/82, 2/8/86. Wyndcliff 28/7/82, 5/7/88. Llansoy 7/7/86. Trelleck Bog 19/7/86.

Magor Marsh 2/7/89.

Sq. **SO** 30,40,50,41. **ST** 48,59.

1011 *Pseudargyrotoza conwagana* Fabr.
Common and widespread in hedgerows and open woods during June and July this species flies freely in sunshine and also comes to light.

Wyndcliff 22/6/71 to m.v.l., 23/6/86. Hael Wds 3/7/71, 12/7/87. Usk, garden m.v.l. trap, frequent, eg. 30/6/71, 8/7/82. Hendre Wds 20/7/79. Slade Wds 21/6/81. Usk, 13/7/82, plentiful by day along old railway line. Llansoy 24/6/87. Ochrwyth 20/6/87. Cwmfelinfach 20/6/87. Llantrisant 27/6/87, (abdt.). Pontllanfraith 2/7/87. Magor Marsh 2/7/89. Clydach 9/7/89.

Sq. **SO** 30,40,50,21,41. **ST** 28,48,19,59.

1013 *Olindia schumacherana* Fabr.
Scarce and local in Monmouthshire, this moth frequents hedge-banks and the margins of woods during June and July.

Usk, garden m.v.l. trap, 10/7/81 (1) (1st v-c. rec.) (GANH), 3/7/83 (1). Hendre Wds 11/7/86 one to m.v.l. Llantrisant 27/6/87, five (♂♂ 2, ♀♀ 3) settled on nettles at edge of wood in afternoon sunshine. Llansoy (July 1988), several by day.

Sq. **SO** 30,40,41. **ST** 39.

1015 *Eulia ministrana* Linn.
Widespread and common in woodland during May and June.

Wye Valley (Barraud 1906, as *Tortrix ministrana*). Coed-y-Bwnydd 5/6/68 (JMC-H 1969).

Slade Wds 3/6/72, plentiful by day. Pont-y-saeson 6/6/72. Trelleck Bog 16/6/73 to m.v.l. Usk, m.v.l. trap, 7/7/77. Hael Wds 27/6/78. St Pierre's Great Wds 17/5/80. Hendre Wds 19/5/80.

Sq. **SO** 30,50,41. **ST** 48,59.

1020 *Cnephasia stephensiana* Doubl.   Grey Tortrix
Occurs locally and sparingly in Monmouthshire.

Usk, garden m.v.l. trap, recorded sporadically eg. 23/7/71, 24/7/73, 21/7/77, 26/7/79. Magor Marsh 13/7/77, 2/7/89.

Sq. **SO** 30. **ST** 48.

1021 *Cnephasia asseclana* D. & S.   Flax Tortrix
 = *interjectana* Haw.
Common and widespread, in gardens, open woodland etc., flying during July and August.

Cwm Tyleri (1,400 ft.) 9/7/73. Magor Marsh 13/7/77, 2/7/89 (melanic f.). Pont-y-saeson 8/7/78. Hendre Wds 20/7/79. Ysguborwen (Llantrisant) 17/7/86.

Sq. **SO** 20,41. **ST** 48,49,59.

1024 *Cnephasia incertana* Treit.   Light Grey Tortrix
Usk, garden m.v.l. trap, 11/7/79. Llansoy 18/6/85, 27/6/86.

Sq. **SO** 30,40.

1025 *Tortricodes alternella* D. & S.
Common in open deciduous woodland during February and March, the males flying readily in sunshine.
   Pont-y-saeson 26/3/68 abundant by day. Redding's Inclosure 7/3/72. Hendre Wds 25/3/72. Usk, garden m.v.l. trap, 25/2/76, 8/3/77 etc.

Sq. **SO** 30,41,51. **ST** 59.

1027 *Neosphaleroptera nubilana* Hb.
One Gwent record to date. Elsewhere in the Principality it appears to have been recorded only from Montgomeryshire.
   Undy 9/7/83 (1) (GANH) (1st v-c. rec.).

Sq. **ST** 48.

1029 *Eana osseana* Scop.
Occurs plentifully in the hilly districts in the north.
   Cwm Tyleri, abundant on 6/8/73 and 13/7/75. Hendre Wds 1/8/80.

Sq. **SO** 20,41.

1030 *Eana incanana* Steph.
One record only to date. In Wales, it was previously known only from Flintshire and Caernarvonshire.
   Slade Wds 21/7/79, one taken by day, (JDB, GANH) (1st v-c. rec.).

Sq. **ST** 48.

1032 *Aleimma loeflingiana* Linn.
Widespread and common among oaks in late June and July.
   Usk, garden m.v.l. trap, 30/6/73 and recorded most years eg. 1979, plentiful 7/7 – 22/7. Trelleck Bog 16/7/79. Hendre Wds 17/7/79, 20/7/79 (abdt.). Slade Wds 23/7/79. Hael Wds 5/7/81, 26/6/86. Magor Marsh 13/7/83. Llansoy 4/7/85. Kilpale 30/6/87 (abdt.). Pontllanfraith 2/7/87. Wyndcliff 5/7/88.

Sq. **SO** 30,40,50,41. **ST** 48,19,49,59.

1033   *Tortrix viridana* Linn.   Green Oak Tortrix
Widespread and common, often abundant, among oaks in June and July. Readily disturbed from foliage by day and at night it comes to light.

Usk (Park Wood) 27/6/67. Usk, garden m.v.l. trap, recorded annually 1966 – 1983 and often plentiful eg. 1979, 28/6 – 27/7. Wyndcliff 13/7/69. Hael Wds 3/7/70. Redding's Inclosure 20/7/71. Pont-y-saeson 18/7/72. Cwm Tyleri 28/6/72. Wentwood 29/6/73. Magor Marsh 3/7/73. St Pierre's Great Wds 13/7/73. Hendre Wds 6/6/78, oaks heavily infested with the black pupae of this species; 20/7/79 abundant to m.v.l. Slade Wds 21/7/79. Llansoy 11/7/86. Trelleck Bog 18/7/86. Llantrisant (Coed-y-prior) 27/6/87 (abdt.). Kilpale 30/6/87. Pontllanfraith 2/7/87. Sugar Loaf (St Mary's Vale) 8/7/87.

Sq. **SO** 20,30,40,50,21,41,51. **ST** 48,19,39,49,59.

1034   *Spatalistis bifasciana* Hb.
"Monmouthshire" (per EFH 1986). No details.

1035   *Croesia bergmanniana* Linn.
Hendre Wds 20/7/79, plfl. to m.v.l. Usk, garden trap, 29/7/79 (1). Slade Wds 28/7/80, 30/6/81, 8/7/83 etc. Llansoy 2/7/86, 14/7/86. Trelleck Bog 24/7/86. Hael Wds 2/8/86. Sugar Loaf (St Mary's Vale) 8/7/87. Wyndcliff 5/7/88. Magor Marsh 2/7/89.

Sq. **SO** 30,40,50,21,41. **ST** 48,59.

1036   *Croesia forsskaleana* Linn.
Common and widespread in gardens and deciduous woodland in July and August.

Usk, garden m.v.l. trap, 19/7/69 and recorded annually until 1983, eg. 1978, frequent 16/7 – 19/8. Ed. 5/7/76. Ld. 1/9/79, but after the hot summer of 1976 a very late specimen was recorded on Nov. 1st. Hael Wds 10/7/73. Magor Marsh 20/8/77. Hendre Wds 17/7/79. Wyndcliff 28/7/82. Slade Wds 23/7/83. Trelleck Bog 17/8/88. Llansoy 21/7/89.

Sq. **SO** 30,40,50,41. **ST** 48,59.

1037   *Croesia holmiana* Linn.
This species, flying in July and August, is locally common in gardens, hedgerows and bushy places and comes readily to light.

Usk, garden m.v.l. trap, 26/7/73 and noted most years, often plentiful eg. 1982, 6/7 – 4/8. Ed. 6/7/82. Ld. 13/8/77. Usk 13/7/82 flying by day on disused railway-line. Llansoy 31/7/86, 18/8/86.

Sq. **SO** 30,40.

1038   *Acleris laterana* Fabr.
      = *latifasciana* Haw.
Common in bushy places from late July to mid-September.

Usk, garden m.v.l. trap, 15/8/66 and recorded most years until 1982 eg. 1979, 27/7 – 17/9. Hael Wds 31/7/82. Wyndcliff 11/8/82. Llansoy 4/8/89.

Sq. **SO** 30,40,50. **ST** 59.

1040 *Acleris caledoniana* Steph.
Usk, 16/10/78, one to garden m.v.l. trap, (GANH) (1st v-c. rec.) det. JDB.

Sq. **SO** 30.

1041 *Acleris sparsana* D. & S.
Usk, one to m.v.l. 18/8/79.

Sq. **SO** 30.

1042 *Acleris rhombana* D. & S.   Rhomboid Tortrix
Flies in gardens and bushy places from Aug. to Oct. and comes to light.

Abertillery dist. 1912 (Rait-Smith 1913, as *Dictyopteryx contaminana* Hb.)

Usk, garden m.v.l. trap, 14/10/78, 21/10/78, 30/8/80 etc. Magor Marsh 13/10/78. Trelleck Bog 9/10/78. Hael Wds 9/10/79. Llansoy 1986.

Sq. **SO** 20,30,40,50. **ST** 48.

1044 *Acleris ferrugana* D. & S.
Occurs among oaks. Bivoltine, the first generation appearing in July and the second emerging in the autumn and seen again flying in the spring after hibernation.

Hendre Wds 20/7/79, 13/7/80. Usk, garden m.v.l. trap, 28/2/80 (2nd generation moth).

Sq. **SO** 30,41.

1045 *Acleris notana* Don.
      = *tripunctana* Hb.
Found in woods containing birches. Bivoltine, flying in July, with a second generation appearing later in the year and overwintering.

Slade Wds 21/7/79 (by day). Usk, garden m.v.l. trap, 1979 21/7 and 8/12. Hendre Wds 13/7/80, 24/4/82.

Sq. **SO** 30,41. **ST** 48.

1047 *Acleris schalleriana* Linn.
Wyndcliff 19/9/90, larval feeding on *Viburnam lantana* (AME in litt.) (1st v-c. rec.).

Sq. **ST** 59.

1048 *Acleris variegana* D. & S.   Garden Rose Tortrix
Common in gardens and hedgerows from July to October and coming to light.
   Usk, garden m.v.l. trap, plentiful most years eg. 1979 (27) 28/7 – 22/10.
   Ed. 26/7/76. Ld. 4/11/78. The form *asperana* Fabr. is often common.
Trelleck Bog 9/10/78 to m.v.l. Llansoy (26), 30/9/86 – 27/10/86.

Sq. **SO** 30,40,50.

1053 *Acleris hastiana* Linn.
A very variable species associated with sallows (*Salix* spp.). Bivoltine, the second generation hibernates and reappears in the spring.
   Usk, garden m.v.l. trap, frequent most years eg. 1978 (19) 5/10 – 11/11, 4/7/82. Pont-y-saeson 18/9/79, 30/9/79. Hael Wds 9/10/79 (5) to m.v.l. Hendre Wds 21/4/80 (brown form). Redding's Inclosure 21/4/81 (d°). Trelleck Bog 17/8/88. Llansoy 17/9/90, 12/10/90.

Sq. **SO** 30,40,50,41,51. **ST** 59

1054 *Acleris cristana* D. & S.
Bradley in *British Tortricoid Moths* (1973 **1**: 206) states that this species occurs "in wooded areas in the southern counties of England north to Herefordshire, Worcestershire and Norfolk but is local and generally uncommon" and is "apparently unknown from Wales".
   Usk, garden m.v.l. trap, 1970 20/2 one, (GANH) (1st v-c. and Welsh rec.). 1972: 23/3 (1). 1973: 5/2 (1), 17/3 (1). 1978: 12/9 f. *sub-fulvovittata* (1).

Sq. **SO** 30.

1055 *Acleris hyemana* Haw.
Abertillery dist. (Rait-Smith 1912, as *Tortrix hyemena*).

Sq. **SO** 20.

1061 *Acleris literana* Linn.
A local woodland species scarce in Monmouthshire. The following records were of moths flying after hibernation.
   Usk, garden m.v.l. trap, 15/4/80. Hendre Wds 21/4/80. Hael Wds 25/4/80.

Sq. **SO** 30,50,41.

1062 *Acleris emargana* Fabr.
Occurs in woods and hedgerows but not common in Monmouthshire. Readily disturbed from bushes by day but comes only sparingly to light.
   Usk, garden m.v.l. trap, 14/10/78, 8/9/82. Usk, several by day along old

railway line 2/9/80. Hael Wds 15/9/80 to m.v.l. Llansoy (1) 14/10/86.

Sq. **SO** 30,40,50.

OLETHREUTINAE

1063   *Celypha striana* D. & S.
Not uncommon from late June to August in rough grassy places.

Usk, garden m.v.l. trap, 21/7/83 and recorded in small numbers most years eg. 1982 (10) 2/7 – 6/9, 1983 (9) 27/6 – 11/8. Cwm Tyleri 12/7/77. Slade Woods 2/8/80. Magor Marsh 24/8/81, plentiful to m.v.l. Llansoy 28/7/84.

Sq. **SO** 20,30,40. **ST** 48.

1066   *Celypha woodiana* Barr.
A rare species, virtually "restricted in Britain to Herefordshire, Worcestershire, Gloucestershire and Monmouthshire". It has been recorded from Tintern (Bradley 1979 **2**: 17).

Sq. **SO** 50

1067a   *Celypha rurestrana* Dup.
Near Tintern 16/6/62, ♂ (1) to m.v.l. (ECP-C) (1st British specimen) but not recognised as such until Heckford found it in N. Devon in 1985 and 1987 and added it to the British list. (Heckford 1988 *Entomologist's Gaz.* **39**: 193).

Sq. **ST** 59.

1076   *Olethreutes lacunana* D. & S.
Very common and widespread, flying from June to August and frequenting hedgerows, gardens, woodland, and bushy places generally.

Usk, garden m.v.l. trap, 6/7/73 and recorded most years and often plentiful eg. 1983, 26/6 – 31/8. Ed. 1/6/81. Ld. 17/9/79. Wyndcliff 2/7/74. Hendre Wds 6/6/78. Pont-y-saeson 8/7/78. Magor Marsh 9/7/78. Hael Wds 26/6/79. Trelleck Bog 16/7/79. Slade Wds 21/7/79. Runston 2/7/80. Redding's Inclosure 24/7/83. Llansoy 28/7/84. Dinham 4/6/87. Ochrwyth 20/6/87. Cwmfelinfach 20/6/87. Pontllanfraith 30/6/88.

Sq. **SO** 30,40,50,41,51. **ST** 28,48,19,49,59.

1079   *Olethreutes bifasciana* Haw.
Scarce and local in Monmouthshire.

Usk, garden m.v.l. trap, 22/6/73 (1) (GANH) (1st v-c. rec.), 18/7/78 (1), 21/7/79 (1). Slade Wds 21/6/81, (2) to m.v.l., 8/7/83 (1), 21/7/83 (1).

Sq. **SO** 30. **ST** 48.

1082   *Hedya pruniana* Hb.   Plum Tortrix
Fairly common in Gwent during June and July; frequenting gardens, hedgerows and bushy places.

Usk, garden m.v.l. trap, 22/6/73 and noted most years eg. 1983, 29/6 – 11/7. Magor Marsh 13/7/77. Pont-y-saeson 8/7/78. Hael Wds 22/6/86. Llansoy 29/6/86. Kilpale 30/6/87. Dixton Bank 18/6/88. Clydach 18/7/89. Slade Wds 20/7/89.

Sq. **SO** 30,40,50,21,51. **ST** 48,49,59.

1083   *Hedya nubiferana* Haw.   Marbled Orchard Tortrix
Common and widespread, frequenting gardens, orchards, hedgerows and woodland during June and July.

Usk, garden m.v.l. trap, 11/7/74 and recorded most years eg. 1981, 24/6 – 2/7. Magor Marsh 13/7/77. Hael Wds 27/6/78. Hendre Wds 20/7/79. Slade Wds 23/7/79. Llansoy 18/6/85. Trelleck Bog 18/7/86. Kilpale 30/6/87. Sugar Loaf (St Mary's Vale) 8/7/87.

Sq. **SO** 30,40,50,21,41. **ST** 48,49.

1084   *Hedya ochroleucana* Fról.
Local and scarce in Monmouthshire.

Usk, garden m.v.l. trap, 18/6/75 (1). Magor Marsh 13/7/77, 9/7/78, 11/7/78 (single moths). Hael Wds 11/8/82 (1). Llansoy 4/7/86 (1).

Sq. **SO** 30,40,50. **ST** 48.

1085   *Hedya atropunctana* Zett.
Bradley in *British Tortricoid Moths* (1979 **2**: 41) writes" . . . Widespread in North Wales, . . . but known only from Monmouthshire (Tintern) in the south." Wyndcliff 10/8/83, one to m.v.l. (GANH). Llansoy, 11/9/89, one.

Sq. **SO** 40. **ST** 59.

1086   *Hedya salicella* Linn.
This species, flying from late June to August, frequents marshes and river valleys but is local and generally scarce in Monmouthshire.

Wye Valley (Bradley et al. 1979 **2**: 42).

Magor Marsh: 1/7/75 (3) to m.v.l.; 1976, abundant 25/6 – 6/7; 1983, 13/7 (abdt.); 2/7/89. Usk, garden m.v.l. trap, 19/7/76, 15/7/79, 8/7/83.

Sq. **SO** 30. **ST** 48,59.

1087   *Orthotaenia undulana* D. & S.
Occurs in open woodland during June and July.

Usk (Park Wood) 10/6, to m.v.l. Usk, garden m.v.l. trap, 29/7/78.

Wentwood, 29/6/73 (abdt.). Slade Wds 21/6/81. Hael Wds 26/6/86. Llansoy 28/6/87.

Sq. **SO** 30,40,50. **ST** 48,49.

1091   *Apotomis lineana* D. & S.
Bradley, 1979 gives "Tintern (Monmouthshire)" in *British Tortricoid Moths* (1979 **2**: 49).

Sq. **ST** 59.

1092   *Apotomis turbidana* Hb.
Locally plentiful in June and July in open woods with birches.
   Tintern dist. (Piffard 1859), as *A. picana* Fröl.
   Wentwood 8/7/68, to m.v.l.; 28/6/69 (abdt.). Hael Wds 3/7/71. Redding's Inclosure 20/7/71. Trelleck Bog 16/7/73. Slade Wds 21/7/73. Hendre Wds 24/6/80. Magor Marsh 21/6/81. Wyndcliff 25/6/86.

Sq. **SO** 50,41,51. **ST** 48,49,59.

1093   *Apotomis betuletana* Haw.
Widespread and very common in open woodland flying among birches from the end of June to early September.
   Wentwood July 1968, 19/8/69. Usk, garden m.v.l. trap, 30/8/69 and noted annually to 1983, sometimes plentiful. Usk (Park Wood) 3/9/69. Hael Wds 3/7/71. Pont-y-saeson 18/7/72. Slade Wds 29/8/72, 21/7/83 (abdt.). Cwm Tyleri 8/8/73. Hendre Wds 4/9/78 (abdt.). Redding's Inclosure 24/7/83. Wyndcliff 10/8/83. Trelleck Bog 19/7/86.

Sq. **SO** 20,30,50,41,51. **ST** 48,49,59.

1094   *Apotomis capreana* Hb.
Local and scarce in open woods, river valleys and in the vicinity of drainage reens in the south of the county.
   Tintern dist. (Piffard 1859).
   Usk, garden trap 30/6/73, 2/8/77, 21/7/79 (all singletons). Undy 13/7/73. Hael Wds, fairly plentiful to m.v.l. on 27/6/78, 4/7/81, & 2/8/86. Slade Wds 15/7/79. Wyndcliff 21/7/87.

Sq. **SO** 30,50. **ST** 48,59.

1095   *Apotomis sororculana* Zett.
Barrett in *The Lepidoptera of the British Islands* (1905 **10**: 364) wrote ". . . no record in Wales." Wye Valley (Barraud 1906, as *Penthina sororculana* Zett.)
   Wyndcliff 24/6/71 (1) to m.v.l. (JMC-H, GANH), 25/6/86 (1) to m.v.l.

(JDB, GANH). Slade Wds 21/7/79, one taken by day (JDB).

Sq. **ST** 48,59.

1096 *Apotomis sauciana* Fröl.
    ssp. *sauciana* Fröl.

Found in the hilly areas of northern Gwent and in the east of the county in open woodland where bilberry (*Vaccinium myrtillus*) grows.

Cwm Tyleri (1,400 ft,) 9/7/73, 12/7/74 (plfl.), 12/7/77. Trelleck Bog 19/7/86 to m.v.l. Trefil 9/7/88. Llansoy 25/6/90.

Sq. **SO** 20,40,50,11.

1097 *Endothenia gentianaeana* Hb.

Found in the south of the county frequenting rough ground and waysides where teasel (*Dipsacus fullonum*) grows.

1976, Undy, bred from seedheads of teasel collected on 24/2/76. 1987, Undy, several bred from teasels collected on 20/2, emerged 10/6 – 7/7 1987.

Sq. **ST** 48.

1099 *Endothenia marginana* Haw.

Tintern 31/5/68, (JMC-H 1969).

Usk, garden m.v.l. trap, 13/5/76 (2). 1976, one male emerged on 23/5/76 from seedheads of teasel (*Dipsacus fullonum*) collected at Magor Marsh (GANH). Slade Wds 2/8/82, one to m.v.l.

Sq. **SO** 30. **ST** 48,59.

1102 *Endothenia nigricostana* Haw.

"Monmouthshire" (EFH, in litt. 1986).

1103 *Endothenia ericetana* Humph. & Westw.

Usk, garden m.v.l. trap, 29/7/82 (1). Hael Wds 31/7/82, one to m.v.l.

Sq. **SO** 30,50.

1106 *Lobesia reliquana* Hb.

Occurs during June in woodland in the east of the county.

Pont-y-saeson 6/6/72 two netted by day (GANH). Wyndcliff 25/6/86, one to m.v.l.. Hael Wds 26/6/86, one to m.v.l.

Sq. **SO** 50. **ST** 59.

1110 *Bactra furfurana* Haw.

Usk, garden m.v.l. trap, 1983 26/6 (1), 30/7 (1).

Sq. **SO** 30.

1111   *Bactra lancealana* Hb.

Common and widespread in marshes and boggy places, especially in the hills of the west and north of the county, often flying by day and occasionally coming to light.

Tintern 31/5/68 (JMC-H 1989).

Usk, garden m.v.l. trap, recorded occasionally eg. 8/8/80, 14/5/82, 8/9/82. Cwm Tyleri 12/7/75 plentiful by day. Slade Wds 2/8/80. Llansoy 26/6/86. Pontllanfraith 2/7/87 and 30/6/88 (plfl). Fforest Coal Pit 15/6/88, (abdt).

Sq. **SO** 20,30,40,22. **ST** 48,19,59.

1112   *Bactra robustana* Christ.

1989, Magor Pill (saltern on Severn Estuary littoral), one to m.v.l. 6/8/89. (GANH) (1st v-c. rec.). Detn. confirmed by JDB.

Sq. **ST** 48.

1113   *Eudemis profundana* D. & S.

Occurs in oakwoods in the east of the county but scarce.

Hael Wds, to m.v.l., 14/8/78 (1), 31/7/82 (3), 11/8/82 (1). Hendre Wds 20/7/79 (1).

Sq. **SO** 50,41.

1115   *Ancylis achatana* D. & S.

Locally common in rough woodland in the eastern half of Monmouthshire during June and July and coming readily to light.

Usk, garden m.v.l. trap, 20/7/78 and noted most years eg. 1979 (11) 11/7 – 28/7, 1983 (9) 6/7 – 15/7. Slade Wds 2/6/81. Hael Wds 22/6/81. Wyndcliff 3/8/81. Llansoy 16/7/86.

Sq. **SO** 30,40,50. **ST** 48,59.

1119   *Ancylis geminana* Don.
      = *biarcuana* Steph.

Very local, flying in damp woods and bogs in May and June.

Trelleck Bog 1973, 22/5 (2), 16/6 plentiful to m.v.l. (GANH). Hael Wds 30/6/76 (MWH pers. comm.).

Sq. **SO** 50.

1119a   *Ancylis diminutana* Haw.

Hendre Wds 2/6/80 (1) (GANH). (1st v-c. rec.).

Sq. **SO** 41.

1119b *Ancylis subarcuana* Dougl.
    = *inornatana* H.-S.
Trelleck Bog 16/6/73 (1) to m.v.l. (GANH) (1st v-c. rec.).

Sq. **SO** 50.

1120 *Ancylis mitterbacheriana* D. & S.
Widespread and not uncommon, flying in deciduous woodland May to July.
    Hael Wds 30/5/78, plentiful to m.v.l., (GANH), (1st v-c. rec.); 26/6/86 etc. Hendre Wds 1980, 19/5,2/6 (plfl.). Wentwood 24/5/80 (GAH). Slade Wds 26/5/80. Usk, flying at dusk 13/7/82.

Sq. **SO** 30,50,41. **ST** 48,49.

1122 *Ancylis obtusana* Haw.
Found in woodland in the east of the county but very scarce.
    Hael Wds 30/6/76 (MWH, in litt.) (1st v-c. and Welsh record). 26/6/86 one to m.v.l. (JDB). Hendre Wds 29/5/82, two netted in daytime (GANH).

Sq. **SO** 50,41.

1126 *Ancylis badiana* D. & S.
Flies from May to August in woodland rides and rough grassy places.
    Tintern 31/5/68 (JMC-H 1969).
    Pont-y-saeson 11/5/68, one by day. Usk, garden m.v.l. trap: 1971, 24/7 (1); 1983 singletons on 18/7, 1/8 and 5/8. St Pierre's Great Wds 19/5/87 (plfl.), flying in afternoon.

Sq. **SO** 30,50. **ST** 59.

1128 *Ancylis myrtillana* Treit.
Common in May and June in the hills of north and west Gwent.
Deri-fach 2/6/68, larvae on bilberry, reared (JMC-H 1969).
1971 Blorenge Mountain 24/6, abundant among bilberry. Pontypool (Mountain Air) 25/5/80. Cwm Tyleri 16/6/81. Usk, garden m.v.l. trap, 31/5/82 (1).

Sq. **SO** 20,30,21. **ST** 29.

1131 *Epinotia subsequana* Haw.
Llandogo 9/5/78, flying amongst stand of giant fir *(Abies grandis)* (1st v-c. and Welsh rec.). Also 16/5, 19/5 (swarming high up around one particular tree) and 26/5. (GANH).

Sq. **SO** 50.

1132 *Epinotia subocellana* Don.
Flying from May to July in damp woodland, bogs etc.
    Trelleck Bog 16/7/79. Hael Wds 24/5/80, 4/7/81. Hendre Wds 29/5/81. Wyndcliff 25/6/86.

Sq. **SO** 50,41. **ST** 59.

1133 *Epinotia bilunana* Haw.
Flies during June and July in woods containing birches.
    Wyndcliff 22/6/71, 24/6/71. Hendre Wds 24/6/79 (2) to m.v.l., 1/7/86. Hael Wds 22/6/81, 26/6/86.

Sq. **SO** 50,41. **ST** 59.

1134 *Epinotia ramella* Linn.
Frequenting open woods and heathy places this moth flies from late June to September and is fairly common in eastern Monmouthshire. The form *costana* Dup. is not uncommon.
    Hael Wds 14/8/78, 9/10/79. Hendre Wds 4/9/78, 27/8/79 (f. *costana*). Slade Wds 21/8/79, 30/6/81 (plentiful to m.v.l.); f. *costana* 15/8/80, 9/9/82, 29/7/86. Kilpale 30/6/87. Trelleck Bog 17/8/88.

Sq. **SO** 50,41. **ST** 48,49.

1135 *Epinotia demarniana* F.v.R.
Bradley in British Tortricoid Moths (1979 **2**:106) writes "A local and rather scarce species in the British Isles."
    Hael Wds 26/6/86 one to m.v.l. (JDB, GANH) (1st v-c. rec.).

Sq. **SO** 50.

1137 *Epinotia tetraquetrana* Haw.
Frequents damp woods, bogs etc. flying in May and June. Fairly common in the eastern half of Monmouthshire.
    Hael Wds 30/5/78, plentiful to m.v.l. Hendre Wds 24/6/79. Wentwood 26/5/80 (plfl.). Slade Wds 3/6/80. Pont-y-saeson 23/6/86. Trelleck Bog 28/5/87, 17/6.87.

Sq. **SO** 50,41. **ST** 48,49,59.

1138 *Epinotia nisella* Cl.
Not uncommon in Gwent, flying in damp woodland in July and August.
    Hael Wds 14/8/78, one (f. *pavonana* Don.) to m.v.l. Slade Wds 15/8/80 (2) to m.v.l., 23/7/83 one (f. *decorana* Hb.). Pontllanfraith 30/6/88 one, by day on damp heath.

Sq. **SO** 50. **ST** 48,19.

1139 *Epinotia tenerana* D. & S.   Nut Bud Moth
Usk, 2/9/80, plentiful by day along old railway line. Slade Wds 30/6/88, one by day.

Sq. **SO** 30. **ST** 48.

1142 *Epinotia tedella* Cl.
Common in plantations of Norway spruce (*Picea abies*) often flying by day and coming to light at night.
    Trelleck Bog 16/6/73, plentiful to m.v.l. Slade Wds 21/7/79, 26/5/80 (abundant by day), Pont-y-saeson 23/6/86.

Sq. **SO** 50. **ST** 48,59.

1147 *Epinotia cruciana* Linn.   Willow Tortrix
Flies in damp woodland during June and July.
    Slade Wds 1980, plentiful by day 24/6 and 2/7 including f. *augustana* Hb. St Pierre's Great Wds 5/7/88. Hael Wds 6/7/88. Wyndcliff 3/7/91.

Sq. **SO** 50. **ST** 48,59.

1150 *Epinotia abbreviana* Fabr.
    = *trimaculana* Don.
Uncommon in Monmouthshire.
    Hendre Wds 20/7/79 to m.v.l. Usk, garden m.v.l. trap, 1/7/82. Llansoy 6/7/87.

Sq. **SO** 30,40,41.

1151 *Epinotia trigonella* Linn.
    = *stroemiana* Fabr.
Slade Wds 9/9/82 plentiful to m.v.l. (GANH) (1st v-c. rec.); 19/8/89 (3) to m.v.l.. Llansoy, 26/8/90 one to u.v.l.
Sq. **SO** 40. **ST** 48.

1153 *Epinotia sordidana* Hb.
Usk, 5/6/68, larvae on alder (JMC-H 1969) (1st v-c. rec.).

Sq. **SO** 30.

1155 *Epinotia brunnichana* Linn.
Widespread in eastern Monmouthshire, it flies in July and August in mixed deciduous woodland and comes to light. Bradley (1979 **2**: 132) states" . . . Generally distributed and common in North Wales; apparently known only from Monmouthshire in the south."
    Piercefield Park 18/7/73 to m.v.l. (GANH) (1st v-c. rec.). Hael Wds

14/8/78 (2), 2/8/81, 31/8/86. Slade Wds 21/8/79 (2), 18/7/81 (by day), 28/7/82 (f. *brunneana* Sheld.), 20/7/89 etc. Hendre Wds 1/8/80. Usk, garden m.v.l. trap, 1981 7/8 (2), 13/8 (1); 1982 13/7, 15/7. Usk (Cwm Cayo) 29/7/82. Trelleck Bog 17/8/88. Redding's Inclosure 2/8/90.

Sq. **SO** 30,50,41,51. **ST** 48,59.

1156   *Epinotia solandriana* Linn.
In Monmouthshire much less frequent than E. *brunnichana*.
    Slade Wds to m.v.l., 28/7/82 f. *parmatana* Hb. (1) and f. *rufana* Sheld. (1); 20/7/89 f. *griseana* Sheld. (1).

Sq. **ST** 48.

1159   *Rhopobota naevana* Hb.   Holly Tortrix
      = *unipunctana* Haw.
Widespread and common in the northern half of the county, flying from June to August in woods, hedgerows and gardens but more especially in the northern hills and moorlands. It is easily disturbed by day and comes to light.
    Blorenge Mountain 24/6/71, massive larval infestation of bilberry (*Vaccinium myrtillus*), reared (GANH). Cwm Tyleri, plentiful to m.v.l. 9/7/73 and 13/7/75. Usk, garden m.v.l. trap, 9/7/75, 1982 1/7 – 7/8. Hael Wds 29/5/82. Trelleck Bog 24/7/86. Llansoy 5/7/88.

Sq. **SO** 20,30,40,50,21.

1162   *Rhopobota myrtillana* Humph. & Westw.
Very scarce in Monmouthshire.
    Trelleck Bog 17/6/87, abundant by day (GANH) (1st v-c. rec.).

Sq. **SO** 50.

1163   *Zeiraphera ratzeburgiana* Ratz.
Slade Wds 15/7/80, one to m.v.l. (GANH) (new v-c. rec.).

Sq. **ST** 48.

1165   *Zeiraphera isertana* Fabr.
Widespread and common in oakwoods.
    Usk, garden m.v.l. trap, 22/7/71, 21/7/77, 4/7/82. Hael Wds 30/8/77. Hendre Wds 20/7/79. Slade Wds 23/7/79. St Pierre's Great Wds 5/7/88.

Sq. **SO** 30,50,41. **ST** 48,59.

1166   *Zeiraphera diniana* Guen.   Larch Tortrix
Found in larch plantations.

Hael Wds 10/7/73. Slade Wds 23/7/79. Usk, garden m.v.l. trap, 27/7/79.

Sq. **SO** 30,50. **ST** 48.

1168   *Gypsonoma sociana* Haw.
Rare in Monmouthshire in damp woodland and marshes.
　　Magor Marsh 8/7/77, one to m.v.l. (GANH det. JDB) (1st v-c. rec.). Hael Wds 24/6/87.

Sq. **SO** 50. **ST** 48.

1169   *Gypsonoma dealbana* Fról.
This moth, flying in woods in July, is widespread and common It is readily disturbed by day and at night comes to light.
　　Usk, garden m.v.l. trap, recorded sporadically eg. 26/8/78, 1982 1/7 – 13/7. Pont-y-saeson 8/7/78. Magor Marsh 11/7/78. Trelleck Bog 16/7/79. Hendre Wds 20/7/79. Slade Wds 21/7/79. Wyndcliff 3/8/81. Hael Wds 6/7/88.

Sq. **SO** 30,50,41. **ST** 48,59.

1174   *Epiblema cynosbatella* Linn.
Common and widespread in gardens, hedgerows and open woodland, flying from late May through June and July.
　　Usk, garden m.v.l. trap, 6/6/79 and recorded most years eg. 1983 (10) 19/6 – 26/7. Ed. 28/5/82. Ld. 26/7/83. Slade Wds 3/6/72. Pont-y-saeson 6/6/72. Hendre Wds 29/5/82. Llansoy 5/6/85, 2/7/85 (10). Wyndcliff 23/6/86.

Sq. **SO** 30,40,41. **ST** 48,59.

1175   *Epiblema uddmanniana* Linn.   Bramble Shoot Moth
Very common and widespread in gardens, hedgerows and open woodland where brambles grow, it flies in late June and July and has been noted in September. It comes readily to light.
　　Tintern dist (Piffard 1859, as *Notocelia uddmanniana* L.). Redding's Inclosure 31/5/68, larvae on *Rubus* (JMC-H 1969).
Usk, garden m.v.l. trap, recorded most years eg. 1981 (9) 21/6 – 26/7; 1982 (6) 4/7 – 23/7, 9/9 (1), 10/9 (1). Wyndcliff 24/6/71. Cwm-mawr 2/7/71. Hael Wds 10/7/73, plentiful to m.v.l. St Pierre's Great Wds 13/7/73. Newport (Coldra) 27/6/75. Magor Marsh 30/6/75. Hendre Wds 22/6/76. Trelleck Bog 22/7/78. Slade Wds 23/7/79. Llansoy 2/7/85. Kilpale 30/6/87. Dixton Bank 18/6/88.

Sq. **SO** 20,30,40,50,41,51. **ST** 38,48,59.

1176 *Epiblema trimaculana* Haw.
Usk, garden m.v.l. trap, 21/7/79 one (JDB). The only Gwent record.

Sq. **SO** 30.

1177 *Epiblema rosaecolana* Doubl.
Not uncommon in Monmouthshire, flying during June and July in open woodland, gardens and hedgerows.

Usk, garden m.v.l. trap, 19/7/79, 19/6/83. Usk, 30/6/82, flying at dusk on old railway line. Llansoy 1986, several, 7/7 – 24/7. Slade Wds 29/7/86. St Pierre's Great Wds 5/7/88.

Sq. **SO** 30,40,41. **ST** 59.

1178 *Epiblema roborana* D. & S.
Flying in July and early August, this species is found in gardens and in hedgerows and open woodland where wild roses grow. It is easily disturbed from rose bushes by day and at night comes to light.

Usk, garden m.v.l. trap, recorded frequently eg. 1971 (July) ; 1977 21/7, 13/8; 1981 8/7 – 18/8. Cwm-mawr 2/7/71. Llangeview 28/7/79, several beaten from hedgerow. Slade Wds 23/7/79, 28/7/82. Llansoy 22/7/87.

Sq. **SO** 30,40,21. **ST** 48.

1180 *Epiblema tetragonana* Steph.
St Pierre's Great Wds 5/7/88 (2) netted in late afternoon (GANH, JDB) (1st v-c. rec.).

Sq. **ST** 59.

1183 *Epiblema foenella* Linn.
Occurs locally in July and the first half of August, frequenting rough ground and situations where mugwort (*Artemesia vulgaris*) grows. It flies in the late afternoon and at night comes to light.

Usk, garden m.v.l. trap, 17/7/73 and noted most years until 1983 eg. 1977 (4) 6/7 – 19/8; 1983 (3) 3/7 – 6/8. Usk (old railway line) 1982, flying around mugwort in afternoon sunshine 8/7, 10/7 etc. (plfl. on 20/7). Slade Wds 20/7/89.

Sq. **SO** 30. **ST** 48.

1184 *Epiblema scutulana* D. & S.
Flies in May and June on rough ground and in marshy places where thistles flourish. Widespread and fairly common in Monmouthshire.

Wye Valley (Barraud 1906, as *Ephippiphora pflugiana*).

Wentwood 21/5/69 (JMD). Hendre Wds (by day) 6/6/78 (2), 16/6/79 (1). Pontllanfraith 25/5/87. Fforest Coal Pit 15/6 (2), by day.

Sq. **SO** 41,22. **ST** 19,49.

1184a *Epiblema cirsiana* Zell.
Found from May to July, frequenting rough ground, waysides and open woods. Widespread and fairly common in Monouthshire.

Magor Marsh 3/8/76, one to m.v.l. Usk, garden m.v.l. trap, 21/7/77 (1). Pont-y-saeson 22/5/79 one by day resting on burdock leaf. Slade Wds 29/5/79 (1). Trelleck (Penorth Mill) 24/5/82, one on herbage by day. Llanllowell 24/6/87 one on roadside herbage.

Sq. **SO** 30,40. **ST** 48,39,59.

1186 *Epiblema sticticana* Fabr.
= *farfarae* Fletch.
Common and widespread in Monmouthshire, flying from late May through June and July, it frequents waste ground, waysides, woodland rides etc. where coltsfoot (*Tussilago farfara*) flourishes.

Slade Wds 15/6/73, 26/5/82. Hendre Wds 6/7/75; 20/7/76, abundant by day (many in cop.). New Church West 12/6/77. Pont-y-saeson 27/5/78, (pltl.). Slade Wds 21/7/79. Trelleck (Fedw Fach) 25/6/88. St Pierre's Great Wds 5/7/88.

Sq. **SO** 40,41. **ST** 48,49,59.

1187 *Epiblema costipunctana* Haw.
Slade Wds 26/5/80, one by day (GANH). (1st v-c. rec.).

Sq. **ST** 48.

1190 *Eucosma aspidiscana* Hb.
Bradley (1979 **2**:179) gives "Monmouthshire".

1197 *Eucosma campoliliana* D. & S.
Rare and local, occurring on the Carboniferous Limestone area in the south-east and on the Severn Estuary littoral. Slade Wds 21/7/83, one to m.v.l. (GANH) (1st v-c. rec.). Kilpale 30/6/87 (1). Uskmouth 16/8/88 (1) to m.v.l.

Sq. **ST** 38,48,49.

1200 *Eucosma hohenwartiana* D. & S.
Locally common in Gwent, flying from June to August, frequenting woodland rides and rough grassy places. The form *fulvana* Steph. occurs.

Magor Marsh, frequent, eg. 8/7/77, 19/6/79. Usk, garden m.v.l. trap, recorded sporadically eg. 6/7/73 (f. *fulvana*), 22/6/74 (f. *fulvana*), 22/6/81, 29/6/83. Slade Wds 2/8/82, 17/7/87. Trelleck Bog 19/7/86.

Sq. **SO** 30,50. **ST** 48.

1201 *Eucosma cana* Haw.
Widespread and common in Monmouthshire frequenting grassy places, rough woodland, waysides etc. and flying from June to August.

Magor Marsh 9/7/74, 19/6/79. Cwm Tyleri 12/7/77. Usk, garden m.v.l. trap 18/7/78; 1983 (5) 8/7 – 19/7. Hendre Wds 20/7/79. Hael Wds 23/7/79, 2/8/86. Slade Wds 21/7/83. Dixton Bank 18/6/86. Llansoy 13/7/86.

Sq. SO 20,30,40,50,41,51. ST 48.

1202 *Eucosma obumbratana* Lien. & Zell.
Slade Wds 23/7/79, one to m.v.l., (GANH) (1st v-c. rec.).

Sq. ST 48.

1205 *Spilonota ocellana* D. & S.   Bud Moth
Common in Gwent, frequenting gardens, hedgerows, open deciduous woods etc. and flying from June to August. Comes readily to light.

Magor Marsh 13/7/77. Usk, garden m.v.l. trap, 11/8/77, 1983 9/7 – 4/8. Hendre Wds 20/7/79. Hael Wds 2/8/81, 26/6/86. Trelleck Bog 12/7/85. Llansoy 12/7/86.

Sq. SO 30,40,50,41. ST 48.

1205a *Spilonota laricana* Hein.
Widespread and common in larch plantations, flying from July to September and coming readily to light.

Usk, garden m.v.l. trap, 21/7/75, 31/7/82. Slade Wds 23/7/79; 28/7/80 (abdt.); 1982, 28/7 (abdt. to m.v.l.), 2/8, 9/9. Trelleck Bog 16/7/79 (plfl.). Llansoy 14/7/86. Ysguborwen (Llantrisant) 17/7/86.

Sq. SO 30,40,50. ST 48,49.

1210 *Rhyacionia buoliana* D. & S.   Pine Shoot Moth
Flies in pine plantations during July and August.

Slade Wds 23/7/79, 21/7/83. Usk, garden m.v.l. trap, 7/8/80, 22/7/82, 31/7/82. Trelleck Bog 17/8/88.

Sq. SO 30,50. ST 48.

1211 *Rhyacionia pinicolana* Doubl.
Occurs in pine plantations in July and August. Bradley 1979 (*British Tortricoid Moths* **2**: 205) writes "In Wales known only with certainty from Monmouthshire (Usk)."
1973, Usk, garden m.v.l. trap, 15/8/73 (GANH) (1st v-c. and Welsh rec.). Slade Wds 2/8/80 (2) to m.v.l., 2/8/82, 21/7/83, 20/7/89. Trelleck Bog 24/7/86.

Sq. SO 30,50. ST 48.

1212   *Rhyacionia pinivorana* Lien. & Zell.   Spotted Shoot Moth
Trelleck Bog 16/7/79, one to m.v.l. Wyndcliff 6/7/81. Usk, garden m.v.l. trap, 14/5/82 (1).

Sq. **SO** 30,50. **ST** 59.

1216   *Enarmonia formosana* Scop.   Cherry-bark Moth
"Monmouthshire" (per EFH).

1217   *Eucosmomorpha albersana* Hb.
Hael Wds 29/5/82, one to m.v.l. (JDB, GANH) (1st v-c. rec.). "This appears to be the only recent record for Wales" (JDB pers. comm.)

Sq. **SO** 50.

1219   *Lathronympha strigana* Fabr.
Common in the eastern half of the county among growths of St John's-wort (*Hypericum* spp.) on waste ground, embankments, open woods etc. from June to August. It flies by day and sometimes comes to light.
    Wyndcliff 24/6/71 to m.v.l. Usk, 23/8/75, plentiful by day along old railway line. Usk, garden m.v.l. trap, 25/6/77, 28/6/79, 31/8/80 etc. Hendre Wds 6/6 and 27/6/78 plentiful by day on open hillside amongst St John's-wort (*H. perforatum* and *H. hirsutum*). Slade Wds 21/7/79. Llangwm-isaf, 1/7/81 flying by day along hedgerow. Llansoy 7/7/86.

Sq. **SO** 30,40,41. **ST** 48,59.

1221   *Strophedra weirana* Doug.
Scarce and local in Monmouthshire.
    Slade Wds 3/6/72 one netted by day (GANH) (1st v-c. and Welsh rec.). Hael Wds, to m.v.l., 22/6/81 (1), 29/5/82 (1), 19/6/88 (1), 22/6/88 (2). Wyndcliff 25/6/86 (1) (JDB). Llansoy 27/6/86 (1),

Sq. **SO** 40,50. **ST** 48,59.

1222   *Strophedra nitidana* Fabr.
Rare in Monmouthshire.
    Hendre Wds 23/5/80, one to m.v.l. (GANH) (1st v-c. rec.).

Sq. **SO** 41.

1223   *Pammene splendidulana* Guen.
Llansoy 22/5/85, ♀ (1) to m.v.l. (GANH) (1st v-c. rec.).

Sq. **SO** 40.

1225 *Pammene obscurana* Steph.
Bradley (1979 2:222) states "unknown in Wales".
    Hael Wds 29/5/82, one to m.v.l. (JDB) (1st v-c. and Welsh rec.).

Sq. **SO** 50.

1228 *Pammene argyrana* Hb.
Hendre Wds 4/6/79, one to m.v.l. (GANH) (1st v-c. rec.). Hael Wds 30/4/93, to m.v.l., ♂ (1) ♀♀ (2).

Sq. **SO** 50,41.

1233 *Pammene aurantiana* Stdgr
Bradley (1979 2:232) states "Its range now extends into Wales where it is known from Usk (Monmouthshire) (Horton 1975: 30) and Dinas Powis (Glamorgan)."
    Usk, garden m.v.l. trap, 25/7/73 one (freshly emerged), (GANH) (1st v-c. and 1st Welsh rec.); 19/8/77 (1); 30/7/83 (1).

Sq. **SO** 30.

1234 *Pammene regiana* Zell.
Usk, garden m.v.l. trap, 18/7/83 (1) (GANH) (1st v-c. rec.).

Sq. **SO** 30.

1236 *Pammene fasciana* Linn.
Usk, garden m.v.l. trap, 21/7/77 (1) (GANH) (1st v-c. rec.). Llansoy 16/7/86 (2), 24/7/86 (1). Wyndcliff 25/6/86.

Sq. **SO** 30,40. **ST** 59.

1236a *Pammene herrichiana* Hein.
Bradley (1979 2: 236) states ". . . *herrichiana* is rare and local being known . . ., and in Wales from St Arvans and Tintern (Monmouthshire)."
    Hael Wds 29/5/82 (1) to m.v.l. (JDB). Wyndcliff 25/6/86 (2) (JDB, GANH).

Sq. **SO** 50. **ST** 59.

1237 *Pammene germmana* Hb.
Bradley (1979 2: 237) states "Unknown from Wales."
    1971, Wyndcliff 24/6 to m.v.l. (JMC-H, GANH) (1st v-c. and Welsh rec.). 1986, Hael Woods 26/6, one to m.v.l. (JDB, GANH).

Sq. **SO** 50. **ST** 59.

1241 *Cydia compositella* Fabr.
Usk, old railway line, one netted by day 25/6/77 (GANH) (1st v-c. rec.). Magor Marsh, 27/5/82, one taken by day (JDB).

Sq. **SO** 30. **ST** 48.

1251 *Cydia jungiella* Cl.
Local and scarce in Monmouthshire.
   Barrett (1907 **11**:235) states "Not recorded in Wales."
   Tintern 31/5/68, by day (JMC-H 1969). (1st v-c. rec.). Magor Marsh 27/5/78, one swept from herbage (GANH); Llangeview 1991, 5/5 (♂ 1), 19/5 (♀♀ 3); 1992, (svl.) 15/5 and 17/5 (GANH).

Sq. **ST** 48,39,59.

1254 *Cydia strobilella* Linn.
Frequents mature plantations of Norway spruce (*Picea abies*) but very local.
   Llandogo 26/5/78, several netted in mid-afternoon (GANH) (1st v-c. rec.).

Sq. **ST** 50.

1255 *Cydia succedana* D. & S.
Slade Wds 15/8/79 (by day); Hael Wds, to m.v.l., 2/8/81, 29/5/82; 1991, Trelleck Common, 20/5, in swarms around flowering gorse bushes, (GANH).

Sq. **SO** 50. **ST** 48.

1256 *Cydia servillana* Dup.
Local and very scarce, frequenting damp woods.
   Bradley (1979 **2**:264) states "In the British Isles known only from England and south-east Wales . . . Llantrisant (Monmouthshire)."
   Llantrisant, 6/6/77, ♂ one netted by day (GANH) (1st v-c. and Welsh rec.). Slade Wds, 3/6/80, ♀ one by day (GANH).

Sq. **ST** 48,39.

1259 *Cydia fagiglandana* Zell.
Flying from June to August this moth is found locally in woods containing beeches.
   Wyndcliff 1971 to m.v.l. 22/6, 24/6 (JMC-H, GANH) (1st v-c. and Welsh records); 3/8/81; 25/6/86 (2) (JDB, GANH); 1/9/89. Hael Wds 22/6/81; 1987, 1/6 (2), 7/7 (1). Clydach 18/7/89 one to m.v.l.

Sq. **SO** 50,21. **ST** 59.

1260 *Cydia splendana* Hb.
This moth, frequents woods and hedgerows containing oaks, and is common and widespread in Monmouthshire during July and August. It flies at dusk and later comes to light.

Usk, garden m.v.l. trap, 31/7/73 and recorded every year until 1983 eg. 1975 21/7 (3), 2/8 (3); 1978 18/7 – 12/9; 4/9/82 (melanic f.). Magor Marsh 4/8/75, 20/8/77. Hendre Wds 27/8/79. Slade Wds 28/8/79. Wyndcliff 21/7/87. Hael Wds 20/8/87. Trelleck Bog 30/8/86. Llansoy 7/8/88.

Sq. **SO** 30,40,50,41. **ST** 48,59.

1261 *Cydia pomonella* Linn.   Codling Moth
Found mainly in gardens and orchards during July and August.

Usk, garden m.v.l. trap, 26/7/79 (2); 1980 8/8 – 27/8; 3/6/82; 1983 9/7 – 7/8. Wyndcliff 25/7/83, 8/8/84. Llansoy 1986, several 25/6 – 28/6.

Sq. **SO** 30,40. **ST** 59.

1267 *Cydia cosmophorana* Treit.
Bradley (1979 **2**:279) states "Local and rather uncommon in England, occurring mainly in the eastern counties . . . Apparently unknown from Wales . . .".

Hendre Wds 29/5/82, one netted during afternoon (JDB, GANH). (1st v-c. and Welsh rec.).

Sq. **SO** 41.

1269 *Cydia conicolana* Heyl.
Bradley (1979 **2**:281) states "In the British Isles this species is known only from East Anglia and the south of England."

Hendre Wds 29/5/79 one to m.v.l. (GANH) (1st v-c. and Welsh rec.); det. JDB, who wrote in litt. "hitherto known from south-east England westwards to Oxfordshire."

Sq. **SO** 41.

1272 *Cydia aurana* Fabr.
Rhadyr (Usk), 15/7/82 , one flying in afternoon sunshine and alighting on hogweed (*Heracleum sphondylium*) flowers on old railway embankment. (GANH) (1st v-c. rec.).

Sq. **SO** 30.

1274 *Dichrorampha alpinana* Treit.
Bradley (1979 **2**:288) gives "Tintern and Usk".

Usk, garden m.v.l. trap, 20/6/77 (1). Usk (old railway line) one swept from herbage 4/6/82.

Sq. **SO** 30. **ST** 59.

1275 *Dichrorampha flavidorsana* Knaggs
Usk, garden m.v.l. trap, 21/6/77 (1) (GANH) (1st v-c. rec.); 25/8/80 (1). Usk, 1982, old railway line, flying by day 1/7 (1), 19/7 (several).

Sq. **SO** 30.

1278 *Dichrorampha sequana* Hb.
Usk, garden m.v.l. trap, 20/6/77 (GANH) (1st v-c. rec.); 1/7/78. Usk, old railway line, 1980 plentiful on tansy (*Tanacetum vulgare*) in afternoon sunshine 24/5 and 25/5; 1982 plentiful on tansy 24/5 – 19/7.

Sq. **SO** 30.

1283 *Dichrorampha montanana* Dup.
Llandogo, on bank of R. Wye, 30/6/76 (MWH in litt.) (1st v-c. rec.).

Sq. **SO** 50.

1284 *Dichrorampha gueneeana* Obraz.
Bradley (1979 **2**: 301) states "In Wales it is known only from Pembrokeshire and Merioneth".
 Usk, disused railway line, one taken by day 22/7/82, (GANH) (1st v-c. rec.).

Sq. **SO** 30.

1286 *Dichrorampha sedatana* Busck
Usk, disused railway line, 1980, plentiful by day on tansy 10/5 (GANH) (1st v-c. rec.) and 11/5; 1982, 14/5 – 31/5 (swarming around tansy flowers in evening sunshine on 14/5 and 20/5 with many in cop.)

Sq. **SO** 30.

1287 *Dichrorampha aeratana* Pier. & Metc.
Slade Wds 21/7/79 one netted by day (JDB) (1st v-c. rec.); 3/6/80 (1); 26/6/86 (1). Usk, disused railway line, 4/6/82 one by day.

Sq. **SO** 30. **ST** 48.

ALUCITIDAE

1288 *Alucita hexadactyla* Linn.   Twenty-plume Moth
This moth frequents gardens and hedges where honeysuckle (*Lonicera*) grows. It flies in August and reappears in May after hibernation.

Usk: 1967: 11/5. 1969: 18/8. 1978: 19/10. 1979: 14/5, 12/8. 1983: 4/8.

Sq. **SO** 30.

PYRALIDAE
CRAMBINAE

1289 *Euchromius ocellea* Haw.
This tropical and sub-tropical species is a rare immigrant and was first recorded in Britain in 1812.
　　　Usk, one in garden m.v.l. trap 14/10/78, (1st v-c. rec.) (Horton 1979). This is the 24th British, and 2nd Welsh record (Skinner 1982).

Sq. **SO** 30.

1290 *Chilo phragmitella* Hb.
Abertillery dist. 1911 (Rait-Smith 1912).

Sq. **SO** 20.

1293 *Chrysoteuchia culmella* Linn.
　　 = *hortuella* Hb.
Abundant in pastures and grassy places during June and July.
　　　Usk, 1971, garden m.v.l. trap, 21/6 (3) and recorded every year to 1983 with maximum numbers usually in first week of July. Ed. 2/6/81. Ld. 15/8/77. Wyndcliff 22/6/71. Hael Wds 3/7/71. Cwm-mawr 2/7/71. Slade Wds 6/7/71. Redding's Inclosure 20/7/71. Magor Marsh 21/7/72. Cwm Tyleri 9/7/73. Wentwood 29/6/73. Undy (Collister Pill) 13/7/73 (abdt.). Hendre Wds 6/6/78. Trelleck Bog 8/7/78. Craig-yr-iar (Llangeview) 28/7/79. Llansoy 26/7/84.

Sq. **SO** 20,30,40,50,41,51. **ST** 38,48,49,59.

1294 *Crambus pascuella* Linn.
Common in bogs, marshes, moist pastures and damp woods during June and July.
　　　Magor Marsh 11/7/69 etc. Gwernesney 13/6/71 (abdt.). Usk 2/8/72. Cwm Tyleri 12/7/77 etc. Hendre Wds 6/6/78, 23/6/79 (abdt.). Pont-y-saeson 3/7/79. Llansoy 1986 23/6 – 18/7. Trelleck Bog 18/7/86.

Sq. **SO** 20,30,40,50,41,51. **ST** 48,59.

1297 *Crambus uliginosellus* Zell.　　　Marsh grass-veneer
Usk dist. 1971, 13/6 and 23/6, abundant in one small isolated bog (GANH) (1st v-c. rec.). 1984, 14/6 continues to be plentiful at this site. To date (1993), this species has not been found elsewhere in Gwent.

Sq. **SO** 40.

1299 *Crambus hamella* Thunb.
Abertillery dist. 1911 (Rait-Smith 1912).

Sq. **SO** 20.

1300 *Crambus pratella* Linn.
= *dumetella* Hb.
Rare, occurring sporadically in eastern half of county.
1973: Usk 8/6, (1) to m.v.l. 1980: Slade Wds, (1) to m.v.l. (GANH). 1986: Wyndcliff, 25/6 (1) to m.v.l. and Pont-y-saeson 26/6 (1) by day (JDB, GANH).

Sq. **SO** 30. **ST** 48,59.

1301 *Crambus lathoniellus* Zinck.
= *nemorella* auctt.
Widespread and fairly common in Monmouthshire.
Wye Valley (Barraud 1906). Abertillery dist. 1911 (Rait-Smith 1912). Tintern 31/5/68 (JMC-H 1969).

Usk, recorded frequently at garden m.v.l. trap eg. 1971, 2/7. 1983, 25/6 – 15/7. Cwm-mawr 2/7/71. Slade Wds 15/6/71, to m.v.l.; 1980 3/6-8/7; 21/6/81. Cwm Tyleri 17/6/73. Hendre Wds 6/6/78, 13/7/80 etc. Magor Marsh 19/6/79. Redding's Inclosure 23/6/81. Llansoy 5/6/85, 12/7/86 etc. Dinham 4/6/87. Pontllanfraith 20/6/87. Dixton Bank 18/6/88. Clydach 15/6/90.

Sq. **SO** 20,30,40,21,41,51. **ST** 48,19,49,59.

1302 *Crambus perlella* Scop.
Flies in June and July in pastures, marshy ground and open woodland but, though widespread in Monmouthshire, is mostly at low density.

The streaked variation f. *warringtonellus* Stainton is not uncommon.

Abertillery dist. 1911 (Rait-Smith 1912). Pontllanfraith, *Crambus perlella* and var. *warringtonellus* 1912 (Rait-Smith 1913).

Usk, garden m.v.l. trap, 1971, 26/7 and 27/7 and in small numbers every year until 1983 when it was frequent. Ed. 25/6/73. Ld. 2/8/80. Usk, 12/7/79 f. *warringtonellus*. Redding's Inclosure 20/7/71, 24/7/83. Magor Marsh, often noted, eg. 21/7/72, 29/7/80, 13/7/82 (abdt.). Undy (Collister Pill), on sea-wall, 13/7/73. Hael Wds 22/6/76. Cwm Tyleri 29/6/76. Hendre Wds 9/7/78, 8/7/79 (f. *warringtonellus*), 17/7/79. Trelleck Bog 22/7/78, 19/7/86 etc. Slade Wds 28/7/81. Wyndcliff 23/6/84. Llansoy 13/7/86. Kilpale 30/6/87.

Sq. **SO** 20,30,40,50,41,51. **ST** 48,49,59.

1303 *Agriphila selasella* Hb.
Abertillery dist. 1911 (Rait-Smith 1912).

Sq. **SO** 20.

1304   *Agriphila straminella* D. & S.
     = *culmella* auctt.
Abundant in grassy places, flying from June to August.

Usk, garden m.v.l. trap, 14/8/66 and noted annually until 1983. Ed. 2/6/71. Ld. 25/8/75. Magor Marsh 1969 1/7 – 5/8; 1983, 13/7 in huge numbers. Wyndcliff 13/7/69, 8/8/84 etc. Hendre Wds 5/7/70, 11/7/86. Hael Wds 5/7/70, 2/8/86. Tintern 1/8/70. Gwehelog (Camp Wood) 28/8/80. Cwm Tyleri 28/7/72, 28/6/86. St Pierre's Great Wds 22/8/72. Piercefield Park 1/7/73. Undy (Collister Pill) 13/7/73 exceptionally abundant. Wentwood 26/7/74. Slade Wds 21/7/79, 22/7/86. Redding's Inclosure 24/7/83. Llansoy 26/7/84, 1986 26/6 – 18/8. Trelleck Bog 18/7/86. Uskmouth 16/8/88.

Sq. **SO** 20,30,40,50,41,51. **ST** 38,48,49,59.

1305   *Agriphila tristella* D. & S.
Common in grassy places in July and August.

Usk, 14/8/66 and every year until 1983 eg. 1983, 24/6 – 16/8. Ed. 24/6/73. Ld. 3/9/79. Magor Marsh 1/7/69. Prescoed 3/8/69. Wyndcliff 13/7/69. Rhyd-y-maen 18/7/70. Cwm Tyleri 28/8/72. Tintern 1/8/70. Redding's Inclosure 20/7/71. Undy (Collister Pill) 13/7/73. Goldcliff 20/8/73. Piercefield Park 18/7/73. Henllys 16/8/73. Hael Wds 17/7/73. Llangybi 26/8/73. Hendre Wds 23/8/77. Slade Wds 2/8/80. Llansoy 26/7/84. Dixton Bank 18/6/88. Uskmouth 16/8/88.

Sq. **SO** 20,30,40,50,42,51. **ST** 38,48,29,39,59.

1306   *Agriphila inquinatella* D. & S.
Cwm Tyleri, 1975 ♀ (1) to m.v.l. 13/7 (GANH).

Sq. **SO** 20.

1307   *Agriphila latistria* Haw.
Very scarce and local and of sporadic occurrence in Monmouthshire, as it appears to be throughout its range in Britain. Beirne in *British Pyralid and Plume Moths* 1952 states that it "has been recorded from all the maritime counties of England south of Lincolnshire and Lancashire and from Monmouth, Ayr, Perth and the Isle of Arran", but is "local and uncommon".

Abertillery dist. 1911 (Rait-Smith 1912).

Usk, garden m.v.l. trap, 1971, Aug. 12, 21, 26; 1973, Aug. 26; 1975, July 28; 1977, Sept. 6. (all singletons). Slade Wds 29/8/72 (3) to m.v.l.

Sq. **SO** 20,30. **ST** 48.

1309   *Agriphila geniculea* Haw.
Local in Monmouthshire but usually plentiful where found. It flies from July to September.

Usk, garden m.v.l. trap, 19/8/66 and recorded annually to 1983 and plentiful most years. Ed. 10/7/71. Ld. 10/9/77. Redding's Inclosure 20/7/71. Magor Marsh 11/8/71, 17/8/76 etc. Cwm Tyleri 5/8/74. Newport Docks 17/8/76. Slade Wds 13/8/83. Llansoy 18/8/86.

Sq. **SO** 20,30,40,51. **ST** 38,48.

1313   *Catoptria pinella* Linn.
Uncommon and sparsely distributed in Monmouthshire.

Usk, garden m.v.l. trap, 30/7/66 and recorded every year until 1983. Ed. 2/7/69. Ld. 20/8/77. Piercefield Park 18/7/73. Cwm Tyleri 1974, 29/7, 23/8. etc. Hael Wds, to m.v.l. 14/8/78 (3), 2/8/86. Hendre Wds 20/7/79. Slade Wds 28/7/80, etc. Llansoy 17/8/86. Redding's Inclosure 2/8/90.

Cwm Clydach 30/8/91 (RS).

Sq. **SO** 20,30,40,50,21,41,51. **ST** 48,59.

1314   *Catoptria margaritella* D. & S.
Rare in Monmouthshire.

Usk, garden m.v.l. trap, 1973 one 10/7 (GANH) (1st v-c. rec.); 1976 2/7 (1).

Sq. **SO** 30.

1316   *Catoptria falsella* D. & S.
Scarce in Monmouthshire, flying during July and August.

Usk, garden m.v.l. trap, 1972 Aug. 3, 5, 8, and several seen most years until 1983. Ed. 29/6/83. Ld. 10/9/82. Hael Wds 10/7/73 (2) to m.v.l. Slade Wds 8/7/83. Llansoy 28/7/84, 17/8/86, 5/8/89; 1990, several to u.v.l. 19/7 – 5/8.

Sq. **SO** 30,40,50. **ST** 48.

SCHOENOBIINAE

1329   *Donacaula forficella* Thunb.
Flies in June and July in reed-beds and reens in S. Gwent. Usk, garden m.v.l. trap, occurred sporadically: 1969, 26/6 (1). 1977, 27/8. 1981, July 11, 25, 27, 30. 1982, 31/7. 1983, 22/6. Magor Marsh, plentiful most years, eg. 1969, 1/7, 11/7 (abdt.), 22/7, 5/8. 1971, 1973, 1976 (25/6 – 3/8), 1978, 1979, 1983. Newport Docks 17/8/76.

Sq. **SO** 30. **ST** 38,48.

SCOPARIINAE

1332   *Scoparia subfusca* Haw.
= *cembrella* auctt.
Occurs sporadically in Monmouthshire.

Abertillery dist. 1914 (Rait-Smith 1915). Usk, garden m.v.l. trap, 1974, 6/7, 8/7. 1976, 21/6. 1977, 18/6. Slade Wds 2/8/80 to m.v.l. Clydach 18/7/89.

Sq. **SO** 20,30,21. **ST** 48.

1333 *Scoparia pyralella* D. & S.
  = *arundinata* Thunb.
Abertillery dist. 1911 (as *Scoparia dubitalis* Hb.) (Rait-Smith 1912).
  Magor Marsh 8/7/77 (GANH). Usk, garden m.v.l. trap, 7/7/80. Dixton Bank 18/6/88 (plfl.).

Sq. **SO** 20,30,51. **ST** 48.

1334 *Scoparia ambigualis* Treit.
Common in woodland in June and July.
  Wye Valley (Barraud 1906).
  Usk, garden m.v.l. trap, 26/5/68 and very common most years eg. 17/6/73, 21/6/74. 1978, abundant from 1/6. 1981, 14/6 etc. Redding's Inclosure 17/6/74. Cwm Tyleri 12/7/77, 8/7/80 (abdt.). Slade Wds 28/5/78. Cwm Coed-y-cerrig 5/6/78. Hael Wds 27/6/78, 14/7/78. Trelleck Bog 8/7/78, 24/7/76. Magor Marsh 21/7/78. Hendre Wds 29/8/78. Llansoy 23/6/86. Dinham 4/6/87. Kilpale 30/6/87.

Sq. **SO** 20,30,40,50,41,51,22. **ST** 48,49.

1334a *Scoparia basistrigalis* Knaggs
Common in deciduous woods.
  Redding's Inclosure, to m.v.l. 7/6/71, 17/6/84. Prescoed 20/6/71 (plfl.). Usk, garden m.v.l. trap, 1973, 18/6, 29/6; 1977, 8/8, 10/8, 5/9; 1982; 1983. 1979, Hael Wds 1979, 26/6, 5/7; 31/7/82. Hendre Wds 20/7/79 (plfl. to m.v.l.) (JDB, GANH); 1/7/86. Slade Wds 1983, July 8,21,23. Llansoy 1986, 28/6 – 12/7.

Sq. **SO** 30,40,50,41,51. **ST** 39,48.

1338 *Dipleurina lacustrata* Panz.
  = *crataegella* auct.
Common in Monmouthshire.
  Usk, garden m.v.l. trap, eg. 21/7/73; 1983, 6/7 – 14/7. Hael Wds 1982, 3/7, 31/7; 2/8/86. Slade Wds 8/7 and 21/7/83. Wyndcliff 25/7 and 10/8/83. Kilpale 30/6/87. Llansoy 20/7/87. Trelleck Bog 25/7/90. Clydach 26/7/90. Redding's Inclosure 2/8/90.

Sq. **SO** 30,40,50,21,51. **ST** 48,49,59.

1339 *Eudonia murana* Curt.
Occurs locally from June to August in the hills of northern Gwent frequenting stone walls and rocks at an altitude of 1,500 ft.

Cwm Tyleri 1973, several to m.v.l. 9/7 (GANH) (1st v-c. rec.); 1974, 29/7, 5/8; 1986, 28/6 (plfl).

Sq. SO 20.

1340 *Eudonia truncicolella* Stt.
Wyndcliff, one to m.v.l. 11/8/82 (GANH).

Sq. ST 59.

1342 *Eudonia angustea* Curt.
Common in Monmouthshire, flying from June to November.

Abertillery dist. 1914, "in hundreds" (Rait-Smith 1915).

Usk, 1966, to m.v.l. 19/8, 7/9. 1974, one indoors 5/1, a very "late" date. Usk, garden m.v.l. trap: 1979, 29/9 – 19/10; 1983, 27/9 (2), 5/11 (1), 6/11 (1). Hael Wds 18/9/82. Llansoy 1986, 22/6, 4/10; 1990, 20/10.

Sq. SO 20,30,40,50.

1343 *Eudonia delunella* Stt.
  – *vandallella* H.-S.
  = *resinea* auct.
Usk, to m.v.l. trap,: 1973, 29/6, 30/6. 1978, 18/7. 1979 21/7, 27/7. 1982, (5) 1/7 – 30/7. Llansoy, 1990, several to u.v.l. 14/7 – 2/8. Redding's Inclosure 2/8/90. Hael Wds 3/8/90. Tintern, 1/7/84 (MJS).

Sq. SO 30,40,50,51.

1344 *Eudonia mercurella* Linn.
Fairly common in Gwent.

Usk, garden m.v.l. trap, recorded most years from 1967 to 1983 eg. 1971, 18/7 (2); 1973, 22/6 – 22/8. Ed. 22/6/73. Ld. 5/9/77. Wyndcliff 17/9/79, 8/8/84, 1/9/89 etc. Hael Wds 5/7/86. Llansoy 1986, 28/6, 17/8.

Sq. SO 30,40,50. ST 59.

NYMPHULINAE

1345 *Elophila nymphaeata* Linn.　　　Brown China-mark
Common near rivers, streams and ponds, flying at dusk from mid-June until early September, and coming readily to light.

Usk, garden m.v.l. trap, recorded most years eg. 1967, Aug. 22, 26, 27, 28; 1979, June 20 (4), July 5, 14, 18; 1983, June 19, 20, 21, 26, July 5, 7, 9, 12, 15, 16. Ed. 15/6/76. Magor Marsh, frequent most years inc. 1970, 29/8;

1976, June 21, 25, July 2, 3, Aug. 3, 8, 9, 21; 1978, 4/9 swarming by day on aquatic vegetation in drainage reen. Llansoy 1986 26/6, 29/6, 11/7. Hael Woods 26/6/86. Hendre Wds 1/7/86. Magor Pill 6/8/89.

Sq. **SO** 30,40,41,50. **ST** 48.

1350 *Nymphula stagnata* Don. Beautiful China-mark
Occurs near rivers and streams but rare in Monmouthshire.

Usk, garden m.v.l. trap, 26/6/76 (1) (GANH) (1st v-c. rec.), 25/7/79 (1), 18/7/83 (1). In the eighteen years from 1966 to 1983 these were the only records of this species at my trap, sited at a distance of 100 yds. from the River Usk.

Sq. **SO** 30.

1348 *Parapoynx stratiotata* Linn. Ringed China-mark
Occurs near rivers, ditches and drainage reens but scarce in Monmouthshire.

Usk, garden m.v.l. trap, a single male on 21/7/69 and single females on 21/7/75, 13/8/78, 13/8/79 and 21/8/82. Magor Marsh 4/8/75 one to m.v.l.

Sq. **SO** 30. **ST** 48.

1354 *Cataclysta lemnata* Linn. Small China-mark
This species, flying from mid-June to late August, frequents still waters such as ponds and drainage reens where there is an abundance of duckweed (*Lemna* spp.).

Usk, garden m.v.l. trap, noted sporadically: 1969, 17/8 ♂, 18/8 ♀ 1970 2/8 ♂ (1), ♀ (1), 3/8 (1). 1971 12/8 ♀ 1975 6/8 ♀ 1976 (3) 5/7 – 31/7. 1980 8/8 (1). 1983 9/8 ♂ Magor Marsh recorded every year from 1968 to 1988 and usually abundant, especially so on 23/8/68, 2/8/70 and 19/6/79. Newport Docks 17/8/76 to m.v.l. Llansoy 2/7/87 ♀ (1). Uskmouth 16/8/88.

Sq. **SO** 30,40. **ST** 38,48.

ACENTROPINAE

1331 *Acentria ephemerella* D. & S. Water Veneer
= *nivea* Ol.

This species with aquatic larvae occurs locally, frequenting ponds, reens etc.

Magor Marsh, to m.v.l. 2/8/75, "in numbers", (MJL in litt.). 1976, swarming at m.v.l. 6/7 (GANH). Wyndcliff 10/8/83 (1) to m.v.l., 7/8/89. Uskmouth 16/8/88. Clydach 18/7/89, swarmed at m.v.l. Slade Wds 20/7/89 (2), 8/8/89. Magor Pill 6/8/89. Llansoy 8/8/90 (1), 25/8/90 (1), 9/7/92 (abdt.).

Sq. **SO** 21,40. **ST** 38,48,59.

EVERGESTINAE

1356   *Evergestis forficalis* Linn.   Garden Pebble
Common in gardens.

Abertillery dist. 1911 (Rait-Smith 1912).

Usk, garden m.v.l. trap, 17/8/66 and recorded annually until 1983 eg. 1967 frequent 2/6 – 6/8. Ed. 31/5/81. Wentwood 19/6/68, 3/7/68. Rhyd-y-maen 18/7/70, 3/9/73. Llansoy 12/6/85, 1986 25/6 – 18/7. Magor Marsh 1/8/88.

Sq. **SO** 20,30,40. **ST** 48,49.

1358   *Evergestis pallidata* Hufn.
Fairly common in damp woods and marshy places.

Usk, garden m.v.l. trap, recorded most years from 1967 to 1983 and sometimes plentiful eg. 1967 10/8, 25/8; 1973 20/7 – 18/8. Redding's Inclosure 20/7/71. Hendre Wds 25/7/78. Hael Wds 31/7/82. Magor Marsh 29/7/80. Llansoy, 1984 26/7; 1986 12/7, 22/7. Wyndcliff 8/8/86.

Sq. **SO** 30,40,50,41,51. **ST** 48,59.

PYRAUSTINAE

1361   *Pyrausta aurata* Scop.
This species is scarce in Monmouthshire, frequenting quarries and woodland rides mainly on the Carboniferous Limestone in the south-east of the county.

Tintern dist. (Piffard 1859). "Monmouthshire" (Barrett 1893 **9**: 170). Caerwent Quarries 31/7/70 (CT in litt.).

Wentwood 11/8/70, flying by day. Slade Wds, by day, 29/5/79 (1) and 26/5/80 (1); at m.v.l. 28/7/81 and 2/8/82; 13/8/83 several by day and at m.v.l. Wyndcliff, at m.v.l. 3/8/81 (1) and a number on 25/7/83, 10/8/83 and 8/8/84. Hael Wds, to m.v.l., 31/7/82 and 5/8/82.

Sq. **SO** 50. **ST** 48,49,59.

1362   *Pyrausta purpuralis* Linn.
More frequent and widespread in Monmouthshire than P. aurata

Usk, occasionally in garden m.v.l. trap: 4/8/66 (1), 19/8/73 (1), 1975 (1), 1980 (2), 1982 (1), 1983 (1). Wentwood 8/7/68, 22/7/68, 11/8/70. Rhyd-y-maen 3/9/73 to m.v.l. Newport Docks 17/8/76 to m.v.l. Graig-yr-iar,

(Llangeview), several flying by day on hilly pasture. Slade Wds, by day 1/6/81, 28/7/81, etc.; to m.v.l. 10/8/81 (10). also 1982, 83 and 84. Wyndcliff 3/8/81, 10/8/83, 27/7/78. Llansoy, to m.v.l., 26/7, 28/7/84; one on lawn in day-time 15/7/90. Trelleck Bog 17/8/88.

Sq. **SO** 30,40,50. **ST** 38,48,49,59.

1365   *Pyrausta cespitalis* D. & S.

This moth, fairly common in Monmouthshire, flies in two broods from late April to early September.

Hadnock Quarries, one by day 27/7/71. Usk, garden m.v.l. trap, 16/8/75 (1), 1976 (6) 2/8 – 27/8, 1979 (1), 1982 (2). Slade Wds 1982, 28/7 (1), 9/9 (1); 1986, 22/7; 1991, 26/4 (1), 14/5 (abdt. on low herbage at edge of forestry roads). Llansoy 25/8/89 (1), 22/8/90 (6). (MEA) : Clydach, 1989 9/7 (1), 21/7 (2).

Sq. **SO** 30,40,21,51. **ST** 48.

1367   *Pyrausta cingulata* Linn.
On Carboniferous Limestone cliffs in the north of the county.
Clydach 1/8/90 (4) to m.v.l. (MEA) (1st v-c. rec).

Sq. **SO** 21.

1369   *Uresiphita polygonalis* D. & S.
   = *limbalis* auct.
A scarce and casual immigrant from southern Europe and the Tropics.

Usk, 1969, one to my garden m.v.l. trap on the night of 13/10 (GANH). This is the only Monmouthshire record to date and is thought to be the first Welsh record (Horton 1970). On the same night another specimen was recorded at m.v.l. in Somerset, at Selworthy near Minehead (Chappel 1970). Hitherto there had been only about a dozen British records of this species.

Sq. **SO** 30.

1373   *Microstega pandalis* Hb.   Bordered Pearl
Pont-y-saeson, 1976, one flying by day 4/6 (REMP).

Sq. **ST** 59.

1374   *Microstega hyalinalis* Hb.
As *Botys hyalinalis*, Pontllanfraith 1912 (Rait-Smith 1913).

Sq. **ST** 19.

1376   *Eurrhypara hortulata* Linn.   Small Magpie
Common in June and July in gardens, hedgerows, waste ground etc.

Croesyceiliog, 1928 (plfl.) (GANH). Usk, m.v.l.trap, 1967, frequent 13/5 – 6/8 and recorded annually until 1983, eg. 22/6/83 (5). Ed. 25/4/69. Ld. 1/9/70, 5/8/78. Pont-y-saeson 5/6/67. Wentwood 3/7/68. Prescoed 5/7/68. Trelleck 9/7/68. Wyndcliff 13/7/69. Rhyd-y-maen 18/7/70. Redding's Inclosure 20/7/71. Magor Marsh 24/6/75. Cwm Tyleri 29/6/76. Hael Wds 27/6/78. Llansoy 4/7/85.

Sq. **SO** 20,30,40,50,51. **ST** 48,39,49,59.

1377 *Perinephila lancealis* D. & S.

Common in Gwent during July and August frequenting moist woods and damp places where hemp agrimony (*Eupatorium cannabinum*) flourishes.

Tintern dist. (Piffard 1859). "Monmouthshire" (Barrett 1893 **9**: 237).

Pont-y-saeson, plentiful at m.v.l. 27/6/67, 18/7/72. Usk, garden m.v.l. trap, 1969 26/6 – 12/7 and recorded most years from 1970 to 1983. Wentwood 28/6/69. Wyndcliff 13/7/69, 25/7/85 etc. Hendre Wds 5/7/70, 11/7/86 etc. Hael Wds 5/7/70. Redding's Inclosure 20/7/71, 23/6/81. Magor Marsh 3/7/76. Slade Wds, very common, 23/7/79, 30/6/84 etc. Trelleck Bog, plentiful, 16/7/79, 24/7/86 etc.

Sq. **SO** 30,50,41,51. **ST** 48,49,59.

1378 *Phlyctaenia coronata* Hufn.
= *sambucalis* D. & S.

Found in hedgerows, waste ground etc. where there is elder (*Sambucus nigra*) growing.

As *Ebulea sambucalis*, Abertillery dist. 1911 (Rait-Smith 1912).

Usk, recorded at m.v.l. most years from 1967 to 1983 eg. 1967, 7/7; 1978 13/7, 20/7, 2/8. Magor Marsh 1969, 1/7, 11/7, 22/7; 1976 3/7, 6/7. Slade Wds 21/7/79. Llansoy 28/6/86, 1/8/90. Hendre Wds 1/7/86.

Sq. **SO** 20,30,40,41. **ST** 48.

1381 *Anania funebris* Ström
= *octomaculata* Linn.

Occurs locally in Monmouthshire, flying by day in rough open woods mainly on Carboniferous Limestone in the south-east. Sadly, this attractive little moth has long since died out at Pont-y-saeson, where it was once so common, due to "coniferisation" of its woodland site. Currently, however, it is still fairly plentiful in at least one western locality.

Pont-y-saeson 1966, 17/5; 1967, abundant, May 9,16, June 5,13,27; 1968, 8/6; 1971, 15/6 (1) (GANH). Wentwood 1968, 19/6, 3/7 (JMD), 1974, 25/6 (CT). Slade Wds 30/5/71, numerous (GANH). Pontllanfraith 1987 20/6 (1), 23/6 (plfl.); 1988 30/6 (1); Fleur-de-lis 17/6/88 (1).

Sq. **ST** 48,19,49,59.

1385 *Ebulea crocealis* Hb.

Local and scarce in Monmouthshire.

Abertillery dist. 1911 (Rait-Smith 1912).

Usk, to garden m.v.l. trap, 1969 31/7, 2/8; 1972 30/8; 1973 30/7; 1975 3/8 (3), 4/8 (1), 5/8 (1); 1978 July 18,23,29, Aug. 3,4 (2); 1981 14/7; 1983 11/7, 16/7. Newport (Coldra) 27/6/75 (2) to m.v.l. Llansoy 23/7/87 (1), 16/7/90 (1).

Sq. **SO** 20,30,40. **ST** 38.

1386 *Opsibotys fuscalis* D. & S.
Local and uncommon in Gwent.

As *Botys fuscalis*, Tintern dist. (Piffard 1859). Abertillery dist. 1911 (Rait-Smith 1912).

Usk, garden m.v.l. trap, 1972 30/7, 3/8; 1980 27/7 (1). Slade Wds 21/7/79, one flying by day amongst cow-wheat (*Melampyrum pratense*). Llansoy, 1986 (5) 13/7 – 18/7.

Sq. **SO** 20,30,40,50. **ST** 48.

1388 *Udea lutealis* Hb.
    = *elutalis* auct.
Common amongst herbage in rough fields, hedgerows, waysides etc. in July and August.

As *Scopula lutealis*, Abertillery dist. 1911 (Rait-Smith 1912).

Usk, garden m.v.l.trap, 13/8/67 and recorded most years until 1983 eg. 1981 26/7 – 21/8 inc. (5) on 7/8. Ed. 7/7/76. Usk, 1982, flying on disused railway line in afternoon sunshine 10/7 etc. Prescoed 3/8/69. Craig-y-dorth 19/8/73. Mynyddislwyn 25/8/73. Runston, flying on roadside banks by day 6/8/74. Magor Marsh 8/8/76 to m.v.l. Slade Wds 11/8/81, 23/7/83. Llansoy 1986 24/7 – 17/8. Magor Pill 19/7/89, 1/8/89 (plfl.).

Sq. **SO** 20,30,40. **ST** 48,19,39,49.

1389 *Udea fulvalis* Hb.
"A casual immigrant which has been taken in Hereford and south Dorset." (Beirne *British Pyralid & Plume Moths* 1952. 143).

Usk, garden m.v.l. trap, 27/7/71 (1) (GANH) (1st v-c. rec.), detn. confirmed by JDB, (in colln. GANH). Several came to the trap in 1971 and in a few previous years but sadly, owing to doubt over their identity they were not recorded.

Sq. **SO** 30.

1390 *Udea prunalis* D. & S.
    = *nivealis* Fabr.
Common in rough woods, hedgerows etc.

As *Scopula prunalis*, Abertillery dist. 1911 (Rait-Smith 1912).

Pont-y-saeson 5/8/67. Usk, 3/8/72; 1973 25/6 – 20/8; and common most years from 1972 to 1983. Hael Wds 17/7/73. Magor Marsh 1/7/73. Llangeview (Allt-y-bela lane) 28/7/79. Slade Wds 21/7/83. Llansoy 1986 5/7 – 25/7.

Sq. **SO** 20,30,40,50. **ST** 48,59.

1392   *Udea olivalis* D. & S.

Widespread and very common in June and July in woods, bushy places and hedgerows.

Abertillery dist. as *Scopula olivalis* 1911 (Rait-Smith 1912).

Usk, garden m.v.l. trap, plentiful 1967 5/6 – 31/7 and annually until 1983. Ed. 3/6/82. Pont-y-saeson 5/6/67. Wyndcliff 6/6/67, 10/7/67, 3/8/81. Wentwood 22/7/68. Magor Marsh 1/7/69. Rhyd-y-maen 6/6/70. Cwm-mawr 2/7/71. Slade Wds 1/6/71, 22/7/86. Hael Wds 4/7/72. Redding's Inclosure 20/7/71, 23/6/81. St Pierre's Great Wds 13/7/73. Piercefield Park 18/7/73. Newport (Coldra) 27/6/75. Hendre Wds 25/6/79. Trelleck Bog 16/7/79. Llansoy 26/7/84. Ysguborwen (Llantrisant) 17/7/86. Kilpale 30/6/87. Dixton Bank 18/6/88.

Sq. **SO** 20,30,40,50,41,51. **ST** 38,48,49,59.

1395   *Udea ferrugalis* Hb.
    = *martialis* Guen.

This migrant species fluctuates widely in numbers from year to year. In some years it is plentiful especially in the summer and autumn and in others very scarce.

Usk, garden m.v.l. trap, 1967 10/8 (1), 25/8 (1); 1969 frequent 3/9 – 28/10; 1970, one only on 22/7; also noted 1971,80 and 82; 1983 11/8 – 3/11. Trostrey Church, at ivy blossom 25/10/69. Llansoy 24/7, 26/7/89.

Sq. **SO** 30,40.

1398   *Nomophila noctuella* D. & S.   Rush Veneer

A common migrant seen most years, sometimes in abundance.

Usk, garden m.v.l. trap, recorded annually 1966 to 1983 except in the years 1967,74,77,79 and 81. 1966 (1) 2/10; 1969 (91) 8/6 – 28/10. Ed. 15/4/80. Ld. 8/11/83. Magor Marsh 1969 22/7, 7/9; 1976 8/8. Rhyd-y-maen 18/7/70. Redding's Inclosure 19/10/71. Llansoy 1986, June 28,29,30, Aug 18.

Sq. **SO** 30,40,51 **ST** 48.

1405   *Pleuroptya ruralis* Scop.   Mother of Pearl

Very common in July and August wherever its larval food-plant the common nettle (*Urtica dioica*) occurs.

Wye Valley, about twenty larvae on nettles 4/6/06 (Barraud 1907). Redding's Inclosure 31/5/68, larvae on *Urtica dioica* (JMC-H 1969).

Croesyceiliog 1928, common (GANH). Usk, garden m.v.l. trap, 21/8/65 and recorded every year until 1983. Very abundant in 1971 reaching maximum numbers from 18/8 to 21/8. Ed. 1/7/76. Ld. 3/10/79. Wentwood 22/7/68. Magor Marsh 22/7/69. Hael Wds 5/7/70. Rhyd-y-maen 18/7/70. Redding's Inclosure 18/7/70. Slade Wds 29/8/72. Newport Docks 17/8/76. Hendre Wds 23/8/77. Llangeview 28/7/79. Wyndcliff 28/7/83. Llansoy

26/7/84. Trelleck Bog 24/7/86.

Sq. **SO** 30,40,50,41,51. **ST** 38,48,39,49,59.

1408   *Palpita unionalis* Hb.
An uncommon immigrant occasionally recorded from the southern maritime counties of Britain.
  Usk 1977, to garden m.v.l. trap, 16/10 ♀ (1) (GANH) (1st v-c rec); 18/10 ♀♀ (2); 20/10 ♀♀ (2), ♂ (1).

Sq. **SO** 30.

[1411   *Leucinodes vagans* Tutt]
As *Aphytoceros vagans* Tutt, a species new to science, said to have been taken at Chepstow, Monmouthshire in October 1888 by Mr J. Mason of Clevedon (Tutt 1890 *Entomologist's Rec. J. Var.* **1**: 203) and was supposed by Tutt to have come from South America or the West Indies.
  Goater (1985) stated "the whereabouts of the type is unknown; the genus is African, whence several similar species have been described."
  However, N.W. Lear (1986) (*Entomologist's Gaz.* **37**: 197) wrote "the type of *Leucinodes vagans* Tutt has been located in the City of Bristol Museum in the C. Bartlett Collection", and that the accompanying data labels indicated that the insect was taken at Tidenham, Glos. about two miles north-east of Chepstow. Tidenham lies about a mile to the east of the River Wye, the county boundary, so it would appear that this record is a West Gloucestershire and not a Monmouthshire one.

PYRALINAE

1413   *Hypsopygia costalis* Fabr.   Gold Triangle
Common in Monmouthshire, at least in the eastern half of the county, inhabiting gardens, woods and rural areas generally, flying from July to early November and often plentiful.
  Brockwells (Caerwent), 18/7/61 (svl.), 5/8/70 (CT in litt.).
  Usk, recorded at m.v.l. trap every year from 1966 to 1983 usually first appearing about July 6 but occasionally much earlier eg. 11/5/67, 23/6/71, 19/6/77. Sometimes quite numerous in October with unusually late records on 7/11/78 and 4/11/83. Prescoed 5/7/68. Magor Marsh 2/8/70, 11/7/78 etc. Hael Wds 12/7/71, 6/8/74. Wyndcliff 21/8/72. Hendre Wds 29/8/78. Pont-y-saeson 18/9/79. Slade Wds 28/7/80 (abdt.). Llansoy 4/7/85 (20) to u.v.l. Trelleck Bog 24/7/86.

Sq. **SO** 30,40,50,41. **ST** 48,39,59.

1415   *Orthopygia glaucinalis* Linn.
Usk, garden m.v.l. trap, recorded most years from 1966 to 1983 but in small

numbers, usually not more than two or three per year eg. 1966 19/8 (1); 1979 single moths on 11/6, 16/10, 17/10; 1983 11/7 (1), 15/7 (3), 16/7 (1), 17/7 (1). Wyndcliff 3/8/81 (1). Slade Wds 23/7/83 (1).

Sq. **SO** 30. **ST** 48,59.

1417   *Pyralis farinalis* Linn.   Meal Moth
Usk, occasionally to m.v.l.: 1968 19/8 (1). 1971 15/9 (1). 1972 20/7 (1). 1973 16/9 (1). 1974 6/7 (1), 8/7 (1). These were the only records from 1966 to 1983. Llansoy 25/8/89 (1), 1/8/90 (1).

Sq. **SO** 30,40.

1421   *Aglossa pinguinalis* Linn.   Large Tabby
Wye Valley (Barraud 1906).
   Usk, one to garden m.v.l. trap, 27/7/72 (GANH). Magor Marsh, 19/7/75 (1), "on wing near shed used as a stable" (MJL in litt.).

Sq. **SO** 30. **ST** 48.

1424   *Endotricha flammealis* D. & S.
Rare in Monmouthshire.
   Usk, garden m.v.l. trap, 25/7/80 (1). Slade Wds 21/7/83 one to m.v.l. (GANH).

Sq. **SO** 30. **ST** 48.

GALLERIINAE

1425   *Galleria mellonella* Linn.   Wax Moth
Llansoy 8/8/90 ♀ (1) to u.v.l (GANH) (1st v-c. rec.).

Sq. **SO** 40.

1426   *Achroia grisella* Fabr.   Lesser Wax Moth
Usk: 9/6/69 imagines emerging from pupae in old honey-comb (JA).
   Wyndcliff: one to m.v.l. 24/6/71 (JMC-H, GANH). Llansoy 1992 31/7 (2).
Sq. **SO** 30, 40. **ST** 59.

1427   *Corcyra cephalonica* Stt.   Rice Moth
Usk: 11/9/82 one in garden m.v.l. trap. (GANH) (det. JDB).

Sq. **SO** 30.

1428   *Aphomia sociella* Linn.   Bee Moth
Common in Monmouthshire.

Usk, garden m.v.l. trap, noted annually from 1969 to 83, eg. 1969: June 26. July 11,19. 1977: July 6, 14, 30 (4), 31. Aug. 13 (2), 15. Ed. 14/6/81.

Ld. 15/8/77. Rhyd-y-maen 18/7/70. Magor Marsh 2/8/70. Hael Wds 10/7/73, 27/6/78, 29/5/82. Redding's Inclosure 14/7/74. Llansoy, 1986, (12) 23/6 to 20/8 and late individuals on 6/10 and 13/10; 1987, (20) 28/6 to 23/7 inc. (5) on 5/7. Hendre Wds 11/7/86.

Sq. **SO** 30,40,50,41,51. **ST** 48.

PHYCITINAE

1433 *Cryptoblabes bistriga* Haw.
1986: Wyndcliff, plfl. to m.v.l. 23/6 and 25/6 (JDB); Hael Wds, 26/6 (JDB); Hendre Wds 11/7/86 (GANH). 1987: Wyndcliff 19/6, Hael Wds 24/6.

Sq. **SO** 50,41. **ST** 59.

1446 *Salebriopsis albicilla* H.-S.
In the British Isles this species occurs only in the lower Wye Valley in Monmouthshire where it was first found in 1964 in woods containing the small-leaved lime (*Tilia cordata*). R.M. Mere 1965, reported its capture at "Tintern" on 23/6/64 (*Entomologist's Gaz.* **16**: 13), while J. Newton (1965 Ibid.) also reported that he and L. Price had taken it there on 19/6/64. They took further specimens in 1967, 68 and 69. (Newton 1985, *The Gloucestershire Naturalist* No **2**: 28).

Since then it has been frequently recorded from the same site eg. 1971, 24/6 (4) (JMC-H, GANH); 1974, 2/7 (2) (GANH); 1986, 25/6 (1) (JDB).

Sq. **ST** 59.

1450 *Metriostola betulae* Goeze
Very local in Monmouthshire flying in open woods where there are scattered birches.

Wentwood, 1968, several at m.v.l. 8/7 (GANH). Slade Wds, 30/6/84. Trelleck Bog 19/7/86 (1).

Sq. **SO** 50. **ST** 48,49.

1451 *Pyla fusca* Haw.
Widespread but infrequent in Monmouthshire occurring in places where heathers grow, including heaths, moorland and also gardens.

Usk, garden m.v.l.trap, 3/8/69 (1), 26/6/77 (1), 1978 (1), 1979 (2). Cwm Tyleri 1973, 9/7 (2) (GANH); 1976, 27/6, 29/6, 8/7; 1977, 9/7, 12/7. Trelleck Bog, 16/7/79 (2). Llansoy 1985, 5/6, 9/7; 1987, 9/7. Sugar Loaf (St Mary's Vale) 8/7/87.

Sq. **SO** 20,30,40,50,21.

1452   *Phycita roborella* D. & S.

Occurs very commonly among oaks in the eastern half of the county during July and August.

Usk, garden m.v.l. trap, 24/7/69 and recorded every year to 1983, often frequent. Ed. 6/7/80. Ld. 6/9/77. Magor Marsh 2/8/70. Redding's Inclosure 27/8/71 (abdt.), 2/8/90. Trelleck Bog 8/7/78. Hael Wds 14/8/78, 2/8/81. Hendre Wds 29/8/78, 4/9/78, 11/7/86. Slade Wds 23/7/79, 23/7/83. Wyndcliff 3/8/81, 10/8/83. Llansoy 26/7/84, 26/7/89, 24/8/90.

Sq. **SO** 30,40,50,41,51. **ST** 48,59.

1454   *Dioryctria abietella* D. & S.

In conifer plantations, but of infrequent occurrence in Gwent though now becoming more widespread. Determinations were confirmed by Messrs S.N.A. Jacobs and M. Shaffer of the BMNH.

Usk, garden m.v.l. trap, 30/7/78 (1) (1st v-c. rec.), 4/8/78 (2), 7/7/81, 14/7/82. Trelleck Bog 16/7/79 (1), 24/7/86 (?). Wyndcliff 3/8/81 (1). Llansoy 17/8/86. Hael Wds 2/8/86 (1), 4/7/87. Slade Wds 20/7/89.

Sq. **SO** 30,40,50. **ST** 48,59.

1436   *Acrobasis repandana* Fabr.
         = *tumidella* Zinck.

Flies during July and August in oakwoods in the east of the county.

Hael Wds: 12/7/71, 10/7/73, 2/8/81, 2/8/86 (plfl.). Hendre Wds, to m.v.l., 17/7/79, 20/7/79 (3). Trelleck Bog 17/8/88.

Sq. **SO** 50,41.

1437   *Acrobasis consociella* Hb.

In oakwoods in eastern half of Monmouthshire but local and scarce.

Hael Wds, 12/7/71, 10/7/73, 2/8/81. Usk, garden m.v.l. trap, 16/7/78, 2/10/79. Slade Wds 23/7/79. These are my only Gwent records in twenty years (GANH).

Sq. **SO** 30,50. **ST** 48.

1438   *Numonia suavella* Zinck.

Scarce and local in the eastern half of Monmouthshire.

Magor Marsh 2/8/70 (1) (GANH). Usk, garden m.v.l. trap, 25/7/73, 31/7/82. Hendre Wds 11/7/86. Llansoy, to u.v.l., 1986, 12/7; 1990, 23/7 (3), and singletons on 1/8, 2/8 and 3/8. Trelleck Bog, to m.v.l., 19/7/86, and in fair numbers on 24/7/86. (GANH).

Sq. **SO** 30,40,50,41. **ST** 48.

1439 *Numonia advenella* Zinck.
Locally common in Monmouthshire.
 Usk, garden m.v.l. trap, 1971, 31/7, 21/8 (2), 24/8, 26/8; 1979, 12/8 (3), 13/8; 1980, several from 6/8 to 6/9. Hael Wds 12/7/71. Slade Wds 28/7/80, 2/8/80 (2). Clydach 18/7/89. Llansoy 1989, to m.v.l., 24/7 (svl.), 28/7, 5/8 (2).

Sq. **SO** 30,40,50,21. **ST** 48.

1440 *Numonia marmorea* Haw.
Rare in Monmouthshire.
 Usk, garden m.v.l. trap, 31/7/71 (1). 1982, single moths on July 8, 11, 14 and 20. (My only Gwent records, GANH).
Sq. **SO** 30.

1458 *Myelois cribrella* Hb.   Thistle Ermine
Locally common in July and August on rough ground and in open woods where thistles flourish.
 Usk, m.v.l. trap, 4/7/67 and noted most years to 1983. Ed.4/6/82. Ld. 21/8/77. Magor Marsh 22/7/69, 25/6/76 (abdt.). Hendre Wds 6/6/78, two freshly emerged moths flying in evening sunshine. Llansoy, 26/7/84, 15/6/88, 11/7/90.

Sq. **SO** 30,40,41. **ST** 48.

1470 *Euzophera pinguis* Haw.
Locally common amongst ash trees (*Fraxinus*), during July and August.
 Usk, garden m.v.l. trap, 25/7/69 and recorded every year from 1969 to 1983. Plentiful some years, eg. 1977, (27) from 21/7 to 26/9. Ed. 13/7/78. Ld. 26/9/77. Magor Marsh 8/8/76, 20/8/77 (GANH). Slade Wds 28/8/79, 6/8/84. Wyndcliff 3/8/81, 8/8/84. Hael Wds 5/8/82. Llansoy 17/8/86, 22/7/89.

Sq. **SO** 30,40,50. **ST** 48,59.

1481 *Homoeosoma sinuella* Fabr.
Local in Gwent, but usually plentiful where found.
 Usk, garden m.v.l. trap, recorded most years eg. 2/7/73; 18/6/75; 1977, 13/6 – 7/7; 1983, 4/7 – 18/7. Magor Marsh 21/6/76, 13/7/83. Slade Wds, flying by day 8/7/80, several to m.v.l. 19/6/85.

Sq. **SO** 30. **ST** 48.

1483 *Phycitodes binaevella* Hb.
Locally common, flying in July.
 Usk, garden m.v.l. trap, 1973, 18/7, 25/7, 1/8. Also noted in 1974, 77,

78, 80, 81, 82, 83. Wentwood 29/6/73. Cwm Tyleri 9/7/73, 1974, and 1977. Llansoy 12/7/86, 13/7/86. Slade Wds 29/7/86, 17/7/87.

Sq. **SO** 20,30,40. **ST** 48,49.

1474   *Ephestia parasitella* Stdgr.
         ssp. ***unicolorella*** Stdgr.

Stated by Beirne (1952) to be widely distributed but very rare in southern England. However, it is very common in the eastern half of Monmouthshire and, to my knowledge, has certainly been so for at least the last twenty five years. (GANH 1990).

Usk, garden m.v.l. trap, 1969, 26/7; 1973, 22/6 (2) etc.; 1979, plentiful 12/7 – 26/7; 1981; 1982 30/5 – 7/7; 1983 20/6   18/7. Tintern 1/7/71 (MWH pers. comm.). Slade Wds 23/7/79. Hael Wds 4/7/81, 26/6/86. Llansoy, 1986 very common 25/6 – 18/7. Wyndcliff 25/6/86. Hendre Wds 1986 1/7, 11/7.

Sq. **SO** 30,40,50,41. **ST** 48,59.

1475   *Ephestia kuehniella* Zell.   Mediterranean Flour Moth
Newchurch West 1978, packet of Quaker Oats heavily infested with larvae. Bred, imagines emerged Sept. and Oct. (GANH).

Sq. **ST** 49.

PTEROPHORIDAE
PLATYPTILIINAE

1495   *Marasmarcha lunaedactyla* Haw.
Magor, 13/7/83, two to m.v.l. (1st v-c. rec.); Slade Wds 21/7 (1) to m.v.l. (GANH) (det. BS).

Sq. **ST** 48.

1498   *Amblyptilia punctidactyla* Haw.
Usk, 21/3/72, Plas Newydd garden, one flying in afternoon sunshine and entering corolla tubes of blue hyacinths (GANH); 26/3/81, one netted in garden.

Sq. **SO** 30.

1501   *Platyptilia gonodactyla* D. & S.
Common on waysides, forestry roads, embankments etc.

Usk, 17/9/66, 20/6/74, 1981 13/6, 14/7, 1/8. Redding's Inclosure 15/6/71 (2) flying by day. Slade Wds, 3/6/80 by day; 1982 to m.v.l. 9/9, 16/9. Hendre Wds 1/8/80. Wyndcliff 15/8/81. Hael Wds 5/7/86 (3) to m.v.l.

Sq. **SO** 30,50,41,51. **ST** 48,59.

1503 *Platyptilia ochrodactyla* D. & S.
Usk, 1977, 30/7 (1) (GANH). 1980, Usk, 25/7 to m.v.l. (1).

Sq. **SO** 30.

1504 *Platyptilia pallidactyla* Haw.
Usk, garden m.v.l. trap, 4/7/67 (1); 1972 21/7 (1), 25/7 (1). Great Barnet's Woods (Mounton), 22/7/76, one flying by day (GANH). Llansoy, one to u.v.l. 28/6/92.

Sq. **SO** 30,40. **ST** 59.

1507 *Stenoptilia zophodactylus* Dup.
 Pont-y-saeson, 1966, three flying in early evening 16/8 (GANH). Usk, 1968, two to m.v.l. 24/9 (REMP).

Sq. **SO** 30. **ST** 59.

1508 *Stenoptilia bipunctidactyla* Scop.
Usk, 1971, one to m.v.l. 22/9 (GANH). Pontllanfraith 4/7/87 (2), 30/6/88 (2), 6/7/88 (frqt.), all by day.

Sq. **SO** 30. **ST** 19.

PTEROPHORINAE

1513 *Pterophorus pentadactyla* Linn.   White Plume Moth
Very common in gardens, waste ground, hedgerows etc. where the greater bindweed (*Calystegia sepium*) flourishes.
 Monmouthshire, "in hundreds" (Ince 1887) (As *Aciptilia pentadactyla*).
 Croesyceiliog, July 1927, plentiful (GANH). Usk, 4/7/67 and most years to 1983 eg. 1971 5/7 – 31/7 and a very late specimen on Sept 9th. Prescoed 5/7/68. Magor Marsh 1975, 1/7, 7/7, 1/9; 25/6/76. Cwm Tyleri 8/7/76 to m.v.l. (AR). Llansoy 4/7/85. Kilpale 30/6/87. Dixton Bank 18/6/88.

Sq. **SO** 20,30,40,51. **ST** 48,39,49.

1517 *Adaina microdactyla* Hb.
Hendre Wds 1982, one netted in daytime 29/5 (JDB) (1st v-c. rec.). Wyndcliff 1986, one to m.v.l. 23/6 (JDB, GANH.).

Sq. **SO** 41. **ST** 59.

1520 *Leioptilus osteodactylus* Zell.
Slade Wds 1980, to m.v.l., 15/7 (1) (GANH) (1st v-c. rec.), 28/7 (1), 2/8 (2). Slade Wds 17/7/87.

Sq. **ST** 48.

1522 *Leioptilus tephradactyla* Hb.
Slade Wds 1980, one netted by day 24/6 (GANH). (1st v-c. rec.). Hael Wds 6/7/88 (JDB).

Sq. **SO** 50. **ST** 48.

1524 *Emmelina monodactyla* Linn.
Common in hedgerows in July and September and again during the spring after hibernation.
    Abertillery dist. 1911 (Rait-Smith 1912).
    Usk, garden m.v.l. trap, recorded most years from 1967 to 1983 eg. 1967: 5/10. 1968: 18/3, 13/4, 23/9, 27/9. 1976: 30/3 – 6/5. 1979: 30/9, 2/10, 8/10 (2). 1982: 10/7, several at dusk on greater bindweed. Trostrey Common 2/10/79. Llansoy 10/9/86, 1/10/86, 27/6/88.

Sq. **SO** 20,30,40.

Bibliography & References

Entomological Journals etc. (abbreviations)
Entom. The Entomologist
Ent. Gaz. Entomologist's Gazette
Ent. Mon. Mag. Entomologist's Monthly Magazine
Ent. Rec. J. Var. Entomologist's Record and Journal of Variation
Ent. Wkly Intell. Entomologist's Weekly Intelligencer
J. Mon. Nat. T. Journal of Monmouthshire Naturalists' Trust
Mag. Nat. Hist. The Magazine of Natural History
MBGBI The Moths and Butterflies of Great Britain and Ireland (Eds. Heath, J.; Emmet, A.M.; et al.) 1976 – 1989
Proc. S. Lond. ent. nat. Hist. Soc. Proceedings of the South London Entomological and Natural History Society
Proc. Trans. Br. Ent. Nat. Hist. Soc. Proceedings and Transactions of the British Entomological and Natural History Society
Rep. Mon. Nat. T. Report of Monmouthshire Naturalists' Trust

ANTHONEY, M.E. *Personal Records of Gwent Lepidoptera 1976 – 1992* (diary).

BARRETT, C.G. *The Lepidoptera of the British Islands.* Vols **1 – 11**. London, 1893 – 1907.

BARRAUD, P.J. *Notes from the Wye Valley (Monmouthshire). Entom.* **39**: 213-214. 1906.

BARRAUD, P.J. *Wye Valley notes. Entom.* **40**: 297-298. 1907.

BEIRNE, B.P. *British Pyralid & Plume Moths.* London. 1952.

BIRD, J.F. *Lepidopterolgical notes from Monmouthshire. Ent. Rec. J. Var.* **17**: 311-315. 1905.

BIRD, J.F. *Butterflies in the Wye Valley during 1906. Ent. Rec. J. Var.* **18**: 277-281. 1906.

BIRD, J.F. *Notes from the Wye Valley: lepidoptera 1906. Ent. Rec. J. Var.* **19**: 59-64. 1907.

BIRD, J.F. *Lepidoptera in Gloucestershire: the Wye Valley in 1909. Ent. Rec. J. Var.* **22**: 4, 37-42. 1910.

BIRD, J.F. Notes from the Wye Valley, lepidoptera in 1911. *Ent. Rec. J. Var.* **24**: 53-59. 1912.

BIRD, J.F. *Lepidoptera in the Wye Valley during 1912. Ent. Rec. J. Var.* **25**: 85-89, 129-132. 1913.

BIRKENHEAD, G.A. *Notes on the Lepidoptera of the Cardiff. district.* In, James, I. (Ed.) Handbook for Cardiff and district. pp. 172-178. British Association, Cardiff. 1891.

BRADLEY, J.D. & MARTIN, E.L. *An Illustrated List of the British. 1956.* Tortricidae (Part 1: *Tortricinae & Sparganothinae*), *Ent. Gaz.* 7: 151-156, Pl. 1-10. 1956.

BRADLEY, J.D. *An Illustrated List of the British Tortricidae,* 1959. (Part 2: *Olethreutinae*), *Ent. Gaz.* 10: 60-80, Pl. 1-19.

BRADLEY, J.D., TREMEWAN, W.G. and SMITH, A. *British Tortricoid Moths,* Part 1, *Cochylidae and Tortricidae: Tortricinae.* The Ray Society, London.

BRADLEY, J.D., TREMEWAN, W.G and SMITH, A. Part 2, *Tortricidae: Olethreutinae.* The Ray Society, London 1979.

BRADLEY, J.D. & FLETCHER, D.S. *A Recorder's Log Book or Label List of British Butterflies and Moths.* London. 1979.

BRADLEY, J.D. & FLETCHER, D.S. *An indexed list of British Butterflies and Moths.* Orpington, Kent. 1986.

BRETHERTON, R.F. *Curtis and Wood on Eriopygodes imbecilla* Fabr. (*Lep.* Noctuidae). *Ent. Rec. J. Var.* 89. 125. 1977.

BUCKLER, (*capture of Lithophane furcifera Hufn. in Monmouthshire*). *Ent. Mon. Mag.* 6: 190. 1870.

CHALMERS-HUNT, J.M. *Notes on Monmouthshire Lepidoptera* mainly extracted from the literature. ms. 28 pp. 1966.

CHALMERS-HUNT, J.M. *Breconshire & Monmouthshire Entomology. Ent. Rec. J. Var.* 81: 39-46. 1969.

CHALMERS-HUNT, J.M. *Local Lists of Lepidoptera or a Bibliographical Catalogue of Local Lists and Regional Accounts of the Butterflies and Moths of the British Isles.* Uffington, Oxfordshire. 1989.

CHAPMAN, T.A. (*re. Cossus cossus Fabr. at Abergavenny*). *Ent. Mon. Mag.* 7: 18. 1871.

CHAPPEL, H.M. *Uresiphita polygonalis in West Somerset. Ent. Rec. J. Var.* 82: 30. 1970.

CHARLES, S.G. *Nature Notes on Monmouth Neighbourhood.* (An article in a booklet guide to Monmouth Town and district which came to my notice in 1943 but which I can no longer trace). (c. 1937).

CLARKSON-WEBB, P. *Lepidoptera (Butterflies in Chepstow dist. 1966). Rep. Mon. Nat. T.* No 3: 10. 1966.

CONWAY, C. *List of Butterflies occurring around Pontnewydd Works,* Monmouthshire with corrections and additions. *Mag. Nat. Hist.* 6: 224-228, 541-544. 1833.

CROWTHER, G.F. *(re. Egira conspicillaris Linn. at Monmouth) Entom.* 72: 186. 1939.

DANIEL, A. *Lepidoptera (Abergavenny, 1964). Rep. Mon. Nat. T.* No. 1: 18. 1964.

DEMUTH, R.P. *Observations on Dr Horton's Note. Ent. Rec. J. Var.* **93**: 22. 1981.

EMMET, A.M. *Addenda and corrigenda to the British list of Lepidoptera. Ent. Gaz.* **38**: 31-52. 1987.

FORD, E.B. *Butterflies.* London. 1945.
FROHAWK, F.W. *British Butterflies.* London. 1934.
FROHAWK, F.W. *Varieties of British Butterflies.* London. 1938.

GOATER, B. *On rearing Eriopygodes imbecilla* (F) (Lepidoptera: Noctuidae). *Ent. Gaz.* **29**: 107. 1978.
GOATER, B. *British Pyralid Moths.* Colchester, Essex. 1986.
GOSS, H. *(re. Callimorpha dominula Linn. at Tintern). Ent. Mon. Mag.* **23**: 219. 1887.
GOSS, H. *(re. C. dominula at Tintern). Ent. Mon. Mag.* **26**: 214. 1890.
GUSTARD S.H. *(re. Utetheisa pulchella Linn. in Monmouthshire). Entom.* **4**: 414. 1871.

HAGGETT, G.M. *Larvae of the British Lepidoptera not figured by Buckler. Proc. Brit. ent. nat. Hist. Soc.* **13**: 100-101, pl. 10 (*Eriopygodes imbecilla* Fab.). 1980.
HALL-SMITH, D.H. *"A Recorder's Log Book or Label List of British Butterflies and Moths".* Index. 59 pp. Leicestershire Museums Service (Publication No 41). 1983.
HEATH, J. and EMMET, A.M. (Eds.) et al., 1976-89. *The Moths and Butterflies of Great Britain and Ireland. (MBGBI)*
Vol. **1**. *Micropterigidae – Heliozelidae.* Heath (Ed.) et al. London 1976.
Vol. **2**. *Cossidae – Heliodinidae.* Heath & Emmet (Eds.) et al. Colchester 1985.
Vol. **7** (1). *Hesperiidae – Nymphalidae.* Emmet (Ed.) et al. Colchester 1989.
Vol **7** (2). *Lasiocampidae – Thyatiridae.* Emmet, Heath et al. Colchester 1991.
Vol. **9**. *Sphingidae – Noctuidae (Hadeninae).* Heath & Emmet (Eds.) et al. London 1979.
Vol. **10**. *Noctuidae (Cuculliinae – Hypeninae), Agaristidae.* Heath & Emmet (Eds.) et al. Colchester 1983.
HORTON, G.A.N. *Provisional list of the moths (macrolepidoptera) of the Usk and Wye Valleys (Monmouthshire). Rep. Mon. Nat. T.* No. **4**: 17-25. 1967.
HORTON, G.A.N. *Migrants 1969 (inc. Uresiphita polygonalis Hübn.). Ent. Rec. J. Var.* **82**: 32. 1970.
HORTON, G.A.N. *Monmouthshire macrolepidoptera (1970). Rep. Mon. Nat. T.* No. **5**: 7-10. 1971.

HORTON, G.A.N. *Monmouthshire Lepidoptera 1970. J. Mon. Nat. T.* Spring 1971: 5-7. 1971.

HORTON, G.A.N. *Monmouthshire macrolepidoptera (1970-71). J. Mon. Nat. T.* Spring 1972: 14-15. Lepidoptera of the Magor Reserve. Ibid. 16-21. 1972.

HORTON, G.A.N. *Egira conspicillaris L. (Lep: Noctuidae) in Monmouthshire. Ent. Rec. J. Var.* **85**: 203. 1973.

HORTON, G.A.N. & HEATH, J. *Eriopygodes imbecilla (Fabricius) (Lep., Noctuidae), A species new to Britain. Ent. Gaz.* **24**: 219 – 222 Pl.8. 1973.

HORTON, G.A.N. *Monmouthshire macrolepidoptera (1972-73). J. Mon. Nat. T.* Spring 1974: 13-14. 1974.

HORTON, G.A.N. *The Butterflies and Moths of Wentwood:* pp. 11. 1974. The Gwent Trust for Nature Conservation (Scientific Committee). Newport, Gwent. Nov. 1974.

HORTON, G.A.N. *Ptycholomoides aeriferanus H.-S. and Pammene aurantiana Staud. in Monmouthshire. Ent. Rec. J. Var.* **87**: 30-31, 1975

HORTON, G.A.N. *Autographa bractea D. & S. in Monmouthshire. Ent. Rec. J. Var.* **88**: 71. 1976.

HORTON, G.A.N. *The discovery of Eriopygodes imbecilla (Fabricius). (Lep.: Noctuidae) as a resident British species.* Ibid. **88**: 246-248. 1976.

HORTON, G.A.N. *Monmouthshire Lepidoptera in 1977. Ent. Rec. J. Var.* **90**: 136. 1978.

HORTON, G.A.N. *Euphydryas aurinia Rott. – Disappearance from Monmouthshire. Ent. Rec. J. Var.* **90**: 246-247. 1978.

HORTON, G.A.N. *Euchromius ocellea Haworth (Lep.: Crambinae) in Monmouthshire. Ent. Rec. J. Var.* **91**: 26. 1979.

HORTON, G.A.N. *Monmouthshire macrolepidoptera: Some Recent Records. Entomologist's Rec. J. Var.* **92**: 150-151. 1980.

HOWARTH, T.G. *South's British Butterflies.* London. 1973.

HUMPHREYS, R.B. *(Lepidoptera, a diary of sundry records mainly from Monmouthshire).* 1931-45.

INCE, C.E.M. *Lepidoptera in Monmouthshire. Entom.* **20**: 236-237. 1887.

JACOBS, S.N.A. *The British Oecophoridae* (Part 1) *and Allied Genera Proc. Trans. Br. Ent. Nat. Hist. Soc.* 1948.

JONES, A.H. *(Conistra rubiginea in Monmouthshire.) Ent. Mon. Mag.* **13**: 162. 1876.

KNIGHT, *(Sundry Monmouthshire Butterfly records). Entom.* **26**: 199. 1893.

LEAR, N.W. *Leucinodes vagans (Tutt) (Lepidoptera: Pyralidae):* location of the type. *Ent. Gaz.* **37**: 197. 1986.

MERE, R.M. *Eupithecia egeneria Herrich-Schäffer (Fletcher's Pug) (Lep., Geometridae) in the British Isles. Ent. Gaz.* **37**: 197. 1962.

MERE, R.M. *Wye Valley, Monmouthshire. Proc. S. Lond. ent. nat. Hist. Soc.,* 1965 (3), **88**. 1965.

MERE, R.M. *Nephopteryx albicilla H.-S. in Monmouthshire – a phycitid moth new to the British Isles. Ent. Gaz.* **16**: 13-14. 1965.

MEYRICK, E. *A Handbook of British Lepidoptera.* London. 1895.

MEYRICK, E. *A Revised Handbook of British Lepidoptera.* London. 1928.

MORRIS, F.O. *A Natural History of British Moths,* Vols. **1 – 4**. London. 1891.

NESBITT, A. *Notes on collecting – Monmouthshire. Ent. Rec. J. Var.* **3**: 132. 1892.

NESBITT, A. *Notes on collecting – Wye Valley (Monmouthshire). Ent. Rec. J. Var.* **3**: 158. 1892a.

NESBITT, A. *Notes on collecting – Wye Valley. Ent. Rec. J. Var.* **3**: 294. 1892b.

NESBITT, A. *Notes on collecting – Wye Valley. Ent. Rec. J. Var.* **4**: 154-155. 1893.

NEWMAN, E. *An illustrated natural history of British Butterflies and Moths.* London. 1871.

NEWTON, J. *(re. Salebriopsis albicilla H.-S. in the Wye Valley). Ent. Gaz.* **16**. 1965.

NEWTON, J. and MEREDITH G.H.J. *Macrolepidoptera in Gloucestershire.* reprinted from *Proc. Cotteswold Naturalists' Field Club* **39** (1982-3). (*Lasiocampa quercus* L. ssp. *callunae* Palm. in Monmouthshire p. 51.). 1984.

NEWTON, J. *(re. Salebriopsis albicilla H.-S. in the Wye Valley). The Gloucestershire Naturalist* **2**: 28. 1985.

PALMER, G.W. *List of butterflies taken or seen near and at Monmouth in 1889 and 1890. Entom.* **23**: 346-347. 1890.

PARRY, *(Endromis versicolora L. in Monmouthshire). Ent. Wkly Intell.* **2**: 43. 1839.

PATTEN, G.L. *Lepidoptera in 1889, Monmouthshire. Entom.* **23**: 69. 1890.

PIFFARD, B. *Captures near Tintern, Monmouthshire.* Ent. Wkly Intell. **6**: 131. 1859.

PONTYPOOL FREE PRESS, *(Death's-head Hawk-moth at Pontypool, quoting an old issue of 1865).* Aug 8 1975.

RAIT-SMITH, W. *The Butterflies of Abertillery, Monmouthshire. Ent. Rec. J. Var.* **18**: 308-311. 1906.

RAIT-SMITH, W. *The season of 1911 in the Abertillery district of Monmouthshire. Ent. Rec. J. Var.* **24**: 133-138, 162-168. 1912.

RAIT-SMITH, W. *The season of 1912 in the Abertillery dist. of Monmouthshire. Ent. Rec. J. Var.* **25**: 158-163, 173-179. 1913.

RAIT-SMITH, W. *Notes on collecting in 1914. Ent. Rec. J. Var.* **27**: 168-173. 1915.

RICHARDSON, A. *Supplement to Donovan's Catalogue of the Macrolepidoptera of Gloucestershire. Proceedings of the Cotteswold Naturalists' Field Club.* 1945.

ROBBINS, S.H. Moths found in a trap at Graig View, Cwmyoy, March and April, October and November 1965. *Rep. Mon. Nat. T.* **3**: 13-14. 1967.

SELLON, E. *Entom.* **5**: 164, (re. *Agrius convolvuli*); *Entom.* 5: 167, (re. *Hyles gallii Rott.*). 1870-71.

SKINNER, B. *The history of Euchromius ocellea (Haworth) (Lep: Pyralidae) in Britain. Ent. Rec. J. Var.* **94**: 139-140. 1982.

SKINNER, B. *Colour Identification Guide to Moths of the British Isles.* London. 1984.

SOUTH, R. *The Butterflies of the British Isles* (Edn. 3. Revised by H.M. Edelsten). London. 1941.

SOUTH, R. *The Moths of the British Isles* (Edn. 4) Vol. **1** (Edited and Revised by H.M. Edelsten and D.S. Fletcher); Vol **2** (Ed. and Rev. by Edelsten, Fletcher and R.J. Collins). London. 1961.

THORNEWILL, C.F. *(Polychrysia moneta Fabr. in Monmouth 1902.). Entom.* **37**: 214. 1904.

TUTT, J.W. *(re Leucinodes vagans Tutt). Ent. Rec. J. Var.* **1**: 203. 1890.

TUTT, J.W., *A Natural History of British Lepidoptera.* **2**: 520. *(re. Eriogaster lanestris Linn. at Abergavenny (Chapman)).* 1899-1906.

WITHERS, B.G. *My experiences with the macrolepidoptera, 1972.* (inc. *Carmelita odontosia* Esp. and *Perizoma taeniatum* Steph. in the Wye Valley). *Ent. Rec. J. Var.* **85**. 168-174, 219-224. 1972.

de WORMS, C.G.M. *Further recent additions to the British macrolepidoptera Ent. Gaz.* **29**: 17-39. (inc. *Eriopygodes imbecilla* Fabr. 19-20. Pl 2.). 1978.

National Grid References

Aberbeeg	SO 20.02	Coed-y-prior (Llantrisant)	ST 38.97
Abercarn	ST 21.95	Coity Mountain	SO 23.07
Aberffrwd	SO 35.09	Coldra (Newport)	ST 35.89
Abergavenny	SO 30.14, 30.13	Craig-y-dorth	SO 48.08
Abernant (Bulmore)	ST 37.91	Craig-y-iar (Llangeview)	SO 40.00
Abertillery	SO 21.04	Craig-y-Master (Llangwm)	ST 49.17
Allt-yr-yn (Newport)	ST 28.88	Crick	ST 48.90
Angidy Valley	SO 50.00, 51.00	Croes Llwyd (Raglan)	SO 39.07
		Croes-Robert (Cwmcarvan)	SO 48.06
Bedwellty	ST 16.98	Croesyceiliog	ST 30.95
Bettws Newydd	SO 36.05	Crosskeys	ST 22.91
Bica Common	ST 46.94	Crumlin	ST 21.98
Bigsweir	SO 53.05	Cwm, The (Shirenewton)	ST 45.93
Bishton	ST 39.87	Cwmcarn	ST 21.92
Black Cliff (Tintern)	ST 53.98	Cwmcarvan	SO 47.07
Black Vein (Crosskeys)	ST 22.91	Cwmcayo (Usk)	SO 37.02
Blackwood (Woodfieldside)	ST 18.97	Cwm Coed-y-cerrig	SO 29.21
Blaenavon Mountain	SO 27.07	Cwmfelinfach	ST 18.91
Blorenge Mountain	SO 26.11, 26.12	Cwm-mawr (Upper Llanover)	SO 28.09
	SO 27.11, 27.12	Cwm Merddog	SO 19.04
British	SO 25.03	Cwm Tyleri	SO 22.06, 22.07
Brockweir	SO 53.01	Cwmyoy	SO 30.22
Brockwells (Caerwent)	ST 47.89		
Brynawel (Sirhowy Valley)	ST 20.91	Darren Wood (Wentwood)	ST 40.94
Bulwark (Chepstow)	ST 53.92	Deri-fach	SO 27.17
		Devauden	ST 48.98
Cadira Beeches (Wentwood)	ST 42.94	Dewstow (Caldicot)	ST 46.88
Cae Cnap (Llangybi)	ST 36.97	Dingestow	SO 45.10
Caerleon	ST 33.91	Dingestow (Treowen)	SO 45.11
Caerwent	ST 46.90	Dinham	ST 47.91
Caldicot (Dewstow)	ST 46.88	Dixton Bank	SO 52.15
Caldicot (The Nedern)	ST 48.89		
Camp Wood (Gwehelog)	SO 37.03	Ebbw Vale	SO 17.09
Candwr Brook	ST 32.94		
Candwr Lane (Llanfrechfa)	ST 32.93	Fedw Fach (Trelleck)	SO 49.06
Carno Reservoir	SO 13.16	Fforest Coal Pit	SO 29.20
Castell-y-bwch	ST 27.92	Five Paths (Wentwood)	ST 43.95
Catbrook	SO 50.02	Fleur-de-lis	ST 15.96
Cefn Ila (Llanbadoc)	SO 36.00	Foresters' Oaks	ST 43.94
Chepstow (Bulwark)	ST 53.92		
Chepstow Park Wood	ST 49.97, 49.98	Gaer-fawr	ST 44.98
	and 50.98	Gaer Hill (Newport)	ST 28.86
Cicelyford	SO 50.03	Gilwern	SO 24.14
Cilfeigan Park	SO 34.00, 35.00	Glascoed	SO 33.01
Clydach	SO 21.12	Goetre (inc. Wern Fawr)	SO 32.05
Clytha	SO 36.09	Goldcliff	ST 37.82
Cockshoots Wood (Usk)	SO 38.01	Graig Syfyrddin	SO 40.20
Coed-y-Bwnydd	SO 36.06	Graig-y-Gaercoed (Usk)	SO 35.02
Coed-y-Fferm (Llangybi)	ST 36.98	Gray Hill	ST 43.93
Coed-y-paen	ST 33.98	Great Barnets Wood	ST 51.93
Coed-y-paen (The Forest)	ST 34.97	Gwehelog	SO 38.04

Gwehelog (Camp Wood)	SO 37.03	Llansoy	SO 44.02
Gwernesney	SO 41.01	Llantarnam (MEA records)	ST 31.92
		(other records)	ST 30.93
Hadnock Quarries	SO 54.15	Llantilio Crossenny	SO 39.14
Hael Woods	SO 52.07, 53.07	Llantrisant	ST 38.97, 39.97
Hendre Woods	SO 46.12, 46.13	d° (Coed-y-prior)	ST 38.97
	and 47.12, 47.13	d° (Nant-y-banw)	ST 40.97
Henllys	ST 26.93	d° (Ysguborwen)	ST 40.95
Highmoor Hill	ST 46.89	Llanvaches	ST 43.91
Hilston Park (Skenfrith)	SO 44.18	Llanvair Discoed	ST 44.92
		Llanvair (Llanishen)	SO 46.04
Itton (The Glyn)	ST 47.96	Llanvapley	SO 36.14
		Llanvetherine	SO 36.17
Jingle Street	SO 47.10	Llanwenarth	SO 27.14
		Llanwern	ST 36.88
Kemeys Commander	SO 34.04	Llwyn-y-celyn Bog	ST 47.94
Kilpale	ST 46.92	Lower Machen (Park Wood)	ST 24.87
Kingcoed	SO 42.05		
Knockall's Inclosure		Maerdy cutting	
(Upper Redbrook, Glos)	SO 54.10	(Llangeview)	SO 40.01
		Magor	ST 42.87
Lady Hill Wood (Usk)	SO 37.02	Magor Marsh	ST 42.86
Langstone	ST 37.89	Magor Pill (sea-wall)	ST 43.84
Lasgarn Wood	SO 27.04	Mathern	ST 52.90
Little Mill	SO 32.02	Mescoed Mawr	ST 27.90
Llanarth	SO 37.10	Monkswood	SO 34.02
Llanbadoc	SO 36.00, 37.00	Monmouth	SO 50.12
Llanbadoc (Cefn Ila)	SO 36.00	Mountain Air (Pontypool)	ST 27.98
Llancayo	SO 37.03	Mynyddislwyn	ST 19.94
Llanddewi Fach	ST 33.96		
Llandegfedd	ST 33.95	Nant Gwyddon	ST 23.95
Llandenny	SO 41.03	Nant-y-banw (Llantrisant)	ST 40.97
Llandenny Walks	SO 39.04	Nant-y-derry	SO 32.06
Llandogo	SO 52.03	Nash	ST 34.84
Llanfaenor	SO 43.16	Nedern, The (Caldicot)	ST 48.89
Llanfihangel Crucorney		Newbridge-on-Usk	ST 38.93
(Pen-y-clawdd Court)	SO 30.20	Newchurch West	ST 42.96
Llanfoist	SO 28.13	Newport:	
Llanfrechfa (Candwr Brook)	ST 32.94	(Allt-yr-yn)	ST 28.88
Llanfrechfa (Candwr Lane)	ST 32.93	(Coldra)	ST 35.89
Llangeview:		(Docks)	ST 31.86
(Allt-y-bela lane)	SO 40.00	(Gaer Hill)	ST 28.86
(Craig-y-iar)	SO 40.00	(St Julian's Wood)	ST 33.89
(Maerdy cutting)	SO 40.01	(Ynysyfro)	ST 28.89
(Flood Route)	ST 39.99		
Llangovan	SO 45.05	Oak Lane (Upper Llanover)	SO 29.09
Llangwm	ST 42.99	Ochrwyth	ST 23.89
Llangwm (Craig-y-Master)	ST 42.97	Old Furnace (Tintern)	SO 51.00
Llangwm-isaf	SO 42.01		
Llangybi	ST 36.96	Panta Arch,	
Llangybi (Cae Cnap)	ST 36.97	(Pont-y-saeson)	ST 50.99
Llangybi (Coed-y-Fferm)	ST 36.98	Pantygelli	SO 29.18
Llanhilleth	SO 22.00	Pant-yr-eos	ST 25.91
Llanishen	SO 47.03	Parc-Seymour	ST 41.91
Llanishen (Llanvair)	SO 46.04	Park Wood (Lower Machen)	ST 24.87
Llanock (= Llanerch) Wood	ST 21.98	Park Wood (Usk)	SO 38.02

Penallt	SO 51.01	Treowen (Dingestow)	SO 45.11
Penorth Mill	SO 47.04	Trewen (Caerwent)	ST 49.90
Penterry	ST 51.98, 51.99	Trinant	SO 20.00
Pentwyn	SO 21.00	Trostrey Church	SO 36.04
Pen-y-clawdd Woods		Trostrey Common	SO 37.04
(Dingestow)	SO 44.08	Trostrey Hill	SO 37.05
Pen-y-clawdd Court		Twmbarlwm Mountain	ST 24.92
(Llanfilhangel Crucorney)	SO 30.20	Twyn-y-Sheriff	SO 40.05
Pen-y-fan	SO 19.00		
Piercefield Park	ST 52.94, 52.95	Undy (Collister Pill)	ST 45.85
Plantations, The		Upper Llanover (Oak Lane)	SO 29.09
(Croesyceiliog)	ST 306.956	Upper Llanover (Cwm-mawr)	SO 28.09
Pontllanfraith	ST 16.96	Upper Redbrook (Glos)	SO 54.10
Pontnewydd	ST 29.96	Usk	SO 37.00, 37.01
Pontypool	SO 29.00		and SO 38.00, 38.01
Pont-y-saeson (Panta Arch)	ST 50.99	Usk Castle	SO 377.011
Porthlong Barn	ST 35.97	Usk (Cockshoots Wood)	SO 36.01
Portskewett	ST 49.88	Usk (Cwmcayo)	SO 37.02
Prescoed	ST 34.98, 34.99	Usk (Flagpole Hill)	SO 38.02
Prioress Mill (Rhadyr)	SO 36.02	Usk (Lady Hill Wood)	SO 37.02
		Usk (Park Wood)	SO 38.02
Raglan	SO 41.07	Usk, Plas Newydd garden	
Ravensnest Wood	ST 50.99	m.v.l. trap	SO 376.011
Redding's Inclosure	SO 53.13, 54.13, 54.14	Uskmouth	ST 34.82
Rhadyr (Usk)	SO 36.01, 36.02		
Rhiwderin	ST 25.87	Wentwood (Cadira Beeches	ST 42.94
Rhyd-y-maen	SO 42.02	d° (Darren Wood)	ST 40.94
Risca	ST 26.90	d° (Five Paths)	ST 43.95
Runston	ST 49.91	d° (Foresters' Oaks)	ST 43.94
		Wern Fawr (Goetre)	SO 32.05
Shirenewton	ST 47.94	Whitebrook	SO 53.06
Skenfrith	SO 45.20	Whitelye	SO 51.01
Slade Woods	ST 45.89, 46.89	Whitson	ST 37.84
Sor Brook (Llanddewi)	ST 32.97	Wolvesnewton	ST 44.09
St Brides Wentlooge	ST 29.82	Wonastow	SO 47.11
St Julian's Wood (Newport)	ST 33.89	Wyesham	SO 51.12
St Mary's Vale		Wyndcliff	ST 52.97
(Sugar Loaf)	SO 28.16		
St Pierre's Great Woods	ST 49.92, 50.92	Ynysddu	ST 17.92
Strawberry Wood (Stanton)	SO 31.21	Ysguborwen (Llantrisant)	ST 40.96
Sudbrook	ST 50.87	Ynysyfro (Newport)	ST 28.89
Sugar Loaf	SO 27.17, 27.18, 27.19		
	and SO 28.17, 28.18, 28.19		
Tal-y-coed	SO 41.15		
Tidenham (Glos)	ST 55.95		
Tintern	SO 52.00		
Tintern Cross			
(= Pont-y-saeson)	SO 50.00		
Tintern (Old Furnace)	SO 51.00		
Tredean Woods	ST 47.99		
Tredunnoc	ST 37.94		
Trefil	SO 11.13		
Trelleck Bog	SO 50.04		
Trelleck Common	SO 50.06, 51.06		
Trelleck (Fedw Fach)	SO 49.06		

Abbreviations of author's names

Barr.	Barrett	Mart.	Martini
Bent.	Bentinck	Metc.	Metcalfe
Bjerk.	Bjerkander	Mill.	Millier
Boisd.	Boisduval	Müll.	Müller
Borkh.	Borkhausen	Nic.	Nicelli
Bours.	Boursin	Nyl.	Nylander
Bradl.	Bradley	Obraz.	Obraztsov
Brem. & Grey	Bremer & Grey	Ol.	Olivier
Bru.	Bruand	Pack.	Packard
Butl.	Butler	Palm.	Palmer
Christ.	Christoph	Panz.	Panzer
Cl.	Clerck	Peters.	Petersen
Cock.	Cockayne	Pier. & Metc.	Pierce & Metcalfe
Curt.	Curtis	Rag.	Ragonot
D. & S.	Denis & Schiffermüller	Ratz.	Ratzeburg
De G.	De Geer	Retz.	Retzius
Don.	Donovan	Roem.	Roemer
Doubl.	Doubleday	Rott.	Rottenburg
Dougl.	Douglas	Scharf.	Scharfenburg
Dup.	Duponchel	Schaw.	Schawerda
Esp.	Esper	Schev.	Scheven
Fabr.	Fabricius	Schr.	Schrank
Fletch.	Fletcher	Schreb.	Schreber
Fol.	Fologne	Scop.	Scopoli
Forst.	Forster	Sheld.	Sheldon
Fröl.	Frölich	Sirc.	Sircom
Fuess.	Fuessly	Sodof.	Sodoffsky
F.v.R.	Fischer von Rösslerstamm	Sorh.	Sorhagen
Geoff.	Geoffroy	Stdgr.	Staudinger
Guen.	Guenée	Steph.	Stephens
H.-S.	Herrich-Schäffer	Stt.	Stainton
Haw.	Haworth	Tengst.	Tengström
Hb.	Hübner	Thunb.	Thunberg
Hein.	Heinemann	Treit.	Treitschke
Heyd.	Heyden	Trem.	Tremewan
Hcyl.	Heylaerts	Ver.	Verity
Hufn.	Hufnagel	Vill.	Villers
Humph. & West.	Humphreys & Westwood	Walk.	Walker
Joann.	Joannis	Wall.	Wallengren
Karsh. & Niel.	Karshalt & Nielsen	Wals.	Walsingham
Klim.	Klimesch	Warr.	Warren
Koll.	Koller	Werneb.	Werneburg
Lien. & Zell.	Lienig & Zeller	Westw.	Westwood
Linn.	Linnaeus	Zell.	Zeller
Mab.	Mabille	Zett.	Zetterstedt
		Zinck.	Zincken

Reference to Plates 23-27

Plate 23

BUTTERFLIES

1. White Letter Hairstreak *Strymonidia w-album Knoch* (under side)
2. Purple Hairstreak *Quercusia quercus* Linn. ♀ (upper side)
3. Purple Hairstreak *Q. quercus* Linn. ♀ (under side)
4. Small Copper *Lycaena phlaeas* Linn.
5. Common Blue *Polyommatus icarus* Rott. ♂
6. Brown argus *Aricia agestis* D.& S. (St Pierre's Great Woods 1968)
7. Holly Blue *Celastrina argiolus* Linn. ♀
8. Wood White *Leptidea sinapis* Linn. (Hendre Woods 1981)
9. Orange-tip *Anthocharis cardamines* Linn. ♂
10. Small Pearl-bordered Fritillary *Boloria selene* D.& S. ab. *flavus-pallidus* (Angidy Valley 1962)
11. Clouded Yellow *Colias croceus* Geoffr. ♀ (Slade Woods 1983)
12. High Brown Fritillary *Argynnis adippe* D.& S. (Angidy Valley 1956)
13. Dark Green Fritillary *Argynnis aglaja* Linn. ♀ (Hendre Woods 1978)
14. Silver-washed Fritillary *Argynnis paphia* Linn. ♂
15. Small Pearl-bordered Fritillary *Boloria selene* D.& S.
16. Marsh Fritillary *Eurodryas aurinia* Rott. (Angidy Valley 1958)
17. White Admiral *Ladoga camilla* Linn. (Angidy Valley 1956)
18. The Ringlet *Aphantopus hyperantus* Linn. (underside)
19. Marbled White *Melanargia galathea* Linn. (Usk 1987)

Plate 24

BUTTERFLIES

1. The Grayling *Hipparchia semele* Linn. ♂ (Wentwood)
2. Red Admiral *Vanessa atalanta* Linn.
3. The Gatekeeper *Pyronia tithonus* Linn. ♀

MOTHS

4. Privet Hawk-moth *Sphinx ligustri* Linn.
5. Cream-spot Tiger *Arctia villica britannica* Ob.
6. Water Ermine *Spilosoma urticae* Esp. (Magor Marsh)
7. Scarlet Tiger *Callimorpha dominula* Linn. (Magor Marsh)

8. Bedstraw Hawk-moth *Hyles gallii* Rott. (Usk 1973)
9. Black Arches *Lymantria monacha* Linn. ♀
10. Lobster Moth *Stauropus fagi* Linn ♂
11. Emperor Moth *Saturnia pavonia* Linn. ♀
12. Lappet *Gastropacha quercifolia* Linn. (Magor Marsh)

Plate 25

MOTHS

1. The Mocha *Cyclophora annulata* Schulze
2. Welsh Wave *Venusia cambrica* Curt. (Trelleck)
3. Pretty Pinion *Perizoma blandiata* D.& S.
4. Devon Carpet *Lampropteryx otregiata* Metc. (north Monmouthshire)
5. Blomer's Rivulet *Discoloxia blomeri* Curt.
6. Galium Carpet *Epirrhoe galiata* D.& S.
7. Cloaked Carpet *Euphyia biangulata* Haw.
8. Grey Mountain Carpet *Entephria caesiata* D.& S.
9. Beautiful Carpet *Mesoleuca albicillata* Linn.
10. Marbled Pug *Eupithecia irriguata* Hb.
11. Bilberry Pug *Chloroclystis debiliata* Hb.
12. Netted Pug *Eupithecia venosata* Fabr.
13. Great Oak Beauty *Boarmia roboraria* D.& S. ab. *infuscata* Stdgr.
14. Orange Moth *Angerona prunaria* Linn. ♂
15. Orange Moth *A. prunaria* Linn. f. *corylaria* Thunb. ♂
16. Peppered Moth *Biston betularia* Linn. ♂
17. Peppered Moth *B. betularia* Linn. f. *carbonaria* Jordan ♂
18. Argent & Sable *Rheumaptera hastata hastata* Linn.
19. Square Spot *Paradarisa consonaria* Hb.
20. Square Spot *P. consonaria* Hb. ab. *waiensis* Richardson (Wentwood)
21. White Satin Moth *Leucoma salicis* Linn. ♂ (Magor Marsh)
22. Small Autumnal Moth *Epirrita filigrammaria* H.-S. (Blorenge)
23. Scallop Shell *Rheumaptera undulata* Linn.
24. Drab Looper *Minoa murinata* Scop.
25. Alder Kitten *Furcula bicuspis* Borkh ♂ (Usk)
26. Small Eggar *Eriogaster lanestris* Linn. ♂ (Usk 1976)
27. Chocolate-tip *Clostera curtula* Linn. ♂ (Usk)
28. Lesser Swallow Prominent *Pheosia gnoma* Esp. ♂
29. Scarce Prominent *Odontosia carmelita* Esp. ♂ (Wye Valley) Page
30. Great Prominent *Peridea anceps* Goeze ♂ (Usk) Page

Plate 26

MOTHS

1. Light Feathered Rustic *Agrotis cinerea* D.& S. ♂ (north Gwent)
2. Light Feathered Rustic *A. cinerea* D.& S. ♀ (north Gwent)
3. Northern Rustic *Standfussiana lucernea* Linn. ♂
4. Barred Chestnut *Diarsia dahlii* Hb. ♀ (Trelleck Bog)
5. Feathered Ranunculus *Eumichtis lichenea lichenea* Hb. (Usk)
6. Dusky Sallow *Eremobia ochroleuca* D.& S.
7. Light Brocade *Lacanobia w-latinum* Hufn.
8. Beautiful Brocade *L. contigua* D.& S.
9. Silver Cloud *Egira conspicillaris* Linn. f. *melaleuca* View. ♂ (Usk)
10. The Silurian *Eriopygodes imbecilla* Fabr. ♂
11. The Silurian *E. imbecilla* Fabr. ♀
12. Glaucous Shears *Papestra biren* Goeze ♂
13. Striped Wainscot *Mythimna pudorina* D.& S. ♂ (Trelleck Bog)
14. Mere Wainscot *Photedes fluxa* Hb. (south Monmouthshire)
15. Obscure Wainscot *Mythimna obsoleta* Hb. ♂ (Magor)
16. Silver Hook *Eustrotia uncula* Cl.
17. White-speck *Mythimna unipuncta* Haw. (Usk, migrant)
18. Southern Wainscot *M. straminea* Treit. (Magor)
19. Large Wainscot *Rhizedra lutosa* Hb. (Magor)
20. The Anomalous *Stilbia anomala* Haw. (Trelleck Bog)
21. Lesser-spotted Pinion *Cosmia affinis* Linn.
22. Double Kidney *Ipimorpha retusa* Linn. (Usk)
23. Small Yellow Underwing *Panemeria tenebrata* Scop.
24. The Miller *Acronicta leporina* Linn f. *grisea* Cochrane
25. Alder Moth *A. alni* Linn.
26. The Coronet *Craniophora ligustri* D.& S.
27. Copper Underwing *Amphipyra pyramidea* Linn.
28. Marbled Green *Cryphia muralis muralis* Forst. (Usk)
29. Marbled Beauty *C. domestica* Hufn.
30. Dusky-lemon Sallow *Xanthia gilvago* D.& S. (Usk)
31. Marbled White Spot *Lithacodia pygarga* Hufn.
32. Red Sword-grass *Xylena vetusta* Hb. (Usk)
33. Gold Spangle *Autographa bractea* D.& S. (Usk)
34. Scarce Silver Y *Syngrapha interrogationis* Linn. (Cwm Tyleri)
35. Scarce Silver-lines *Bena prasinana* Linn.

Plate 27

MOTHS

1. Red-necked Footman *Atolmis rubricollis* Linn. (Wye Valley)
2. Rosy Footman *Miltochrista miniata* Forst. (Hael Wds)
3. Hoary Footman *Eilema caniola* Hb. (Lower Wye Valley)
4. Orange Footman *Eilema sororcula* Hufn. (Redding's Inclosure)
5. Peach Blossom *Thyatira batis* Linn. (Hael Wds)
6. Figure of Eighty *Tethea ocularis* Linn. (Usk)
7. Poplar Lutestring *Tethea or* D.& S. (Hael Wds)
8. Oak Lutestring *Cymatophorima diluta* D.& S. (Tintern)
9. Satin Lutestring *Tetheella fluctuosa* Hb. (Wentwood)
10. Barred Hook-tip *Drepana cultraria* Fabr. ♂ (Wyndcliff)
11. Oak Hook-tip *Drepana binaria* Hufn. ♂ (Usk)
12. Scarce Hook-tip *Sabra harpagula* Esp. (Wye Valley)
13. Great Brocade *Eurois occulta* Linn. (Usk 1982)
14. White-marked *Cerastis leucographa* D.& S. (Usk)
15. Chamomile Shark *Cucullia chamomillae* D.& S. (Usk)
16. Large Ranunculus *Polymixis flavicincta* D.& S. (Usk)
17. Grey Shoulder-knot *Lithophane ornitopus* Hufn. (Llansoy)
18. The Sprawler *Brachionycha sphinx* Hufn. (Usk)
19. Pale Pinion *Lithophane socia* Hufn. (Usk)
20. Blair's Shoulder-knot *Lithophane leautieri* Boisd. (Llansoy)
21. Tawny Pinion *Lithophane semibrunnea* Haw. (Usk)
22. Grey Chi *Antitype chi* Linn. (Usk)
23. Brindled Green *Dryobotodes eremita* Fabr. Llansoy
24. The Suspected *Parastichtis suspecta* Hb. (Trelleck Bog)
25. Bordered Sallow *Pyrrhia umbra* Hufn. (Magor sea-wall)
26. Beautiful Snout *Hypena crassalis* Fabr. ♂ (Trelleck)
27. Beautiful Snout *H. crassalis* Fabr. ♀ (Wentwood)
28. Buttoned Snout *Hypena rostralis* Linn. (Usk)
29. Cream-bordered Green Pea *Earias clorana* Linn. (Usk)

INDEX of ENGLISH NAMES

Alder Kitten, 18, **142**, *plate 25*
Alder Moth, 17, **194**, *plate 26*
Angle Shades, **199**
Annulet, **137**
Anomalous, 21, **212**
Antler Moth, **175**
Argent & Sable, **97**, *plate 25*
Ash Bud Moth, **247**
August Thorn, **122**
Autumn Green Carpet, **92**
Autumnal Moth, 22, **100**, *plate 25*
Autumnal Rustic, **163**

Barred Carpet, 18, **101**
Barred Chestnut, 19, **165**, *plate 26*
Barred Fruit-tree Tortrix, **265**
Barred Hook-tip, 17, **69**, *plate 27*
Barred Red, **137**
Barred Rivulet, **101**
Barred Sallow, **192**
Barred Straw, **91**
Barred Umber, **120**
Barred Yellow, **93**
Beaded Chestnut, **191**
Beautiful Brocade, 19, **172**, *plate 26*
Beautiful Carpet, **88**, *plate 25*
Beautiful China-mark, **298**
Beautiful Golden Y, **217**
Beautiful Hook-tip, **221**
Beautiful Snout, 17, **222**, *plate 27*
Beautiful Yellow Underwing, 21, **170**
Bedstraw Hawk-moth, 23, **140**, *plate 24*
Bee Moth, **305**
Berger's Clouded Yellow, **33**
Bilberry Pug, **111**, *plate 25*
Bilberry Tortrix, **267**
Birch Mocha, **77**
Bird-cherry Ermine, **245**
Black Arches, 18, **149**, *plate 24*
Black Rustic, **184**
Blackneck, **220**
Black-veined White, 20, 24, **34**
Blair's Shoulder-knot, **185**
Bleached Pug, **107**
Blomer's Rivulet, 18, **114**

Blood-vein, **78**
Blossom Underwing, **177**
Blotched Emerald, **76**
Blue-bordered Carpet, **93**
Bordered Beauty, **122**
Bordered Gothic, **171**
Bordered Pearl, **300**
Bordered Pug, **108**
Bordered Sallow, 21, **213**, *plate 27*
Bordered White, **134**
Brick, **189**
Bright-line Brown-eye, **173**
Brimstone, **34**
Brimstone Moth, **121**
Brindled Beauty, **127**
Brindled Green, **187**, *plate 27*
Brindled Pug, **109**
Brindled White-spot, 18, **133**
Broad-barred White, **174**
Broad-bordered Bee Hawk-moth, **139**
Broad-bordered Yellow Underwing, **162**
Broken-barred Carpet, **94**
Broom Moth, **174**
Broom-tip, **112**
Brown Argus, **41**, *plate 23*
Brown China-mark, 20, **297**
Brown House-moth, **253**
Brown Oak Tortrix, **266**
Brown Rustic, **198**
Brown Scallop, **98**
Brown Silver-line, **120**
Brown-line Bright-eye, **180**
Bud Moth, **286**
Buff Arches, **71**
Buff Ermine, **155**
Buff Footman, **152**
Buff-tip, **141**
Bulrush Wainscot, 20, **209**
Burnet Companion, **220**
Burnished Brass, **215**
Buttoned Snout, **223**, *plate 27*

Cabbage Moth, **171**
Camberwell Beauty, 22, **47**
Campion, **174**
Canary-shouldered Thorn, **123**
Centre-barred Sallow, **192**
Chalk Hill Blue, 23, **42**
Chamomile Shark, **182**, *plate 27*
Chequered Fruit-tree Tortrix, **264**

Cherry-bark Moth, **287**
Chestnut, **188**
Chevron, **90**
Chimney Sweeper, 19, **113**
Chinese Character, **70**
Chocolate-tip, **147**, *plate 25*
Cinnabar, **156**
Clay, **180**
Clay Triple-lines, **78**
Cloaked Carpet, **99**, *plate 25*
Cloaked Minor, **206**
Clouded Border, **118**
Clouded Brindle, **203**
Clouded Buff, 17, **154**
Clouded Drab, **179**
Clouded Magpie, **117**
Clouded Silver, **136**
Clouded Yellow, 21, 22, **33**, *plate 23*
Clouded-bordered Brindle, **202**
Cocksfoot Moth, **224**
Codling Moth, **290**
Comma, **48**, *plate 80*
Common Blue, 21, **41**, *plate 23*
Common Carpet, **86**
Common Emerald, **76**
Common Footman, **152**
Common Heath, **134**
Common Lutestring, **73**
Common Marbled Carpet, **92**
Common Pug, **107**
Common Quaker, **178**
Common Rustic, **206**
Common Swift, **62**
Common Wainscot, **181**
Common Wave, **135**
Common White Wave, **134**
Conformist, 24, 26, **185**
Confused, 21, **203**
Convolvulus Hawk-moth, 23, **137**
Copper Underwing, **197**, *plate 26*
Cork Moth, **235**
Coronet, 18, **196**, *plate 26*
Coxcomb Prominent, **145**
Cream-bordered Green Pea, 24, **214**, *plate 27*
Cream-spot Tiger, 18, **154**, *plate 24*
Crescent, 20, **209**
Crescent Striped, 20, **202**
Crimson Speckled, 23, **153**
Currant Clearwing, **65**
Currant Pug, **107**

Cyclamen Tortrix, **267**

Dark Arches, **201**
Dark Brocade, **187**
Dark Chestnut, **189**
Dark Dagger, **195**
Dark Fruit-tree Tortrix, **265**
Dark Green Fritillary, 26, **51**, *plate 23*
Dark Marbled Carpet, **92**
Dark Spectacle, **218**
Dark Sword-grass, 23, **159**
Dark Umber, **98**
Dark-barred Twin-spot Carpet, **83**
Death's-head Hawk-moth, 23, **138**, **316**
December Moth, **65**
Devon Carpet, 16, **89**, *plate 25*
Diamond-back Moth, 23, **248**
Dingy Footman, 20, **151**
Dingy Shears, **200**
Dingy Shell, **114**
Dingy Skipper, **32**
Dog's Tooth, 21, **173**
Dot Moth, **171**
Dotted Border, **129**
Dotted Chestnut, **189**
Dotted Clay, **167**
Double Dart, **163**
Double Kidney, **200**, *plate 26*
Double Line, 24, 25, **180**
Double Lobed, 20, **204**
Double Square-spot, **166**
Double-striped Pug, **112**
Drab Looper, **115**, *plate 25*
Drinker, **67**
Dun-bar, **201**
Dusky Brocade, **203**
Dusky Sallow, 21, **207**, *plate 26*
Dusky Thorn, **123**
Dusky-lemon Sallow, 17, **193**, *plate 26*
Dwarf Pug, **110**

Ear Moth, **208**
Early Grey, **186**
Early Moth, **136**
Early Thorn, **124**
Early Tooth-striped, **116**
Elephant Hawk-moth, **140**
Emperor Moth, 21, **68**, *plate 24*
Engrailed, **132**
European Map, **49**
Eyed Hawk-moth, **139**

False Mocha, 77
Fan-foot, 224
Feathered Gothic, 176
Feathered Ranunculus, 188, *plate 26*
Feathered Thorn, 126
Fern, 97
Festoon, 64
Figure of Eight, 147, *plate 27*
Figure of Eighty, 72, *plate 27*
Five-spot Burnet, 64
Flame, 160
Flame Carpet, 82
Flame Shoulder, 160
Flax Tortrix, 269
Fletcher's Pug, 18, 26, 105, 316
Flounced Chestnut, 190
Flounced Rustic, 208
Forester, 63
Four-dotted Footman, 17, 141
Fox Moth, 21, 67
Foxglove Pug, 104
Freyer's Pug, 106
Frosted Green, 74
Frosted Orange, 209

Galium Carpet, 21, 87, *plate 25*
Garden Carpet, 84
Garden Dart, 158
Garden Pebble, 299
Garden Tiger, 153
Gatekeeper, 21, 56, *plate 24*
Gem, 23, 82
Ghost Moth, 61
Glaucous Shears, 21, 173, *plate 26*
Goat Moth, 63
Gold Spangle, 218, *plate 26*
Gold Spot, 20, 216
Gold Swift, 61
Gold Triangle, 304
Golden Plusia, 216
Golden-rod Pug, 109
Gothic, 168
Grass Emerald, 75
Grass Rivulet, 102
Grayling, 21, 26, 55, *plate 24*
Great Oak Beauty, 18, 132, *plate 25 & 17*
Great Prominent, 18, 144, *plate 25*
Green Arches, 169
Green Carpet, 95
Green Hairstreak, 19, 37
Green Oak Tortrix, 271

Green Pug, 111
Green Silver-lines, 214
Green-brindled Crescent, 186
Green-veined White, 36, 155
Grey Arches, 171
Grey Birch, 133
Grey Chi, 188, *plate 27*
Grey Dagger, 195
Grey Mountain Carpet, 22, 87, *plate 25*
Grey Pine Carpet, 94
Grey Pug, 108
Grey Shoulder-knot, 184, *plate 27*
Grey Tortrix, 269
Grizzled Skipper, 17, 32

Haworth's Pug, 103
Hawthorn Moth, 247
Heart & Club, 158
Heart & Dart, 159
Heath Rustic, 168
Hebrew Character, 179
Hedge Brown, 21, 56, *plate 24*
Hedge Rustic, 176
Herald, 221
High Brown Fritillary, 26, 50, *plate 23*
Hoary Footman, 18, 152, *plate 27*
Holly Blue, 42, *plate 23*
Holly Tortrix, 282
Honeysuckle Moth, 247
Humming-bird Hawk-moth, 23, 140

Ingrailed Clay, 164
Iron Prominent, 143

Kentish Glory, 24, 69
Knot Grass, 196

Lackey, 66
Lappet, 25, 68, *plate 24*
Larch Pug, 110
Larch Tortrix, 282
Large Emerald, 75
Large Fruit-tree Tortrix, 265
Large Ranunculus, 187, *plate 27*
Large Skipper, 31
Large Tabby, 305
Large Tortoiseshell, 23, 46
Large Wainscot, 20, 210, *plate 26*
Large White, 35
Large Yellow Underwing, 161
Latticed Heath, 119

Lead Belle, **85**
Lead-coloured Drab, **178**
Lead-coloured Pug, 19, **103**
Least Black Arches, **157**
Least Carpet, 23, **79**
Least Yellow Underwing, **162**
Leopard Moth, **62**
Lesser Broad-bordered Yellow Underwing, **162**
Lesser Swallow Prominent, **144**, *plate 25*
Lesser Treble-bar, **113**
Lesser Wax Moth, **305**
Lesser Yellow Underwing, **161**
Lesser-spotted Pinion, **200**, *plate 26*
Light Arches, **202**
Light Brocade, 17, **172**, *plate 26*
Light Emerald, **136**
Light Feathered Rustic, 22, **158**, *plate 26*
Light Grey Tortrix, **270**
Light Knot Grass, 22, **195**
Light Orange Underwing, **74**
Lilac Beauty, 17, **122**
Lime Hawk-moth, **138**
Lime-speck Pug, **105**
Ling Pug, 21, **106**
Little Emerald, **76**
Little Thorn, 17, **120**
Lobster Moth, 17, **143**, *plate 24*
Lunar Hornet Moth, **65**
Lunar Marbled Brown, **147**
Lunar Thorn, **124**
Lunar Underwing, **192**
Lunar-spotted Pinion, **201**
Lychnis, **174**

Magpie, **117**
Maiden's Blush, **77**
Mallow, **87**
Maple Prominent, **146**
Map-winged Swift, 21, **62**
Marbled Beauty, **196**, *plate 26*
Marbled Brown, **146**
Marbled Green, **197**, *plate 26*
Marbled Minor, **204**
Marbled Orchard Tortrix, **275**
Marbled Pug, 18, **104**, *plate 25*
Marbled White, 17, 21, **55**, *plate 23 & 14*
Marbled White Spot, **213**, *plate 26*
March Moth, **75**
Marsh Fritillary, 18, 19, **52**, **53**, **64**, *plate 23*
Marsh Grass-veneer, 19, **292**

Marsh Oblique-barred, 19, **223**
Marsh Pug, **105**
May Highflyer, **96**
Mazarine Blue, 24, **42**
Meadow Brown, **57**
Meal Moth, **305**
Mediterranean Flour Moth, **309**
Mere Wainscot, **207**, *plate 26*
Merveille du Jour, **187**
Middle-barred Minor, **205**
Miller, **194**, *plate 26*
Minor Shoulder-knot, **183**
Mocha, 18, **77**, *plate 25*
Mother of Pearl, **303**
Mother Shipton, **220**
Mottled Beauty, **131**
Mottled Grey, **95**
Mottled Pug, **104**
Mottled Rustic, **212**
Mottled Umber, **129**
Mouse Moth, **198**
Mullein, **183**
Muslin Footman, 18, **150**
Muslin Moth, **155**

Narrow-bordered Bee Hawk-moth, **139**
Narrow-bordered Five-spot Burnet, **64**, *plate 19*
Narrow-winged Pug, **109**
Neglected Rustic, 19, **167**
Netted Pug, **105**, *plate 25*
Northern Drab, **177**
Northern Eggar, 21, **66**, **67**
Northern Rustic, 22, **161**, *plate 26*
Northern Spinach, 21, **90**
Northern Winter Moth, **100**
November Moth, **99**
Nut Bud Moth, **281**
Nutmeg, 21, **170**
Nut-tree Tussock, **215**

Oak Beauty, **127**
Oak Eggar, **66**
Oak Hook-tip, **69**, *plate 27*
Oak Lutestring, 18, **73**, *plate 27*
Oak Nycteoline, **214**
Oak-tree Pug, **110**
Oblique Carpet, 22, **82**
Obscure Wainscot, 20, **182**, *plate 26*
Ochreous Pug, **109**
Old Lady, **198**

Olive, **200**
Orange Footman, 18, **151**, *plate 27*
Orange Moth, **126**, *plate 25*
Orange Sallow, **192**
Orange Swift, **61**
Orange Underwing, **74**
Orange-tip, **36**, *plate 23*
Orchard Ermine, **245**

Painted Lady, 22, **45**
Pale Brindled Beauty, **127**
Pale Clouded Yellow, **33**
Pale Eggar, **65**
Pale Mottled Willow, **212**
Pale November Moth, **99**
Pale Pinion, **184**, *plate 27*
Pale Prominent, **146**
Pale-shouldered Brocade, **172**
Pale Tussock, **148**
Parsnip Moth, **255**
Pauper Pug, **105**
Peach Blossom, **71**, *plate 27*
Peacock, **47**
Peacock Moth, 18, **118**
Pearl-bordered Fritillary, 17, **50**, *plate 23*
Pearly Underwing, 23, **164**
Pebble Hook-tip, **70**
Pebble Prominent, **144**
Peppered Moth, **128**, *plate 25*
Phoenix, **90**
Pine Beauty, **176**
Pine Carpet, **93**
Pine Shoot Moth, **286**
Pinion-streaked Snout, **223**
Pink-barred Sallow, **193**
Plain Golden Y, **217**
Plain Pug, **108**
Plain Wave, **81**
Plum Tortrix, **275**
Poplar Grey, **194**
Poplar Hawk-moth, **139**
Poplar Kitten, **143**
Poplar Lutestring, **72**, *plate 27*
Powdered Quaker, 20, **178**
Pretty Chalk Carpet, **97**
Pretty Pinion, **102**, *plate 25*
Privet Hawk-moth, 17, **138**, *plate 24*
Purple Bar, **89**
Purple Clay, **165**
Purple Emperor, 16, 24, **44**
Purple Hairstreak, 17, **38**, *plate 23*

Purple Thorn, **124**
Puss Moth, **142**

Queen of Spain Fritillary, 22, **50**

Red Admiral, 22, **44**, *plate 24*
Red Chestnut, **169**
Red Sword-grass, **186**, *plate 26*
Red Twin-spot Carpet, **83**
Red Underwing, **219**
Red-barred Tortrix, **268**
Red-Green Carpet, **92**
Red-line Quaker, **190**
Red-necked Footman, 18, **150**, *plate 27*
Rhomboid Tortrix, **272**
Riband Wave, **81**
Rice Moth, **305**
Ringed China-mark, 20, **298**
Ringlet, **58**, *plate 23*
Rivulet, **101**
Rosy Footman, 18, **150**, *plate 27*
Rosy Minor, **206**
Rosy Rustic, **208**
Round-winged Muslin, 20, **150**
Royal Mantle, 23, **86**
Ruby Tiger, **156**
Ruddy Carpet, 18, **85**
Rufous Minor, **205**
Rush Veneer, 23, **303**
Rustic Shoulder-knot, **204**

Sallow, 17, **193**
Sallow Kitten, **143**
Saltern Ear, 21, **208**
Sandy Carpet, **102**
Satellite, **188**
Satin Beauty, 17, **130**
Satin Lutestring, 17, **72**, *plate 27*
Satin Wave, **80**
Scalloped Hazel, **125**
Scalloped Hook-tip, **69**
Scalloped Oak, **125**
Scallop Shell, **98**, *plate 25*
Scarce Burnished Brass, **216**
Scarce Footman, **152**
Scarce Hook-tip, 18, 26, **70**, *plate 27*
Scarce Prominent, 17, **146**, *plate 25*
Scarce Silver-lines, 18, **214**, *plate 26*
Scarce Silver Y, 22, **218**, *plate 26*
Scarce Tissue, **98**
Scarce Umber, **129**

Scarlet Tiger, 20, 26, **156**, *plate 24*
Scorched Carpet, **118**
Scorched Wing, **121**
September Thorn, **123**
Seraphim, **116**
Setaceous Hebrew Character, **166**
Shaded Broad-bar, **84**
Shark, **183**
Sharp-angled Carpet, **99**
Sharp-angled Peacock, **118**
Shears, **170**
Short-cloaked Moth, **157**
Shoulder Stripe, **88**
Shoulder-striped Wainscot, **182**
Shuttle-shaped Dart, **159**
Silurian, The, 22, **175**, *plate 11, 12, 13 & 26*
Silver Cloud, **176**, *plate 26*
Silver Hook, 19, **213**, *plate 26*
Silver Y, 22, 23, **216**
Silver-ground Carpet, **83**
Silver-striped Hawk-moth, **141**
Silver-studded Blue, **40**
Silver-washed Fritillary, 18, **51**, **52**, *plate 23*
Silvery Arches, **170**
Single-dotted Wave, **80**
Six-spot Burnet, **63**, *plate 19*
Six-striped Rustic, **167**
Skin Moth, **236**
Slender Brindle, 17, **204**
Slender Pug, **103**
Small Angle Shades, **199**
Small Argent & Sable, **86**
Small Autumnal Moth, 22, **100**, *plate 25*
Small Black Arches, **157**
Small Blood-vein, **78**
Small Brindled Beauty, **127**
Small China-mark, 20, **298**
Small Clouded Brindle, **203**
Small Copper, 21, **40**, *plate 23*
Small Dotted Buff, **207**
Small Dusty Wave, **80**
Small Eggar, **66**, *plate 25*
Small Elephant Hawk-moth, **141**
Small Emerald, **76**
Small Engrailed, **132**
Small Fan-foot, **224**
Small Fan-footed Wave, **79**
Small Heath, 21, **58**
Small Magpie, **300**
Small Mottled Willow, 23, **211**

Small Pearl-bordered Fritillary, 17, **49**, *plate 22, 23*
Small Phoenix, **91**
Small Purple-barred, **221**
Small Quaker, **177**
Small Rivulet, **101**
Small Rufous, 20, **210**
Small Scallop, **81**
Small Seraphim, 20, **116**
Small Skipper, **31**
Small Square-spot, **165**
Small Tortoiseshell, **46**
Small Wainscot, **207**
Small Waved Umber, 17, **96**
Small White, **35**
Small White Wave, **114**
Small Yellow Underwing, **212**, *plate 26*
Small Yellow Wave, **115**
Smoky Wainscot, **181**
Smoky Wave, 21, **79**
Snout, **222**
Southern Wainscot, 20, **181**, *plate 26*
Speckled Wood, **53**
Speckled Yellow, 17, **122**
Spectacle, **219**
Spinach, **91**
Spindle Ermine, **246**
Spotted Shoot Moth, **287**
Sprawler, 17, **183**, *plate 27*
Spring Usher, **128**
Spruce Carpet, **94**
Square Spot, **133**, *plate 25*
Square-spot Dart, 23, **158**
Square-spot Rustic, **168**
Stout Dart, 23, **163**
Straw Dot, **221**
Straw Underwing, **199**
Streak, **112**
Streamer, **88**
Striped Hawk-moth, 23, 25, **140**
Striped Twin-spot Carpet, 21, **89**
Striped Wainscot, 19, **181**, *plate 26*
Suspected, 19, **191**, *plate 27*
Svensson's Copper Underwing, **197**
Swallow Prominent, **145**
Swallow-tailed Moth, **125**
Sword-grass, **186**
Sycamore, **194**

Tawny-barred Angle, **119**
Tawny Marbled Minor, **205**

Tawny Pinion, **184**, *plate 27*
Tawny Speckled Pug, **108**
Tawny-barred Angle, **119**
Thistle Ermine, **308**
Thyme Pug, 22, **109**, *plate 17*
Timothy Tortrix, **267**
Tissue, **98**
Toadflax Pug, **104**
Treble Brown Spot, **80**
Treble Lines, **210**
Treble-bar, **113**
Triple-spotted Clay, **166**
True Lover's Knot, 21, **163**
Turnip Moth, **158**
Twenty-plume Moth, **291**
Twin-spot Carpet, 21, **103**
Twin-spotted Quaker, **179**
Twin-spotted Wainscot, 20, **209**
Uncertain, The, **211**

Vapourer, **148**
Variegated Golden Tortrix, **266**
Vestal, 23, **81**
Vine's Rustic, **211**
The V-Moth, **119**
The V-Pug, **111**

Wall, **54**
Water Carpet, **88**
Water Ermine, 20, **155**, *plate 24*
Water Veneer, 20, **298**
Waved Black, **222**
Waved Carpet, 18, **115**
Waved Umber, **130**
Wax Moth, **305**
Welsh Wave, 19, **114**, *plate 25*
White Admiral, 18, **43**, *plate 7, 21, 23*
White Ermine, **154**
White-letter Hairstreak, 17, **38**
White Plume Moth, **310**
White Satin Moth, **149**, *plate 25*
White-banded Carpet, 23, **97**
White-line Snout, **223**
White-marked, 17, **169**, *plate 27*
White-pinion Spotted, **135**
White-Shouldered House-moth, **253**
White-speck, 23, **182**, *plate 26*
White-spotted Pug, **107**
Willow Beauty, **130**
Willow Tortrix, **281**
Winter Moth, **100**

Wood Tiger, 21, **153**, *plate 14*
Wood White, 18, **33**, *plate 23*
Wormwood Pug, **106**

Yellow-barred Brindle, **116**
Yellow Horned, **74**
Yellow-line Quaker, **190**
Yellow Shell, **87**
Yellow-tail, **149**

INDEX of SCIENTIFIC NAMES

(Synonyms are given in italics. In the case of homonyms the generic name follows the specific name.)

abbreviana, **281**
abbreviata, **109**
abietella, **307**
Abraxas, **117**
Abrostola, **218, 219**
abruptaria, **130**
absinthiata, **106**
Acasis, **116**
Acentria, 20, **298**
aceris, Acronicta, **194**
achatana, **278**
Acherontia, **138**
Achlya, **74**
Achroia, **305**
Acleris, **271, 272, 273**
Acompsia, **260**
Acrobasis, **307**
Acrocercops, **239**
Acrolepia, **248**
ACROLEPIINAE, **248**
Acronicta, 17, 22, **194, 195, 196**, *plate 26*
ACRONICTINAE, **194**
acuminatella, **259**
Adaina, **310**
Adela, **233, 234**
ADELINAE, **233**
adippe, 50, *plate 23*
Adscita, **63**
adscitella, **253**
adusta, **187**
adustata, **118**
advenaria, 17, **120**
advenella, **308**
aegeria, **53**
aeneella, **230**
aeratana, **291**
aeriferanus, **267**, 315
aescularia, **75**
aestivaria, **76**
Aethalura, **133**
Aethes, **263**
aethiops, Xenolechia, **258**

affinis, Cosmia, **200**, *plate 26*
affinitata, **101**
Agapeta, **263**
agathina, **168**
agestis, **41**, *plate 23*
Aglais, **46**
aglaja, **51**, *plate 23*
Aglossa, **305**
Agonopterix, **256, 257**
Agriopis, **128**
Agriphila, **293, 294**
Agrius, 23, **137**, 317
Agrochola, **189, 190, 191**
Agrotis, 22, 23, **158, 159**, *plate 26*
Alabonia, **254**
albedinella, **237**
albersana, 18, **287**
albicilla, 18, 26, **88, 306**, 316, *plate 25*
albicillata, **88**, *plate 25*
albicosta, **251**
albifasciella, **226**
albifrontella, **252**
albipunctata, **77**
albipunctella, **235**
albistria, **245**
albulata, Asthena, **114**
albulata, Perizoma, **102**
alchemillata, **101**
alchimiella, **238**
Alcis, **131**
Aleimma, **270**
alismana, 20, **262**
Allophyes, **186**
alni, 17, **194**, *plate 26*
alniaria, **123**
alnifoliella, **242**
alpinana, **290**
alsines, **211**
Alsophila, **75**
alstroemeriana, **256**
alternaria, **118**
alternata, **86**
alternella, **270**
alticolella, **251**
Alucita, **291**
ALUCITIDAE, **291**
ambigua, **211, 296**
ambigualis, **296**
Amblyptilia, **309**
Amphipoea, 21, **208**
Amphipyra, 197, **198**, *plate 26*

AMPHIPYRINAE, **197**
Anacampsis, **260**
Anania, 19, **301**
Anaplectoides, **169**
Anarta, 21, **170**, *plate 17*
anceps, Peridea, 18, **144**, *plate 25*
Ancylis, 19, 22, **278**, **279**
angelicella, **257**
Angerona, **126**, *plate 25*
anglicella, **239**
angulifasciella, **226**
angustana, **264**
angustea, **297**
angusticollella, **232**
angustoriana, **268**
annulata, 18, **77**, *plate 25*
anomala, 21, **212**, *plate 26*
anomalella, **230**
Anthocharis, **36**, *plate 23*
Anthophila, **243**
Anticlea, **88**
antiopa, **47**
antiqua, **148**
Antispila, **234**
Antitype, **188**, *plate 27*
Apamea, 17, 20, 21, **201**, **202**, **203**, **204**
Apatura, 24, **44**
Apeira, 17, **122**
Aphantopus, **58**, *plate 23*
Aphelia, **267**
Aphomia, **305**
apicipunctella, **252**
Aplocera, **113**
Apocheima, **127**
Apoda, **64**
Aporia, 20, 24, **34**
Aporophyla, **184**
Apotomis, **276**, **277**
applana, **256**
aprilina, **187**
Araschnia, **49**
arcella, Nemapogon, **235**
Archanara, 20, **209**, **210**
ARCHIEARINAE, **74**
Archiearis, **74**
Archips, **265**, **266**
Arctia, 18, **153**, **154**, *plate 24*
ARCTIIDAE, 150
arcuatella, **226**
ardeaepennella, **250**
arenella, **256**

areola, **186**
argentella, Elachista, **253**
argentipedella, **226**
argiolus, **42**, *plate 23*
argus, **40**, **41**
Argynnis, **50**, **51**, *plate 23*
argyrana, **288**
Argyresthia, **244**, **245**
ARGYRESTHIINAE, **244**
argyropeza, **226**
Aricia, **41**, *plate 23*
aruncella, **225**
arundinata, **296**
aspidiscana, **285**
Aspilapteryx, **238**
assimilata, **107**
Asthena, **114**
atalanta, **44**, *plate 24*
Atethmia, **192**
Atolmis, 18, **150**, *plate 27*
atomaria, **134**
atrata, 19, **113**
atrella, **257**
atricapitana, **264**
atricapitella, **229**
atricollis, **226**
atricomella, **252**
atropos, 23, **138**
atropunctana, **275**
aucupariae, **230**
augur, **163**
aurago, **192**
aurana, **290**
aurantiana, **288**, 315
aurantiaria, **129**
aurata, **299**
aureatella, **225**
aurella, **227**
aurinia, **52**, 315
auroguttella, **238**
Autographa, 23, **216**, **217**, **218**, 315, *plate 26*
autumnata, **100**
avellana, **64**
avellanella, **255**
aversata, **81**
Axylia, **160**

Bactra, 21, **227**, **278**
badiana, **279**
badiata, **88**

334

badiella, **255**
badiipennella, **250**
baja, **167**
bajularia, **76**
basistrigalis, **296**
Batia, **253**
batis, **71**, *plate 27*
Batrachedra, **261**
BATRACHEDRINAE, **261**
Bedellia, **237**
BEDELLIINAE, **237**
bembeciformis, **65**
Bena, 18, **214**, *plate 26*
berbera, **197**
bergmanniana, **271**
betulae, Heliozela, **234**
betulae, Metriostola, **306**
betulae, Parornix, **239**
betulae, Thecla, 24, **38**
betularia, **128**, *plate 25*
betuletana, **276**
betulicola, **231**
biangulata, **99**, *plate 25*
bicolorata, **174**
bicruris, **174**
bicuspis, 18, **142**, *plate 25*
bidentata, **125**
bifasciana, Olethreutes, **274**
bifasciana, Spatalistis, **271**
bifaciata, **101**
bifida, **143**
bilineata, **87**
bilunana, **280**
bimaculata, Lomographa, **135**
binaevella, **308**
binaria, **69**, *plate 27*
bipunctidactyla, **310**
biren, 21, **173**, *plate 26*
Biselachista, **253**
biselata, **79**
Biston, **127**, **128**, **plate 25**
bistortata, **132**
bistriga, **306**
bistrigella, **232**
blancardella, **240**
blanda, **211**
blandella, Brachmia, **260**
blandiata, **102**, *plate 25*
BLASTOBASIDAE, 261
Blastobasis, **261**
BLASTODACNINAE, **262**

blattariella, **260**
blomeri, 18, **114**, *plate 25*
Boarmia, 18, **132**, *plate 25*
Boloria, **49**, **50**, *plate 23*
boleti, **235**
Brachmia, **260**
Brachionycha, **183**
Brachylomia, **183**
bractea, **218**, **315**, *plate 26*
bractella, 18, 19, **254**
brassicae, Mamestra, **171**
brassicae, Pieris, **35**
brockeella, **244**
brongniardella, **239**
brumata, **100**
brunnea, **165**
brunnichana, **281**, **282**
Bryotropha, **259**
BUCCULATRIGINAE, **237**
Bucculatrix, **237**
bucephala, **141**
buoliana, **286**
Bupalus, **134**

Cabera, **134**, **135**
caerulcocephala, **147**
caesiata, 22, **87**, *plate 25*
caesiella, **247**
caespititiella, **251**
cagnagella, **246**
caja, **153**
c-album, **48**
caledoniana, **272**
Callimorpha, 20, 26, **156**, 314, *plate 24*
Callistege, **220**
Callophrys, **37**
Caloptilia, **238**
calthella, **225**
Calybites, **238**
cambrica, 19, **114**, *plate 25*
camilla, **43**, *plate 23*
Campaea, **136**
campoliliana, **285**
Camptogramma, **87**
cana, **286**
caniola, 18, **152**, *plate 27*
capreana, **276**
capucina, **145**, **186**, **187**
Caradrina, **212**
carbonaria, **128**, *plate 25*
Carcina, **254**

cardamines, **36**, *plate 23*
cardui, **45**
carmelita, 17, **146**, **317**, *plate 25*
carpinata, **116**
Caryocolum, **259**
casta, **235**
castanea, 19, **167**
Cataclysta, 20
Catarhoe, 18, 23, **85**, **86**
Catocala, **219**
CATOCALINAE, **219**
Catoptria, **295**
cavella, **241**
Celaena, 20, **209**
Celastrina, **42**, *plate 23*
celerio, **141**
Celypha, 18, **274**
cembrella, **295**
CEMIOSTOMINAE, **236**
centaureata, **105**
centrago, **192**
cephalonica, **305**
Cepphis, 17, **120**
Ceramica, **174**
Cerapteryx, **175**
cerasana, **265**
cerasicolella, **241**, **250**
cerasivorella, **250**
Cerastis, 17, **169**, *plate 27*
Cerura, **142**
cerusella, **252**
cervinalis, **98**
cespitalis, **300**
cespitis, **176**
chaerophyllella, **249**
chamomillae, **182**, *plate 27*
Charanyca, **210**
chenopodiata, **84**
Chesias, **112**
chi, **188**, *plate 27*
Chilo, **292**
CHIMABACHINAE, **255**
CHLOEPHORINAE, **214**
Chloroclysta, **92**
Chloroclystis, **111**, *plate 25*
chlorosata, **120**
choragella, **235**
CHOREUTIDAE, **243**
Choristoneura, **266**
christyi, **99**
chrysitis, **215**

chryson, **216**
chrysoprasaria, **76**
Chrysoteuchia, **292**
Cidaria, **93**
ciliella, **256**
Cilix, **70**
cinerea, 22, **158**, *plate 26*
cinerella, **260**
cingulata, Pyrausta, 22, **300**, *plate 17*
cinnamomeana, **265**
circellaris, **189**
citrago, **192**
citrata, **92**
clathrata, **119**
clavaria, **87**
clavipalpis, **212**
clavis, **158**
clematella, **235**
Clepsis, **267**
clerkella, **237**
cloacella, **235**
clorana, 24, **214**, *plate 27*
Clostera, **147**, *plate 25*
Cnephasia, **269**, **270**
cnicana, **263**
c-nigrum, **166**
COCHYLIDAE, 313
Cochylidia, **264**
Cochylis, **263**, **264**
Coenobia, 20, **210**
Coenonympha, **58**
Coleophora, **249**, **250**, **251**
COLEOPHORIDAE, **249**
COLIADINAE, **33**
Colias, **33**, *plate 23*
Colocasia, **215**
Colostygia, **95**
Colotois, **126**
combinella, **246**
comes, **161**
Comibaena, **76**
comma, Mythimna, **182**
complana, **152**
compositella, **289**
confusalis, **157**
confusella, **231**
conicolana, **290**
conigera, **180**
Conistra, **188**, **189**, 315
conjugella, **245**
consimilana, **267**

consociella, **307**
consonaria, **133**, *plate 25*
conspicillaris, **176**, 313, 315, *plate 26*
contigua, 19, **172**, *plate 26*
continuella, **228**
convolvuli, 23, **137**, **317**
conwagana, **269**
Corcyra, **305**
coridon, 23, **42**
coronata, **301**
corylana, **264**
corylata, **94**
coryli, Colocasia, **215**
coryli, Phyllonorycter, **241**
corylifoliella, **241**
Cosmia, **200**, **201**, *plate 26*
cosmophorana, **290**
COSMOPTERIGIDAE, **262**
Cosmorhoe, **89**
COSSIDAE, **62**, **314**
COSSINAE, **63**
Cossus, **63**, **313**
costaestrigalis, **223**
costalis, **304**
costella, **259**
costipunctana, **285**
costosa, **257**
CRAMBINAE, **292**, **315**
Crambus, 19, **292**, **293**
Craniophora, 18, **196**, *plate 26*
crassalis, 17, **222**, *plate 27*
crataegana, **266**
crataegella, Dipleurina, **296**
crataegella, Scythropia, **247**
crataegella, Stigmella, **230**
crataegi, Aporia, 20, 24, **34**
crataegi, Bucculatrix, **237**
crataegi, Trichiura, **65**
crenata, Apamea, **202**
crepuscularia, **132**
cribrella, **308**
cristana, **273**
Crocallis, **125**
croceus, **33**, *plate 23*
croesella, **234**
Croesia, **271**
cruciana, **281**
cruda, **177**
Cryphia, **196**, **197**, *plate 26*
Cryptoblabes, **306**
cuculata, 23, **86**

cucullatella, **157**
Cucullia, **182**, **183**, *plate 27*
cucullina, **146**
CUCULLIINAE, **182**
culmella, Agriphila, **294**
culmella, Chrysoteuchia, **292**
cultraria, 17, **69**, *plate 27*
currucipennella, **250**
curtula, **147**, *plate 25*
Cyaniris, 24, **42**
Cybosia, 17, **151**
Cyclophora, 18, **77**, **78**, *plate 25*
Cydia, **289**, **290**
Cymatophorima, 18, **73**, *plate 27*
cynosbatella, **283**
Cynthia, **45**
cytisella, **257**

dahlii, 19, **165**, *plate 26*
Dasychira, **148**
daucella, **255**
dealbana, **283**
debiliata, **111**, *plate 25*
decimalis, **176**
defoliaria, **129**
degeerella, **233**
Deilephila, **140**, **141**
Deileptenia, 17, **130**
demarniana, **280**
dentaria, **124**
dentella, Ypsolopha, **247**
deplana, **152**
Depressaria, **255**, **256**
DEPRESSARIINAE, **255**
derivata, **88**
designata, **82**
devoniella, **239**
Diachrysia, **215**, **216**
Diacrisia, 17, **154**
Diaphora, **155**
Diarsia, 19, **164**, **165**, *plate 26*
Dicestra, 21, **170**
Dichomeris, **260**
Dichonia, **187**
Dichrorampha, **290**, **291**
didymata, 21, **96**, **103**
diffinis, Teleiopsis, **259**
Diloba, **147**
diluta, 18, **73**, *plate 27*
dilutata, **99**
dimidiata, **80**

diniana, **282**
Dioryctria, **307**
Diplodoma, **234**
Discoloxia, 18, **114**, *plate 25*
DISMORPHIINAE, **33**
dissoluta, 20, **210**
distentella, **242**
distinctaria, 22, **109**
distinguenda, **231**
ditrapezium, **166**
Ditula, **268**
Diurnea, **255**
dodecella, **258**
dodonaea, Drymonia, **146**
dodonaea, Tischeria, **232**
dodoneata, **110**
dolabraria, **121**
domestica, Bryotropha, **259**
domestica, Cryphia, **196**, *plate 26*
dominula, 20, 26, **156**, **314**, *plate 24*
Donacaula, **295**
Drepana, 17, **69**, **70**, *plate 27*
DREPANIDAE, **69**
dromedarius, **143**
Drymonia, **146**, **147**
Dryobotodes, **187**, *plate 27*
dubitana, **264**
dubitata, **98**
dulcella, **227**
duplaris, **73**

Eana, **270**
Earias, 24, **214**, *plate 27*
Ebulea, **301**
Ecliptopera, **91**
Ectoedemia, **226**, **227**
Ectropis, 18, **132**
efformata, **113**
egeneria, 26, **105**, 316
Egira, **176**, 313, 315, plate 26
Eidophasia, **248**
Eilema, 18, 20, **151**, **152**, *plate 27*
ekebladella, **231**
Elachista, **252**, **253**
ELACHISTIDAE, **252**
Electrophaes, **94**
Eligmodonta, **144**
elinguaria, **125**
elongella, Caloptilia, **238**
elpenor, **140**
elutalis, **302**

emargana, **273**
emarginata, **81**
Ematurga, **134**
Emmelina, **311**
Enargia, **200**
Enarmonia, **287**
Endothenia, **277**
Endotricha, **305**
ENDROMIDAE, **69**
Endromis, 24, **69**, **316**
Endrosis, **253**
ENNOMINAE, **117**
Ennomos, **122**, **123**
Entephria, 22, **87**, *plate 25*
Epagoge, **268**
Epermenia, **249**
EPERMENIIDAE, **248**
Ephestia, **309**
Epiblema, **283**, **284**, **285**
epilobiella, **261**
epilobiella, **262**
Epinotia, **279**, **280**, **281**, **282**
Epione, **122**
Epirrhoe, 21, **86**, **87**, *plate 25*
Epirrita, 22, **99**, **100**, *plate 25*
epomidion, **203**
Erannis, **129**
eremita, **187**, *plate 27*
Eremobia, 21, **207**, *plate 26*
ericetana, **277**
ericetella, **259**
Eriocrania, **225**
ERIOCRANIIDAE, **225**
Eriogaster, **66**, 317, *plate 25*
Eriopygodes, 21, 22, **175**, 313, 314, 315, 317, *plate 26*
erosaria, **123**
erxlebella, **246**
Erynnis, **32**
Esperia, **253**
Ethmia, **257**
ETHMIIDAE, **257**
Euchoeca, **114**
Euclidia, **220**
Eucosma, **285**
Eucosmomorpha, 18, **287**
Eudemis, **278**
Eudonia, 22, **297**
Eulamprotes, **257**
Eulia, **269**
Eulithis, 21, **90**, **91**, **96**

Eumichtis, **188**, *plate 26*
euphrosyne, **50**
Eupithecia, 18, 19, 21, 22, 26, **103-110**, 316, *plate 25*
Euplexia, **199**
Eupoecilia, **264**
Euproctis, **149**
Eupsilia, **188**
Eurodryas, **52**, *plate 23*
Eurois, 24, **168**, *plate 27*
Eurrhypara, **300**
Eustrotia, 19, **213**, *plate 26*
Euxoa, 23, **158**
Euzophera, **308**
EVERGESTINAE, **299**
Evergestis, **299**
evonymella, **245**
exanthemata, **135**
exclamationis, **159**
exigua, 23, **211**
exiguata, **104**
Exoteleia, **258**
expallidata, **107**
exsoleta, **186**
extersaria, 18, **133**
Eyphyia, **99**

fabriciana, **243**
fagana, **214**
fagella, **255**
fagi, Stauropus, 17, **143**, *plate 24*
fagiglandana, **289**
fagivora, **239**
Falcaria, **69**
falcataria, **70**
falsella, **295**
Falseuncaria, **264**
farfarae, **285**
farinalis, **305**
fasciana, **288**
fasciaria, **137**
fasciuncula, **205**
ferrago, **180**
ferrugalis, 23, **303**
ferrugana, **272**
ferrugata, **83**
ferruginea, **198**
festaliella, **249**
festucae, 20, **216**
filigrammaria, 22, **100**, *plate 25*
filipendula, Zygaena, **63**, **64**

fimbriata, **162**
finitimella, **239**
firmata, **93**
flammea, Panolis, **176**
flammealis, **305**
flammeolaria, **115**
flavago, **209**
flavicincta, **187**, *plate 27*
flavicornis, **74**
flavidorsana, **291**
flavifrontella, **254**
flavipennella, **249**
flavofasciata, **102**
flexula, **221**
floslactata, **78**
fluctuata, **84**
fluctuosa, 17, **72**, *plate 27*
fluxa, **207**, *plate 26*
foenella, **284**
Fomoria, **227**
fonterella
forficalis, **299**
forficella, 20, **295**
formosana, **287**
formosanus, **267**
forsskaleana, **271**
forsterana, **268**
fraxinella, **247**
fuciformis, **139**
fucosa, 21, **208**
fugitivella, **258**
fuliginaria, **222**
fuliginosa, **156**
fulvalis, 23, **302**
fulvata, **93**
fulvescens, **202**
fulviguttella, **248**
fulvimitrella, **236**
funebris, 19, **301**
furcata, 21, **96**
furcifera, 24, 26, **185**, 313
Furcula, 18, **142**, *plate 25*
furcula, **142**
furfurana, **277**
furuncula, **206**
furva, 21, **203**
fusca, Pyla, **306**
fuscalis, **302**
fuscantaria, **123**
fuscocuprella, **250**
fusconebulosa, 21, **62**

339

fuscoviridella, **244**

galathea, **55**, *plate 23*
galiata, 21, **87**, *plate 25*
Galleria, **305**
gallii, 23, **140**, 317, *plate 24*
gamma, 23, **216**
Gastropacha, 25, **68**, *plate 24*
gei, **227**
Gelechia, **259**
GELECHIIDAE, 257
GELECHIINAE, **257**
geminana, 19, **278**
geminipuncta, 20, **209**
geniculea, **294**
geniculella, **243**
gentianaeana, **277**
geoffrella, **254**
Geometra, **75**
GEOMETRIDAE, 74, 316
GEOMETRINAE, **75**
germmana, **288**
gerningana, 22, **268**
gerronella, **260**
gibbosella, **260**
gilvago, 17, **193**, *plate 26*
gilvicomana, **262**
glareosa, **163**
glaucata, **70**
glaucicolella, **251**
glaucinalis, **304**
glutinosae, **231**
glyphica, **220**
GLYPHIPTERIGIDAE, 244
Glyphipterix, **244**
gnoma, **144**, *plate 25*
gnophos, **137**
goedartella, **244**
Gonepteryx, **34**
gonodactyla, **309**
goossensiata, **106**
Gortyna, **209**
gothica, **179**
gracilis, 20, **178**
GRACILLARIIDAE, 238
GRACILLARIINAE, **238**
graminis, **175**
Graphiphora, **163**
griseata, Timandra, **78**
grisella, **305**
griseola, 20, **151**

grossulariata, **117**
grotiana, **268**
gryphipennnella, **249**
gueneeana, **291**
Gymnoscelis, **112**
Gypsonoma, **283**

Habrosyne, **71**
Hada, **170**
Hadena, **174**
HADENINAE, **170**, **314**
halterata, **116**
hamana, **263**
hamella, **293**
hammoniella, **234**
harpagula, 18, **26**, 70, *plate 27*
harrisella, **240**
hastata, **97**, *plate 25*
hastiana, **273**
haworthi, **225**
haworthiata, **103**
hebenstreitella, **266**
Hecatera, **174**
hecta, **61**
Hedya, 20, **275**
heegeriella, **240**
Heliophobus, **171**
Heliozela, **234**
HELIOZELIDAE, 234, 314
hellerella, **262**
helvola, **190**
hemargyrella, **229**
Hemaris, **139**
Hemistola, **76**
Hemithea, **76**
heparana, **265**
hepariella, **246**
hepatica, **170**
HEPIALIDAE, 61
Hepialus, 21, **61**, **62**
heracliana, Agonopterix, **256**
heracliana, Depressaria, **255**
heringi, **227**
herminata, **234**
Herminia, **224**
HESPERIIDAE, 31, 314
HESPERIINAE, **31**
hexadactyla, **291**
Hipparchia, **55**, *plate 24*
Hippotion, **141**
hirtaria, **127**

hispidaria, **127**
Hofmannophila, **253**
holmiana, **271**
Homoeosoma, **308**
Hoplodrina, **211**
Horisme, 17, **96**, **97**
hortuella, **292**
hortulata, **300**
hohenwartiana, **285**
humilis, **252**
humuli, **61**
hyale, **33**
hyalinalis, **300**
hybnerella, **230**
hybridella, **264**
Hydraecia, **208**
Hydrelia, 18, **115**
Hydriomena, 21, **96**
hyemana, **273**
Hylaea, **137**
Hyles, 23, 25, **140**, 317, *plate 24*
Hypatima, **260**
Hypena, 17, **222**, **223**, *plate 27*
HYPENINAE, **222**, 314
hypenodes, 19, **223**
hyperantus, **58**, *plate 23*
Hypsopygia, **304**
Hysterophora, **262**

icarus, **41**, *plate 23*
icterata, **108**
icteritia, **193**
Idaea, 23, **79**, **80**, **81**
imbecilla, 21, 22, **175**, 313, 314, 315, 317, *plate 26*
imitaria, **78**
impluviata, **96**
impura, **181**
Inachis, **47**
incanana, **270**
incerta, **179**
incertana, **270**
Incurvaria, **232**
INCURVARIIDAE, **232**
INCURVARIINAE, **232**
indigata, **109**
inquinatella, **294**
interjecta, **162**
interjectana, **269**
interrogationis, 22, **218**, *plate 26*
intricata, **106**

io, **47**
Ipimorpha, **200**, *plate 26*
ipsilon, 23, **159**
iris, 24, **44**
irriguata, 18, **104**, *plate 25*
isertana, **282**

jacobaeae, **156**
janthina, **162**
Jodis, **76**
josephinae, **254**
jota, **217**
junglella, **289**
junoniella, **241**
jurtina, **57**

kleemannella, **243**
kuehniella, **309**

Lacanobia, 17, 19, **172**, **173**, *plate 26*
lacertinaria, **69**
lactearia, **76**
lacunana, **274**
Ladoga, **43**, *plate 23*
laevigatella, **244**
Lampronia, **232**
Lampropteryx, 16, **88**, **89**, *plate 25*
lancealana, **278**
lancealis, **301**
lanestris, **66**, **317**, *plate 25*
langiella, **261**
Laothoe, **139**
lapponica, **231**
Larentia, **87**
LARENTIINAE, **82**
laricella, **250**
lariciata, **110**
Lasiocampa, 21, **66**, 316
LASIOCAMPIDAE, **65**, 314
Lasiommata, **54**
Laspeyria, **221**
laterana, **271**
lathonia, **50**
Lathronympha, **287**
latifasciana, **271**
latistria, **294**
latruncula, **205**
lautella, **242**
leautieri, **185**, *plate 27*
lecheana, **267**
legatella, **112**

Leioptilus, **310**, **311**
lemnata, 20, **298**
leporina, **194**, *plate 26*
Leptidea, **33**, *plate 23*
leucatella, **258**
Leucinodes, **141**, **304**, 315, 317
leucographa, 17, **169**, *plate 27*
Leucoma, **149**, *plate 25*
leucophaearia, **128**
Leucoptera, **236**, **237**
leucostigma, 20, **209**
leuwenhoekella, **262**
levana, **49**
libatrix, **221**
lichenea, **188**, *plate 26*
Ligdia, **118**
ligula, **189**
ligustri, Craniophora, 18, **196**, *plate 26*
ligustri, Sphinx, 17, **138**, *plate 24*
limacodes, **64**
LIMACODIDAE, **64**
limbalis, **300**
limosipennella, **250**
linariata, **104**
lineana, **276**
linearia, **78**
lineata, 23, 25, **140**
literana, **273**
literosa, **206**, **257**
Lithacodia, **213**, *plate 26*
lithargyrinella, **250**
LITHOCOLLETINAE, **240**
Lithophane, 24, 26, **184**, **185**, **313**, *plate 27*
LITHOSIINAE, 18, **150**
lithoxylaea, **202**
litura, **191**
liturata, **119**
liturella, **256**
literosa, **206**
Lobesia, **277**
Lobophora, **116**
locupletella, **261**
loeflingiana, **270**
Lomaspilis, **118**
Lomographa, **135**, **136**
lonicerae, **63**, **64**
lota, **190**
Lozotaenia, **268**
Lozotaeniodes, **267**
lubricipeda, **154**
lucella, **247**

lucernea, 22, **161**, *plate 26*
lucidella, **258**
lucipara, **199**
luctuata, 23, **97**
luculella, **259**
lunaedactyla, **309**
lunosa, **192**
lunularia, **124**
Luperina, **208**
lupulinus, **62**
luridata, **85**
lurideola, **152**
lutarea, Metriotes, **249**
lutarea, Paraswammerdamia, **246**
luteella, **231**
luteolata, **121**
luteum, **155**
luticomella, **252**
lutipennella, **249**
lutosa, 20, **210**, *plate 26*
lutulentella, **258**
luzella, **232**
Lycaena, **40**, *plate 23*
LYCAENIDAE, **37**
LYCAENINAE, **40**
lychnidis, **191**
Lycia, **127**
Lycophotia, 21, **163**
Lygephila, **220**
Lymantria, 18, **149**, *plate 24*
LYMANTRIIDAE, **148**
Lyonetia, **237**
LYONETIIDAE, **236**
LYONETIINAE, **237**
Lysandra, 23, **42**

macilenta, **190**
MACROGLOSSINAE, **139**
Macroglossum, 23, **140**
Macrothylacia, 21, **67**
macularia, 17, **122**
maculosana, **262**
maestingella, **241**
magdalenae, **230**
Malacosoma, **66**
malella, **230**
malvae, **32**
Mamestra, **171**
Maniola, **57**
Marasmarcha, **309**
margaritata, **136**

margaritella, **295**
marginana, **277**
marginaria, **129**
marginata, **118**
marginea, **232**
marginella, **260**
marginicolella, **228**
marmorea, **308**
masculella, **232**
matura, **199**
maura, **198**
mediofasciella, **226**
megacephala, **194**
Meganola, **157**
megera, **54**
Melanargia, **55**, *plate 23*
Melanchra, **171**
Melanthia, **97**
mellinata, **91**
mellonella, **305**
mendica, Diaphora, **155**
mendica, Diarsia, **164**
Menophra, **130**
menyanthidis, 22, **195**
mercurella, **297**
Mesapamea, **206**
Mesoleuca, **88**, *plate 25*
Mesoligia, **206**
mesomella, 17, **151**
messaniella, **240**
messingiella, **248**
metallica, **233**
metaxella, **233**
meticulosa, **199**
Metriostola, **306**
Metriotes, **249**
mi, **220**
miata, **92**
micacea, **208**
microdactyla, **310**
MICROPTERIGIDAE, **225**, 314
Micropterix, **225**
Microstega, **300**
microtheriella, **231**
Miltochrista, 18, **150**, *plate 27*
Mimas, **138**
miniata, 18, **150**, *plate 27*
minima, **207**
miniosa, **177**
ministrana, **269**
Minoa, **115**, *plate 25*

Mirificarma, **259**
miscella, **261**
mitterbacheriana, **279**
Mompha, **261**, 262
MOMPHIDAE, **261**
MOMPHINAE, **261**
monacha, 18, **149**, *plate 24*
moneta, **216**, 317
Monochroa, **257**, 258
monodactyla, **311**
monoglypha, **201**
Monopis, **236**
montanana, **291**
montanata, **83**
Mormo, **198**
Morophaga, **235**
morpheus, **212**
mouffetella, **258**
mucronata, **85**
muelleriella, **240**
mulinella, **259**
multistrigaria, **95**
munda, **179**
mundana, 18, **150**
muralis, **196**, 197, *plate 26*
murana, 22, **297**
murinata, **115**, *plate 25*
musculana, **266**
Myelois, **308**
myllerana, **243**
myrtillana, Ancylis, 22, **279**
myrtillana, Rhopobota, 19, **282**
myrtillella, **228**
myrtilli, 21, **170**
Mythimna, 19, 23, 24, **180-182**, *plate 26*

Naenia, **168**
naevana, **282**
nana, Cochylis, **264**
nana, Hada, **170**
nanata, **109**
napi, **36**
Nebula, **89**
nebulata, **114**
nebulosa, **171**
Nemapogon, **235**
NEMAPOGONINAE, **235**
Nematopogon, **233**
Nemophora, **233**
nemoralis, Herminia, **224**
nemorella, Crambus, **293**

Neofaculta, **259**
Neosphaleroptera, **270**
NEPTICULIDAE, **226**
nervosa, Agonopterix, **257**
nervosa, Depressaria, **255**
neustria, **66**
nicellii, **242**
Niditinea, **236**
nigra, Aporophyla, **184**
nigricomella, **237**
nigricostana, **277**
nisella, **280**
nitidana, **287**
nivea, 20, **298**
Noctua, **161**, **162**
noctuella, 23, **303**
NOCTUIDAE, **158**, 313, 314, 315
NOCTUINAE, **158**
Nola, **157**
NOLIDAE, **157**
Nomophila, 23, **303**
Nonagria, 20, **209**
notana, **272**
notha, **74**
Notodonta, **143**
NOTODONTIDAE, **141**
nubiferana, **275**
nubilana, **270**
Nudaria, 18, **150**
Numonia, **307**
nupta, **219**
Nycteola, **214**
nylandriella auct., **230**
nylandriella Tengst., **230**
nymphaeata, **297**
NYMPHALIDAE, **43**, 314
Nymphalis, 23, **46**, 47
Nymphula, 20, **298**
NYMPHULINAE, **297**

obelisca, 23, **158**
obeliscata, **94**
oblonga, 20, **202**
obscurana, **288**
obscuratus, **137**
obsoleta, 20, **182**, *plate 26*
obstipata, 23, **82**
obtusana, **279**
obumbratana, **286**
occulta, 24, **168**, *plate 27*
ocellana, Agonopterix, **256**

ocellana, Spilonota, **286**
ocellata, Cosmorhoe, **89**
ocellata, Smerinthus, **139**
ocellea, 23, **292**, **315**
Ochlodes, **31**
ochraceella, Mompha, **261**
ochrodactyla, **310**
ochroleuca, 21, **207**, *plate 26*
ochroleucana, **275**
Ochropacha, **73**
ocularis, **72**, *plate 27*
oculea, **208**
Odezia, 19, **113**
Odontopera, **125**
Odontosia, 17, **146**, 317, *plate 25*
Oecophora, 18, 19, **254**
OECOPHORIDAE, **253**, **315**
OECOPHORINAE, **253**
Oegoconia, **260**
oehlmanniella, **232**
OENOCHROMINAE, **75**
oleracea, **173**
Olethreutes, **274**
OLETHREUTINAE, **274**, 313
Oligia, **204**, **205**
Olindia, **269**
olivalis, **303**
Omphaloscelis, **192**
Operophtera, **100**
OPHIDERINAE, **220**
ophiogramma, 20, **204**
opima, **177**
Opisthograptis, **121**
Opsibotys, **302**
or, **72**
Orgyia, **148**
ornitopus, **184**, *plate 27*
Orthonama, 20, 23, **82**
Orthopygia, **304**
Orthosia, 20, **177**, **178**, **179**
Orthotaenia, **275**
osseana, **270**
osteodactylus, **310**
otregiata, 16, **89**, *plate 25*
Ourapteryx, **125**
oxyacanthae, Allophyes, **186**
oxyacanthae, Phyllonorycter, **240**
oxyacanthella, **230**

padella, **245**
paleana, **267**

pallens, **181**
palliatella, **251**
pallidactyla, **310**
pallidata, **299**
palpina, **146**
Palpita, 23, **304**
Paltodora, **257**
Pammene, **287**, **288**, 315
pamphilus, **58**
Pancalia, **262**
pandalis, **300**
Pandemis, **264**, **265**
Panemeria, **212**, *plate 26*
Panolis, **176**
PANTHEINAE, **215**
panzerella, **233**
Papestra, 21, **173**, *plate 26*
paphia, **51**, *plate 23*
papilionaria, **75**
Parachronistis, **258**
Paradiarsia, **163**
Parapoynx, 20, **298**
Pararge, **53**
Parascotia, **222**
Parasemia, **153**
parasitella, Ephestia, **309**
parasitella, Triaxomera, **236**
Parastichtis, 19, **191**, *plate 27*
Paraswammerdamia, **246**
parenthesella, **247**
paripennella, **251**
Parornix, **239**
parthenias, **74**
pascuella, **292**
pastinacella, **255**
pastinum, **220**
pavonia, 21, **68**, *plate 24*
pectinataria, **95**
pectinea, **232**
pennaria, **126**
pentadactyla, **310**
Peribatodes, **130**
Peridea, 18, **144**, *plate 25*
Peridroma, 23, **164**
Perizoma, 18, 21, **96**, **101**, **102**, **103**, 317, *plate 25*
perlella, **293**
perpygmaeella, **229**
persicariae, **171**
Petrophora, **120**
petryi, **234**

Phalera, **141**
Phalonidia, **262**
Phaulernis, **248**
Pheosia, **144**, **145**, *plate 25*
Philedone, 22, **268**
Philereme, **98**
Philudoria, **67**
phlaeas, **40**, *plate 23*
Phlogophora, **199**
Photedes, **207**, *plate 26*
Phragmatobia, **156**
phragmitella, Chilo, **292**
phryganella, **255**
Phtheochroa, **262**
Phycita, **307**
PHYCITINAE, 18, **306**
Phycitodes, **308**
PHYLLOCNISTINAE, **243**
Phyllocnistis, **243**
Phyllonorycter, **240-243**
Phylloporia, **232**
Phytometra, **221**
picarella, **235**
piercella, **236**
PIERIDAE, **33**
PIERINAE, **34**
Pieris, **35**, **36**
pilosaria, **127**
pinella, **295**
pinguinalis, **305**
pinguis, **308**
piniaria, **134**
pinicolana, **286**
pinicolella, **241**, **261**
pinivorana, **287**
pisi, **174**
plagiata, **113**
plagicolella, **228**
Plagodis, **120**, **121**
plantaginis, 21, **153**
Platyptilia, **309**, **310**
PLATYPTILIINAE, **309**
Plebejus, **40**
plecta, **160**
Plemyria, **93**
Pleuroptya, **303**
plumbella, **246**
plumbeolata, 19, **103**
Plusia, 20, **216**
PLUSIINAE, **215**
Plutella, 23, **248**

PLUTELLINAE, **247**
poae, **252**
podana, **265**
Poecilocampa, **65**
Polia, **170**, **171**
polychloros, 23, **46**
Polychrysia, **216**
Polygonia, **48**
Polymixis, **187**, *plate 27*
Polyommatus, **41**, *plate 23*
Polyploca, **74**
pomonella, **290**
populata, 21, **90**, **96**
populeti, **178**
populi, Laothoe, **139**
populi, Poecilocampa, **65**
porata, **77**
porcellus, **141**
porphyrea, 21, **163**
porrectella, **248**
potatoria, **67**
prasina, 18, **169**
prasinana, **214**, *plate 26*
pratella, **293**
Prays, **247**
primaria, **136**
proboscidalis, **222**
procellata, **97**
Prochoreutis, **243**
PROCRIDINAE, **63**
profundana, **278**
pronuba, **161**
propinquella, Agonopterix, **256**
propinquella, Mompha, **261**
proximella, **258**
pruinata, **75**
prunalis, **302**
prunaria, **126**, *plate 25*
prunata, **90**
pruni, **40**
pruniana, **275**
pruniella, **245**
Pseudargyrotoza, **269**
Pseudoips, **214**
Pseudopanthera, 17, **122**
pseudospretella, **253**
Pseudoswammerdamia, **246**
Pseudatemelia, **254**
Pseudoterpna, **75**
psi, **195**
Psoricoptera, **260**

Psyche, **235**
PSYCHIDAE, **234**
PSYCHINAE, **235**
Psychoides, **235**
Pterapherapteryx, 20, **116**
PTEROPHORIDAE, **309**
PTEROPHORINAE, **310**
Pterophorus, **310**
Pterostoma, **146**
Ptilodon, **145**
Ptilodontella, **146**
Ptycholoma, **267**
Ptycholomoides, **267**, 315
pudibunda, **148**
pudorina, 19, **181**, *plate 26*
pulchella, Elachista, **104**, **252**
pulchella, Utetheisa, 23, **153**, 314
pulchellata, **104**
pulcherrimella, **256**
pulchrina, **217**
pulveraria, **120**
pulverella, **257**
pulverosella, **226**
punctaria, **77**
punctidactyla, **309**
punctulata, **133**
purpuralis, Pyrausta, **299**
pusaria, **134**
pustulata, **76**
puta, **159**
putris, **160**
pygarga, **213**, *plate 26*
pygmaeata, **105**
pygmaeella, Argyresthia, **245**
pygmaeella, Stigmella, **229**
pygmeana, **248**
pygmina, **207**
Pyla, **306**
pyralella, **296**
pyraliata, **91**
PYRALIDAE, **292**, 315, 317
pyralina, **201**
Pyralis, **305**
pyramidea, **197**, *plate 26*
Pyrausta, 22, **299**, **300**
PYRAUSTINAE, **299**
pyrella, **246**
PYRGINAE, **32**
Pyrgus, **32**
pyrina, **62**
pyritoides, **71**

Pyronia, **56**, *plate 24*
Pyrrhia, 21, **213**, *plate 27*

quadripuncta, **260**
quercana, **254**
quercifolia, 25, **68**, *plate 24*
quercifoliae, **227**
quercifoliella, **240**
quercinaria, **122**
quercus, Lasiocampa, 21, **66**, 316
quercus, Quercusia, **38**, **66**, *plate 23*
Quercusia, **38**, *plate 23*
quinqueguttella, **242**

rajella, **242**
ramella, **280**
rapae, **35**
raschkiella, **261**
ratzeburgiana, **282**
ravida, 23, **163**
reaumurella, **233**
rectangulata, **111**
Recurvaria, **258**
regiana, **288**
reliquana, **277**
remissa, **203**
repandana, **307**
repandaria, **122**
repandata, **131**
resinea, Eudonia, **297**
resplendella, **234**
reticulata, **171**
retinella, **245**
retusa, **200**, *plate 26*
revayana, **214**
revinctella, **253**
rhamni, **34**
Rheumaptera, 97, **98**, *plate 25*
Rhizedra, 20, **210**, *plate 26*
Rhodometra, 23, **81**
rhombana, **272**
rhomboidaria, **130**
rhomboidella, **260**
Rhopobota, 19, **282**
Rhyacionia, **286**, **287**
Rhynchopacha, **258**
ribeata, 17, **130**
ridens, **74**
rivata, **86**
Rivula, **221**
rivularis, **174**

roborana, **284**
roboraria, 18, **132**, *plate 25*
roborella, Phycita, **307**
robustana, 21, **278**
robustella, **238**
Roeslerstammia, **246**
rosaecolana, **284**
rostralis, **223**, *plate 27*
rubi, Callophrys, **37**, **85**
rubi, Diarsia, **165**
rubi, Macrothylacia, 21, **67**
rubidata, **18**
rubiginata, Plemyria, **93**
rubiginea, **189**
rubricollis, 18, **150**, *plate 27*
rubricosa, **169**
rufa, **20**
rufata, **112**
rufescens, **260**
ruficapitella, **229**
ruficiliana, **264**
ruficornis, **147**
rufifasciata, **112**
rufimitrella, **234**
rufocinerea, **252**
rugosana, **262**
rumicis, **196**
rupicapraria, **136**
rupicola, **264**
ruralis, **303**
Rusina, **198**
rusticella, **236**

Sabre, 18, **26**, *plate 27*
sacraria, 23, **81**
Salebriopsis, 18, **26**, **306**, 316
salicata, 21, **89**
salicella, Hedya, 20, **275**
salicis, Leucoma, **149**, *plate 25*
salicis, Stigmella, **228**
sambucaria, **125**
sannio, 17, **154**
sarcitrella, **253**
SARROTHRIPINAE, **214**
Saturnia, 21, **68**, *plate 24*
SATURNIIDAE, 68
SATYRIDAE, 53
saucia, 23, **164**
sauciana, **277**
scabiosella, **233**
scabrella, **247**

347

SCARDIINAE, **235**
schalleriana, **272**
SCHOENOBIINAE, **295**
Schrankia, **223**
schreberella, **242**
Schreckensteinia, **249**
SCHRECKENSTEINIIDAE, 249
schumacherana, **269**
schwarziellus, **233**
Scoliopteryx, **221**
scolopacina, 17, **204**
Scoparia, **295**, **296**
scopariella, Agonopterix, **257**
scopariella, Phyllonorycter, **241**
SCOPARIINAE, **295**
Scopula, 21, **78**, **79**, **302**, **303**
scoticella, **239**
Scotopteryx, **84**, **85**
Scrobipalpa, **259**
scutulana, **284**
Scythropia, **247**
secalis, **206**
sedatana, **291**
segetum, **158**
selasella, **293**
selene, **49**, *plate 23*
Selenia, **124**
semele, **55**, *plate 24*
semiargus, 24, **42**
semibrunnea, **184**, *plate 27*
semifascia, **238**
semifulvella, **236**
Semioscopis, **255**
Semiothisa, 18, **118**, **119**
senex, 20, **150**
septembrella, **227**
sequana, **291**
sequella, **248**
serella, **228**
seriata, 80
sericealis, 221
sericella, **234**
serratella, **249**
servillana, **289**
Sesia, 65
SESIIDAE, 64
SESSIINAE, **65**
sexalata, 20, **116**
sexstrigata, **167**
silaceata, **91**
similis, Bryotropha, **259**

similis, Euproctis, **149**
simpliciata, **108**
simpliciella, **244**
sinapis, **33**, *plate 23*
sinuella, Homoeosoma, **308**
siterata, **92**
smeathmanniana, **263**
Smerinthus, **139**
socia, **184**, *plate 27*
sociana, **283**
sociella, **305**
solandriana, **282**
somnulentella, **237**
sorbi, Stigmella, **228**
sorbiella, **245**
sordens, **204**
sordidana, **281**
sororcula, 18, **151**, *plate 27*
sororculana, **276**
sororculella, **259**
spadicearia, **83**
Spaelotis, 23, **163**
Spargania, 23, **97**
sparsana, **272**
spartifoliella, **237**
speciosa, **228**
spectrana, **267**
SPHINGIDAE, 137, **314**
SPHINGINAE, **137**
Sphinx, 17, **138**, *plate 24*
sphinx, 17, **183**, *plate 27*
Spilonota, **286**
Spilosoma, 20, **154**, **155**, *plate 24*
spinolella, **241**
splendana, **290**
splendidissimella, **227**
splendidulana, **287**
Spodoptera, 23, **211**
stabilis, **178**
stagnata, **298**
Standfussiana, 22, **161**, *plate 26*
statices, **63**
Stauropus, 17, **143**, *plate 24*
steinkellneriana, **255**
stellatarum, 23, **140**
Stenoptilia, **310**
stephensiana, **269**
STERRHINAE, **77**
stettinensis, **242**
stigmatella, **238**
Stigmella, **227-231**

Stilbia, 21, **212**, *plate 26*
straminata, **81**
straminea, Mythimna, 20, **181**, *plate 26*
straminella, **294**
strataria, **127**
stratiotata, 20, **298**
Striatipennella, **251**
striana, **274**
strigana, **287**
strigilis, **204**
strigula, **157**
strobilella, **289**
stroemiana, **281**
Strophedra, **287**
Strymonidia, *plate 23*
suasa, 21, **173**
suavella, **307**
subbimaculella, **227**
subfusca, **295**
subfuscata, **108**
subocellana, Epinotia, **280**
subochreella, **255**
subpropinquella, **256**
subpurpurella, **225**
subsequana, **279**
subsericeata, **80**
subtusa, **200**
succedana, **289**
succenturiata, **108**
suffumata, **88**
sulphurella, **253**
suspecta, 19, **191**, *plate 27*
swammerdamella, **223**
Swammerdamia, **246**
sylvata, Abraxas, **117**
sylvata, Hydrelia, 18, **115**
sylvaticella, **251**
sylvella, Phyllonorycta, **243**
sylvella, Ypsolopha, **247**
sylvestris, 31, **106**, **249**, **267**
sylvina, **61**
SYMMOCINAE, **260**
Synanthedon, **65**
Syndemis, **266**
Syngrapha, 22, **218**, *plate 26*
syringaria, 17, **122**
syringella, **238**

taenialis, **223**
taeniatum, 18, **317**
tages, **32**

Taleporia, **234**
TALEPORIINAE, **234**
tantillaria, **110**
tarsipennalis, **224**
tedella, **281**
TEICHOBIINAE, **235**
Telechrysis, **253**
Teleiodes, **258**, **259**
Teliopsis, **259**
temerata, **136**
tenebrata, **212**, *plate 26*
tenebrella, **257**
tenerana, **281**
tenerella, **240**
tenuiata, **103**
tephradactyla, **311**
ternata, 21, **79**
tersata, **97**
tesserana, **263**
testacea, **208**
testata, **90**
Tethea, **72**, *plate 27*
Tetheella, 17, **72**, *plate 27*
tetragonana, **284**
tetralunaria, **124**
tetraquetrana, **280**
thalassina, **172**
Thalpophila, **199**
Thecla, 24, **38**
THECLINAE, **37**
Thera, **93**, **94**
Theria, **136**
Tholera, **176**
thoracella, **237**
thrasonella, **244**
Thumatha, 20
thunbergella, **225**
Thyatira, **71**, *plate 27*
THYATIRIDAE, **71**, **314**
Thymelicus, **31**
tiliae, Mimas, **138**
tiliae, Stigmella, **229**
Timandra, **78**
Tinea, **236**
TINEIDAE, 235
TINEINAE, **236**
tipuliformis, **65**
Tischeria, **231**, **232**
TISCHERIIDAE, 231
tithonus, 56, *plate 24*
tityrella, **229**

349

tityus, 139
togata, 193
torquillella, 239
TORTRICIDAE, **262**, 313
TORTRICINAE, **264**, 313
Tortricodes, **270**
Tortrix, **269**
tragopoginis, **198**
transversa, **188**
transversata, **98**
trapeziella, **253**
trapezina, **201**
tremula, **145**
triangulum, **166**
Triaxomera **236**
Trichiura, **65**
Trichopteryx, **116**
tricolorella, **259**
tridens, Acronicta, **195**
trifasciella, **243**
trifolii, Dicestra, 21, **170**
trifolii, Zygaena, **64**
trigemina, **218**
trigeminata, **80**
trigonella, **281**
trigrammica, **210**
trimaculana, Epiblema, **284**
trimaculana, Epinotia, **281**
tringipennella, **238**
trinotella, **236**
Triphosa, **98**
triplasia, **219**
tripuncta, **253**
tripunctana, 107, **272**
tripunctaria, **107**
trisignaria, **106**
tristata, **86**
tristella, **294**
tristrigella, **242**
trochilella, **251**
troglodytella, **251**
truncata, **92**
truncicolella, **297**
tubulosa, **234**
tumidella, **307**
tunbergella, **225**
turbidana, Apotomis, **276**
turca, 24, **180**
turfosalis, **223**
typhae, 20, **209**

typica, **168**
Tyria, **156**

uddmanniana, **283**
Udea, 23, **302**, **303**
uliginosellus, 19, **292**
ulmella, Bucculatrix, **237**
ulmifoliella, **242**
ulmivora, **229**
umbra, 21, **213**, *plate 27*
umbratica, **183**
unangulata, **99**
unanimis, **203**
uncula, 19, **213**, *plate 26*
undulana, **275**
undulata, **98**, *plate 25*
unimaculella, **225**
unionalis, 23, **304**
unipuncta, 23, **182**, *plate 26*
unipunctana, **282**
unipunctella, **243**
unitella, **253**
Uresiphita, 23, **300**, 313, 314
urticae, Aglais, **46**
urticae, Spilosoma, 20, **155**, *plate 24*
ustella, **248**
Utetheisa, 23, **253**, 314

vaccinii, **188**
vagans, **141**, **304**, 315, 317
vandaliella, **297**
Vanessa, **44**, *plate 24*
variata, **94**
variegana, **273**
v-ata, **111**
venata, **31**
venosata, **105**, *plate 25*
Venusia, 19, **114**, *plate 25*
verbasci, **183**
verhuella, **235**
versicolor, **205**
versicolora, 24, **69**, 316
vetulata, **98**
vetusta, **186**, *plate 26*
viburnana, **267**
villica, 18, **154**, *plate 24*
viminalis, **183**
viminetella, **250**
vinula, **142**
viretata, **116**

virgaureata, **109**
viridana, **271**
viridaria, **221**
viscerella, **230**
vitalbata, 17, **96**
vitella, **248**
vittata, 20, **82**
vulgana, **268**
vulgata, **107**
vulgella, **258**
vulpinaria, **79**

wailesella, **236**
w-album, **38**, **40**, *plate 23*
wauaria, **119**
weaverella, **236**
weirana, **287**
williana, **263**
w-latinum, 17, **172**, *plate 26*
wolffiella, **235**
woodiana, **274**

Xanthia, 17, **192**, **193**, *plate 26*
xanthographa, **168**
Xanthorhoe, **82**, **83**, **84**

xenolechia, **258**
Xestia, 19, **166**, **167**, **168**
Xylena, **186**, *plate 26*
Xylocampa, **186**
xylosteana, **266**
xylostella, 23, **248**

Yponomeuta, **245**, **246**
YPONOMEUTIDAE, 244
YPONOMEUTINAE, **245**
ypsillon, **200**
Ypsolopha, **247**, **248**

Zeiraphera, **282**
Zelleria, **246**
Zeuzera, **62**
ZEUZERINAE, **62**
ziczac, **144**
zoegana, **263**
zophodactylus, **310**
Zygaena, **63**, **64**
ZYGAENIDAE, 63
ZYGAENINAE, **63**

351